Instrumental Methods of Organic Functional Group Analysis

Instrumental Methods of Organic Functional Group Analysis

EDITED BY

SIDNEY SIGGIA

CHEMISTRY DEPARTMENT
UNIVERSITY OF MASSACHUSETTS,
AMHERST, MASSACHUSETTS

WILEY-INTERSCIENCE
a Division of John Wiley & Sons, Inc.
New York • London • Sydney • Toronto

CHEMISTRY

Library of Congress Catalog Card Number: 77 168642

ISBN 0 471 79110 5

Printed in the United States of America.

10 9 8 7 6 5 4 3 2

Contributing Authors

NUCLEAR MAGNETIC RESONANCE METHODS

Harry Agahigian
Baron Consulting Company
Orange, Connecticut

GAS CHROMATOGRAPHIC METHODS

Morton Beroza and
May N. Inscoe
Pesticide Chemistry Division
U.S. Department of
Agriculture
Beltsville, Maryland

RADIOCHEMICAL METHODS

Donald R. Campbell
The General Tire and
Rubber Company
Corporate Research and
Development
Akron, Ohio

ABSORPTION SPECTROPHOTOMETRIC METHODS

J. Gordon Hanna
Chief Chemist
The Connecticut Agricultural
Experiment Station
New Haven, Connecticut

AUTOMATIC WET CHEMICAL ANALYSIS

Robert A. Hofstader and
W. K. Robbins
Esso Research and
Engineering Company
Linden, New Jersey

ELECTROANALYTICAL METHODS

Alan F. Krivis
Department of Chemistry
University of Akron
Akron, Ohio

Preface

This book is meant as a supplement to my earlier book, *Quantitative Organic Analysis via Functional Groups*. The latter covered the wet methods; this book is concerned with the instrumental methods. As a matter of fact, I inherently object to the delineations of "wet" and "instrumental," since the buret and the balance are instruments and hence all analysis is instrumental. Let us just say, therefore, that this book picks up where the other leaves off and covers approaches that the earlier volume does not discuss.

Both this book and its predecessor rely heavily on the chemistry of the functional groups. The underlying philosophy in both books is the same and is twofold: (1) the reactions of the functional group are quite selective and thus provide the discrimination which is necessary for good analysis; (2) very often the group cannot be measured directly and must be converted to another form first. The use of reactions opens up the possibility of measuring the products of a reaction and/or the reagent consumed by the reaction.

This book is divided into chapters by the functional group involved. This was done because it was believed that the users of this book will be interested in the measurement of a particular functional group or groups and that all methods for each group should be found together. This approach proved difficult in a book composed of contributed sections because there are no specialists in the measurement of specific groups; the specialists exist only in the measuring techniques. It was necessary, therefore, that I solicit contributions from these instrumental experts and then reorganize the contributions by functional groups.

The format of this book is essentially the same as that of my earlier book. The coverage is of the more general methods for the various groups; specific methods for individual compounds are treated only if the compound is very widely encountered. Sufficient details are given for the most widely used methods so that these methods can, in many cases, be applied directly from this textbook. In some cases, however, referral to the original source will be needed for details of the instrumentation used. My earlier book covered the use of only the common laboratory measuring devices; this one includes some specialized instruments.

Chapter 19 on automated wet methods is included because these techniques must be considered in discussing functional group analysis. Their inclusion is justified in this text by the fact that special instrumentation is used to accomplish the automation.

It is assumed that the reader knows the basic use of the required instrumentation or has access to the necessary expertise. It is not possible to describe the operation of all the devices used in the analyses discussed in this book. Included, however, is an Appendix which describes some general gas chromatographic techniques that apply to the broad topic of functional group analysis and not to any specific group. Also included in the Appendix are literature references to the basic instrumental techniques and apparatus.

I should like to thank the contributing authors for their diligence, promptness, and effective presentations. I am also grateful to Mrs. Young-Ja Chon, who handled the secretarial work and to L. R. Whitlock who helped with the proofreading.

SIDNEY SIGGIA

Amherst, Massachusetts
April 1971

Contents

Instrumental Methods of Organic Functional Group Analysis

1 Hydroxyl groups

I. ABSORPTION SPECTROPHOTOMETRIC METHODS*

A. Direct Infrared Determination of Alcohols and Phenols

Although hydroxyl groups show strong absorptions in the infrared region, attempts to develop general quantitative procedures based on direct infrared measurements often are hampered by the tendency of the groups to form hydrogen bonds among themselves and with other polar groups. The intensities of the absorptions normally depend on the degree of association in the system. Procedures specifying conditions to control the amount of association are successful in specific cases. The conditions are designed either to promote essentially complete association or complete dissociation, or to maintain the ratio of associated to unassociated species in both the sample system and the standard reference system. Intermolecular bonding is concentration dependent and extensive dilution with nonpolar solvents favors dissociation, whereas the use of polar solvents such as pyridine promotes essentially complete association. Intramolecular bonding, on the other hand, is not affected by dilution with a nonpolar solvent.

References to some typical infrared determinations of alcohols and phenols are given in Table 1-1.

B. Direct Ultraviolet Measurement of Phenols

Although the hydroxyl group does not absorb in the ultraviolet region, phenols are measurable by ultraviolet spectrometry because of their resonating structures. The adherence of phenol concentration to Beer's law is illustrated in Figure 1-1, prepared from data presented by Schmauch and Grubb (1).

* Written by J. Gordon Hanna.

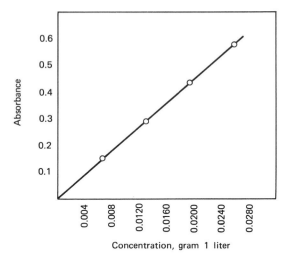

Figure 1-1.   Adherence of phenol to Beer's law in ultraviolet (1).

C.   Measurement after Chemical Reaction

1. Benzoate Esters of Primary and Secondary Alcohols.   In pyridine medium 3,5-dinitrobenzoyl chloride and the active hydrogen atoms of primary and secondary alcohols react rapidly to form benzoate esters. Johnson and Critchfield (2) isolated the dinitro products by extraction, treated them with acetone in the presence of strong alkali, and measured the intense color of the quinoid ions formed. Spectrometric values of these colored products were then related to the amounts of primary or secondary alcohol originally present.

METHOD FROM JOHNSON AND CRITCHFIELD (2)

Reagents.   3,5-Dinitrobenzoyl chloride solution. Dissolve 1 gram of reagent-grade 3,5-dinitrobenzoyl chloride in 10 ml of redistilled pyridine. Use a hot-water bath to maintain solution. Prepare the solution fresh immediately before use.

Procedure.   Prepare a pyridine (redistilled) solution of the sample to contain 2 to 50 γ of hydroxyl per milliliter. Transfer 2 ml of this solution to a 100-ml glass-stoppered graduated cylinder and add 1 ml of the 3,5-dinitrobenzoyl chloride solution. For a blank add 1 ml of the 3,5-dinitrobenzoyl chloride solution to 2 ml of pyridine in another 100-ml graduated cylinder. Allow the sample and the blank to stand for 15 minutes at room temperature and then

Table 1-1. Typical Direct Infrared Determinations of Alcohols and Phenols

Hydroxyl compound	Sample system	Wavelength, μ	Reference
Aliphatic primary alcohols	Carbon tetrachloride or tetrachloro-methylene	1.4	1
Aliphatic unbranched primary alcohols	Chloroform	1.4	2
Alcohols	Carbon tetrachloride	2.719–2.761	3
Methanol		9.84	4
Ethanol	Carbon tetrachloride	11.37	5
Polyglycols	Quinoline	3.05	6
Polypropylene glycols	Liquid	2.84	7
Polyesters and polyethers	Chloroform-carbon tetrachloride	2.6–3.2	8
Oxidized fatty esters	Carbon tetrachloride	2.76	9
Alkyd resins	Dichloromethane	2.85	10
Epoxy resins	Pyridine	3.08	11
Residual hydroxyl in cellulose acetate	Pyrrole	1.445	12
Phenols	Carbon tetrachloride	2.7–2.84	13
Steroids	Pyridine	3.05	14

[1] R. O. Crisler and A. M. Burrill, *Anal. Chem.*, **31**, 2055 (1959).

[2] E. Shauenstein and H. Puchner, *Monatshefte*, **93**, 253 (1962).

[3] E. L. Saier and R. H. Hughes, *Anal. Chem.*, **30**, 513 (1958).

[4] T. Oba, *Esei Shikensho Hokoku*, **76**, 53 (1958).

[5] F. A. Bronan, D. E. Broadlick, and J. E. Hamilton, *J. Pharm. Sci.*, **51**, 242 (1962).

[6] J. G. Hendrickson, *Anal. Chem.*, **36**, 126 (1964).

[7] E. A. Burns and R. F. Muraca, *Anal. Chem.*, **31**, 397 (1959).

[8] C. L. Hilton, *Anal. Chem.*, **31**, 1610 (1959).

[9] N. H. E. Ahlers and N. G. McTaggert, *Analyst*, **79**, 70 (1954).

[10] J. F. Murphy, *Appl. Spectry.*, **16**, 139 (1962).

[11] M. R. Adams, *Anal. Chem.*, **36**, 1688 (1964).

[12] J. A. Mitchell, C. D. Bockman, Jr., and A. V. Lee, *Anal. Chem.*, **29**, 499 (1957).

[13] R. F. Goddu, *Anal. Chem.*, **30**, 2009 (1958).

[14] P. Kabasakalian, E. R. Townley, and M. D. Yudis, *Anal. Chem.*, **31**, 375 (1959).

add 25 ml of 2 N hydrochloric acid. From a pipet add 20 ml of n-hexane to each. Stopper the cylinders and shake vigorously for 30 seconds. Allow the phases to separate completely; then pipet 2 ml of the top layers of each into separate 25-ml graduated cylinders. Avoid getting any precipitate in the pipet. Add 10 ml of acetone and 0.3 ml of 2 N aqueous sodium hydroxide to each, shake well, and let stand 3 to 5 minutes. Determine the absorbance of the solution at 525 mμ on the basis of a reading of zero for the blank. If a turbidity forms in the colored solution, add a pinch of sodium chloride crystals and mix before determining the absorbance.

Calibration Curve.   Prepare a calibration curve with absolute ethanol as the sample and plot absorbance versus micrograms of hydroxyl. Beer's law is followed for concentrations up to 100 γ of hydroxyl.

Pyridine has the twofold advantage of being an excellent solvent for the alcohol and of forcing the reaction to completion by absorbing the hydrogen chloride that is released. The latter function contributes to the fact that the reaction is complete in only 15 minutes at room temperature for all the C_1 to C_{20} primary and secondary alcohols tested.

Data obtained by Johnson and Critchfield (2) are shown in Tables 1–2 and 1–3.

Table 1-2.   Lower Limit of Determination of Alcohols by the 3,5-Dinitrobenzoyl Chloride Method (2)

Alcohol	Lower limit, ppm[a]
Methanol	2.8
Ethanol	4.1
2-Propanol	5.3
Butanol	6.5
2-Butanol	6.5
1-Pentanol	7.7
Hexanol	8.9
2-Ethylhexanol	11.4
5-Ethyl-2-heptanol	12.6
2-Eicosanol (mixed isomers)	26.1

[a] One-gram samples and an absorbance of 0.01.

Dihydroxy compounds, including ethylene glycol, diethylene glycol, 2,2-dimethyl-1,3-butanediol, and 2,2-dimethyl-1,3-propanediol, cannot be determined by this method. Ethylene glycol and diethylene glycol produce no color, and only about 50% reaction is realized with the other compounds. In contrast to ethylene glycol, its monomethyl ether, methyl Cellosolve, reacts quantitatively and can be determined.

Table 1-3.   Recovery of Alcohol from Known Mixtures by the 3,5-Dinitrobenzoyl Chloride Method (2)

	Alcohol, % by weight		
Mixture	Added	Found	Recovered
Hexanol in methyl isobutyl ketone	0.189	0.177	98.6
Butanol in butyl ether	0.144	0.144	100.0
Ethanol in diethyl butyral	0.200	0.200	100.0
Ethanol in pentanedione	50.00	47.00	94.0
Ethanol in acetaldehyde	0.405	0.413	102.0
Ethanol in vinyl ethyl ether	0.033	0.034	103.0
Ethanol in ethyl acrylate	0.098	0.096	98.2
2-Propanol in isopropyl ether	0.824	0.800	97.0
Methanol in acetone	0.041	0.040	97.5
Ethanol in butylamine	48.00	49.90	104.0
Ethanol in water	0.600	0.600	100.0
Average:			99.5
Average deviation:			±2.2

Although phenol and 1-naphthol respond quantitatively, difficulties with the color stabilities are encountered. Hydroquinone fails to respond, but its monomethyl ether reacts quantitatively.

Water and primary and secondary amines interfere because 3,5-dinitrobenzoyl chloride reacts with each preferentially to alcohols. The products of these reactions, however, are not generally soluble in hexane, and their interference usually can be overcome by the addition of sufficient excess reagent.

Scoggins (3) substituted a direct ultraviolet absorption measurement of the alkyl nitrobenzoate products for the formation and measurement of the colored products required by the procedure of Johnson and Critchfield (2). Scoggins used p-nitrobenzoyl chloride as the esterification reagent for primary and secondary alcohols through C_{12} in hydrocarbons. He separated the esters from excess reagent and read the absorbance at 253 mμ.

METHOD FROM SCOGGINS (3)

Reagent.   Dissolve 1 gram of p-nitrobenzoyl chloride in 25 ml of redistilled pyridine. Prepare fresh daily.

Procedure. Transfer 1 ml of liquid hydrocarbon sample to a 125-ml separatory funnel (Teflon stopcock) and add 2 ml of pyridine (redistilled). Then add 1 ml of *p*-nitrobenzoyl chloride solution to the separatory funnel and shake to mix. Allow the mixture to react at room temperature for 30 minutes. Add 25 ml of cyclohexane to the separatory funnel, mix well, and wash the mixture with 10 ml of 2 *M* potassium hydroxide. Allow the phases to separate and discard the lower phase. Wash the cyclohexane with two 10-ml portions of 2 *M* hydrochloric acid and follow this with two 10-ml alkali washes. Finally wash the material with 10 ml of 2 *M* hydrochloric acid and allow the phases to separate. Measure the absorbance of the cyclohexane phase in a 1-cm cell at 253 mμ versus a blank treated in the same manner. Determine the alcohol concentration from a previously prepared calibration curve.

Calibration Curve. Treat 1-ml aliquots of cyclohexane-alcohol blends containing from 25 to 300 μg of alcohol per milliliter of solution in the manner given in the procedure. Plot the total absorbance values versus the micrograms of alcohol. Beer's law is followed for concentrations up to 300 μg of alcohol per 25 ml of solution.

The results of the analysis of synthetic blends of ethanol in cyclohexane are shown in Table 1–4. In the concentration range of 25 to 200 ppm the average recovery was 100.4%. The application of the method to higher alcohols is shown in Table 1–5. The blends were prepared from commercial alcohols without further purification. Results given for these blends are based on the ethanol calibration. Therefore, errors are magnified by the alcohol/ethanol ratio.

Table 1-4. Analysis of Ethanol-Cyclohexane Blends by the *p*-Nitrobenzoyl Chloride Method (3)

	Ethanol, ppm		Standard deviation
Added	Recovered (average)[a]		
34.0	37.0 (6)		4.7
78.0	75.9 (7)		2.1
131.8	127.0 (6)		5.3
196.2	194.1 (6)		7.4
206.1	206.1 (6)		8.1
212.2	212.4 (6)		4.2
25.2	25.9 (3)		2.0

[a] Figures in parentheses represent numbers of determinations.

Table 1-5. Analysis of Higher Alcohol Blends by the p-Nitrobenzoyl Chloride Method (3)

Alcohol	Alcohol, ppm		Standard deviation	Alcohol recovery, %
	Added	Recovered[a]		
Butanol-1	131.7	123.6	3.5	94
3-Methylbutanol-1	58.8	61.7	2.9	105
Hexanol-1	156.1	165.1	4.5	106
Cyclohexanol	114.5	115.5	7.1	101
Butanol-2	97.4	95.0	1.4	98
Propanol-4	66.2	68.0	3.2	103
Heptanol-1	155.4	160.8	11.4	103
Octanol-1	128.6	129.6	4.5	101
Dodecanol	155.5	164.4	3.1	106

[a] Average of four determinations.

2. Oxidation of Secondary Alcohols to Ketones. Critchfield and Hutchinson (4) selectively oxidized secondary alcohols with potassium dichromate to form ketones and determined the amount of ketone formed by a 2,4-dinitrophenylhydrazine colorimetric method. Because excess dichromate ion oxidizes 2,4-dinitrophenylhydrazine and the methanol used as a solvent, the excess must be reduced chemically. It was found that 50% hypophosphorous acid is suitable for this purpose and has no effect on the ketones. It is necessary to neutralize the acidity of the reduced reaction mixture before the 2,4-dinitrophenylhydrazine reaction if any color is to be obtained.

METHOD FROM CRITCHFIELD AND HUTCHINSON (4)

Reagents. Acid potassium dichromate, approximately 0.3 N. Dissolve 15 grams of reagent-grade potassium dichromate in 500 ml of distilled water. Slowly add 360 ml of concentrated sulfuric acid and cool to prevent excessive heating. Dilute to 1 liter with distilled water and mix.

Hypophosphorous acid, 50% solution.

Carbonyl-free methanol. Reflux 3 gal of methanol containing 50 grams of 2,4-dinitrophenylhydrazine and 15 ml of concentrated hydrochloric acid for 4 hours and collect the distillate until the head temperature reaches 64.8°C.

Pyridine stabilizer. Pyridine-water solution (80:20, v/v).

2,4-Dinitrophenylhydrazine reagent. Suspend 0.05 gram of reagent-grade 2,4-dinitrophenylhydrazine crystals in 25 ml of carbonyl-free methanol. Add 2 ml of concentrated hydrochloric acid and mix to effect solution. Dilute to 50 ml with carbonyl-free methanol.

Optimum Oxidation Time for Pure Alcohols.   Pipet 15.0 ml of potassium dichromate reagent into each of eight 250-ml flasks. Reserve four flasks for blank determinations. Introduce 2.0 to 3.0 meq of the alcohol to be oxidized into each of the other flasks. If an aliquot of a dilution of the sample is used, introduce the same volume of solvent into the blank flasks. Allow one sample and one blank flask to stand for 30, 60, 90, and 120 minutes, respectively. Add 100 ml of distilled water and 10 ml of 15% potassium iodide to the sample and the blank flasks. Titrate immediately with standard 0.1 N sodium thiosulfate to a greenish yellow. Add a few milliliters of a 1% starch solution and continue to titrate to the disappearance of the starch-iodide color. Take the optimum oxidation time as the time at which no further dichromate is consumed.

Calibration Curve.   Add 50 ml of distilled water (or acetonitrile if specified in Table 1–6) to a 100-ml volumetric flask. Transfer 100 times the maximum sample specified in Table 1–6 to the volumetric flask, dilute to volume with the appropriate solvent, and mix. Transfer 5.0-, 10.0-, 15.0-, and 20.0-ml aliquots of the dilution into separate 100-ml volumetric flasks and dilute to volume with the solvent. Use 5-ml aliquots in place of the sample in the following procedure to determine the absorbance produced by each of these standards.

Procedure.   Pipet 15.0 ml of the potassium dichromate reagent into each of three 50-ml volumetric flasks. Reserve one of the flasks for the blank determination. Transfer sufficient sample that does not exceed the secondary alcohol content specified in Table 1–6 into each of the other flasks. Prepare a dilution in water or in redistilled acetonitrile if the optimum sample size is too small to be weighed accurately. The total alcohol content of the sample should not exceed 3.8 meq as determined by the volumetric procedure previously described. If the sample is weighed directly and is insoluble in the reagent, add sufficient redistilled acetonitrile to effect solution. Add the same volume to the blank. Allow the samples and the blanks to stand at 0°C (unless another temperature is specified in Table 1–6) until the alcohols are completely oxidized. The optimum times for secondary alcohols are listed in Table 1–6; the optimum times for primary alcohols can be determined by the volumetric procedure.

Immerse the flasks in an ice-water bath and pipet 1.0 ml of the hypophosphorous acid into each flask. Swirl the flasks during the additions.

**Table 1-6.   Reaction Conditions for the Determination of
Secondary Alcohols (4)**

Compound	Secondary alcohol, mg (maximum)	Oxidation time, minutes
2-Butanol	2.92	60–120[a]
3-Heptanol	6.16[b]	5–60
4-Heptanol	5.36[b]	5–75
2,5-Hexanediol	1.47	30–60
2-Hexanol	3.63	5–60
2-Propanol	1.96	5–60[a]
Isopropanolamine	2.17	120–210[a, c]
2-Octanol	4.06[b]	10–60
4-Octanol	5.47[b]	30–100
3-Pentanol	6.66	5–60

[a] At room temperature.

[b] Dilute in acetonitrile or use as cosolvent to effect solution in oxidation step.

[c] Allow 17 hours for quantitative reaction with 2,4-dinitrophenylhydrazine.

Remove the flasks from the bath and immerse them in a water bath at room temperature for 15 minutes. Dilute the contents of the flasks to volume with carbonyl-free methanol and mix. Transfer a 3-ml aliquot of each dilution to separate 50-ml volumetric flasks. Immerse the flasks in an ice-water bath and pipet 3.0 ml of 4 N potassium hydroxide into each flask. Remove the flasks from the bath and allow the contents to warm to room temperature. Pipet 3.0 ml of the 2,4-dinitrophenylhydrazine reagent into each flask, mix, and allow the flasks to stand at room temperature for 30 minutes unless otherwise specified in Table 1–6. Pipet 15 ml of the pyridine stabilizer into each flask. Transfer 3.0 ml of freshly prepared 10% methanolic potassium hydroxide into each flask and mix. Allow the flasks to stand at room temperature for 5 minutes and then filter the solutions through No. 40 Whatman filter paper. Collect the filtrates in separate 25-ml glass-stoppered graduated cylinders and allow the cylinders to stand for 10 minutes. Measure the absorbance of the sample versus the blank at 480 mμ in 1-cm cells. Read the concentration of the secondary alcohol from the calibration curve.

The optimum oxidation time selected for the determination of secondary alcohols in the presence of primary alcohols will depend on the time necessary to oxidize both types of alcohols to acids. In general the sample should not consume more than 85% of the dichromate ion.

Usually a separate calibration curve is necessary for each secondary alcohol determined. Ketones that contain a methyl group adjacent to the carbonyl group generally give the most sensitive color reactions, and the same effect was noted for the corresponding alcohols. The most sensitive color reaction was obtained from 2,5-hexanediol, the ketone of which contains a methyl group adjacent to each carbonyl.

The method is not successful for diols in which the hydroxyls are separated by fewer than four carbon atoms. Also, it is not applicable to cyclic secondary alcohols or to highly branched aliphatic alcohols that are oxidized to acids, not ketones. Most ketones interfere quantitatively. Any compound that is oxidized to a carbonyl compound and resists further oxidation will interfere in the method.

3. *Formation of Alkyl Halides of Alcohols.* Scoggins and Miller (5) used the reaction of tertiary alcohols and hydriodic acid to form tertiary alkyl iodides and water as the basis for a method to determine tertiary alcohols:

$$
\begin{array}{ccc}
\text{R}' & & \text{R}' \\
| & & | \\
\text{R—C—OH} + \text{HI} \rightleftharpoons \text{R—C—I} + \text{H}_2\text{O} & & (1) \\
| & & | \\
\text{R}'' & & \text{R}''
\end{array}
$$

Cyclohexane is used to remove the tertiary alkyl iodide as it forms and thereby force the reaction to completion. The organic iodide absorption in the ultraviolet region is then measured.

An analysis time of 15 minutes is claimed, with an average recovery of 100.1% reported for synthetic blends of 2-methyl-2-propanol in the 50- to 2000-ppm range. The ultraviolet absorbance of tertiary alkyl iodides is stable for several days.

High results are caused by propanone or other compounds that contain the CH_3CO grouping or form it by oxidation if the hydriodic acid contains free iodine. High concentrations of olefins interfere seriously, but this difficulty can be avoided by prior hydrogenation of the sample over platinum catalyst at room temperature and atmospheric pressure. Aromatic compounds if present interfere because of their ultraviolet absorptions.

METHOD FROM SCOGGINS AND MILLER (5)

Procedure.   Transfer a quantity of hydrocarbon sample containing not more than 0.08 mM of tertiary alcohol to a 40-ml screw-cap vial equipped with a polyethylene gasket in the cap and dilute to 25 ml with cyclohexane (for aqueous solutions weigh 100 to 200 mg of sample into the vial and add 25 ml of cyclohexane). Add 3 ml of hydriodic acid and shake for 3 minutes. Use either commercial hydriodic acid, 50 to 55%, or iodine-free hydriodic acid, 72%. Transfer the contents of the vial to a separatory funnel. Mix thoroughly, allow the phases to separate, and discard the aqueous phase. Add 10 ml of 1 M sodium hydroxide and 3 drops of hydrogen peroxide and shake until the iodine color disappears. Transfer a portion of the hydrocarbon phase to a 1-cm cell and measure the absorbance versus a reagent blank. Scan from 300 to 240 mμ. Use a previously prepared calibration curve to convert the absorbance to tertiary alcohol concentration.

To prepare the calibration curve, dilute from 0 to 0.1 mM of the appropriate tertiary alcohol to 25 ml with cyclohexane and proceed as described above. Plot the absorbance versus the millimoles of tertiary alcohol.

Analytical data obtained by the original authors are shown in Table 1-7.

4. Determination of Alcohols as Ferric Hydroxamate Esters after Acetylation. The ferric hydroxamate method for esters (see Chapter 3, Section I.C) can be used to determine trace amounts of alcohols after acetylation (6). The alcohol-containing sample is first acetylated under mild

Table 1-7. Analysis of Tertiary Alcohol-Cyclohexane Blends[a] (5)

Alcohol	Concentration, ppm Added	Found	Standard deviation[b]	Maximum mμ[c]	Molar absorptivity, liters/mole cm[d]
2-Methyl-2-propanol	489.2	457.1	25.1 (4)	269	591
2-Methyl-2-butanol[e]	424.6	380.9	23.8 (5)	267	602
2-Methyl-2-butyne-2-ol[f]	561.6	560.0	... (2)	...	...
2-Methyl-2-pentanol	571.2	560.7	19.7 (5)	268	613
3-Ethyl-3-pentanol	195.5	212.2	1.5 (4)	266	689
2,4-Dimethyl-4-hexanol	459.1	460.3	4.6 (5)	267	636
4-Methyl-4-heptanol	671.2	653.7	7.8 (3)	269	618
4-Methyl-4-octanol	740.4	763.8	8.1 (3)	269	654
4-Methyl-4-nonanol	214.6	188.4	7.4 (5)	267	585

[a] 2,4-Dimethyl-4-hexanol calibration used for analysis.
[b] Figures in parentheses represent numbers of determinations.
[c] Wavelength of maximum absorbance of corresponding alkyl iodide.
[d] Molar absorptivity of corresponding alkyl iodide.
[e] Freshly distilled before standard solutions prepared.
[f] Hydrogenated before analysis.

buffered acid-catalyzed conditions in pyridine at room temperature to an acetate. Then the excess anhydride is hydrolyzed with a minimum of water at room temperature. The acetate is converted with hydroxylamine to the anion of the corresponding hydroxamic acid in basic solution, and ferric perchlorate is added to the acidified solution to form the purple ferric chelate (7). The procedure of Gutnikov and Schenk (6) follows.

METHOD FROM GUTNIKOV AND SCHENK (6)

Reagents.   Acetylating reagent. Pipet 1 ml of 72% perchloric acid into a 50-ml volumetric flask and immerse in ice water. Slowly pipet 20 ml of pyridine down the sides into the flask and dilute to volume with acetic anhydride. Allow to come to room temperature before making the final dilution to volume with acetic anhydride. Prepare fresh daily.

Hydroxylamine perchlorate, 1.4 M. Dissolve 195 grams of sodium perchlorate in 400 ml of absolute methanol with heat. Heat and stir to dissolve 105 grams of reagent-grade hydroxylamine hydrochloride in 550 ml of absolute methanol. Stir during the slow addition of the methanolic sodium perchlorate to the latter solution. Add 50 ml of benzene and allow to come to room temperature. Then chill the solution and the precipitate for 1 hour in an ice bath. Filter the sodium chloride on a sintered glass filter and store the solution in a polyethylene bottle.

Stock ferric perchlorate solution. Dissolve 50 grams of anhydrous ferric perchlorate in 400 ml of absolute ethanol, add 40 ml of 72% perchloric acid. and dilute exactly to 500 ml with ethanol.

Ferric perchlorate reagent. Add 100 ml of the stock ferric perchlorate solution to about 1.7 liters of absolute ethanol. Slowly, with cooling, add 140 ml of 72% perchloric acid in small portions. Allow to cool and dilute to 2 liters with absolute ethanol.

Procedure.   Weigh the sample into a 10-ml or larger volumetric flask so that the solution contains 0.005 to 0.07 mM of hydroxyl per milliliter when diluted to volume with the acetylating reagent. Allow to stand 5 minutes or longer for hindered compounds. The ratio of the sample to the total volume at this point should be no more than 1:20 or 0.5 ml of sample to 10 ml total volume.

Pipet exactly 1 ml of this solution into a 50-ml volumetric flask (for the blank, use 1 ml of the acetylating reagent), add 0.5 ml of hydrolysis reagent (4:1, pyridine-water), and allow to stand at room temperature for 10 minutes. Pipet 3 ml of the 1.4 M hydroxylammonium perchlorate reagent into the flask and then add 10 ml of 2.5 M methanolic sodium hydroxide solution. After 20 minutes, add about 20 to 23 ml of a ferric perchlorate-acetone mixture, made for each sample from 2.5 ml of acetone and 35 ml of ferric perchlorate reagent. Allow to stand for 5 minutes. Dilute to volume with the same ferric perchlorate-acetone mixture, mix thoroughly, and read, at 524 mμ in 1-cm cells, the absorbance of the sample versus a blank treated in the same manner.

Weigh samples (each no more than 2 ml) containing low concentrations of hydroxyl groups directly into a 50-ml volumetric flask. Add exactly 1 ml of the acetylating reagent, allow to react for 10 minutes or longer, and then proceed with 0.5 ml of the hydrolysis reagent and the other reagents as directed.

Substances such as sugars that are insoluble in the acetylating reagent may be first dissolved in a minimum amount of water and then treated with the acetylating reagent as described.

Alcohols were determined in the presence of more than a hundredfold excess of an amine or water or in the presence of chloroform or tributyl phosphate. Beers' law is obeyed over the concentration range of 1.0 to 14 $\times$ 10^{-4} M calculated as the hydroxyl group in the final solution measured. The method is applicable to primary and secondary alcohols as well as to *tert*-butyl alcohol, polyhydroxyl compounds, sugars, mercaptans, and unhindered phenols. Primary and secondary amines that are not sterically hindered consume acetic anhydride preferentially to form substituted amides that react much more slowly than esters with the alkaline hydroxylamine reagent. Corrections are usually unnecessary for mixtures containing more than 10 meq % of alcohol. Aldehydes and ketones at the same concentrations as the hydroxyl groups interfere, probably because of acetylation of their enolate forms.

5. *Reaction of Alcohols and Isocyanates.* Phenyl isocyanate and most hydroxyl compounds, including sterically hindered and tertiary alcohols, react rapidly to form carbanilates (N-phenyl carbamates):

$$\text{ROH} + \text{C}_6\text{H}_5\text{NCO} \rightarrow \text{C}_6\text{H}_5\text{NHCOOR} \tag{2}$$

The carbanilates absorb strongly in the ultraviolet region. Malm and his coworkers (8) based a procedure for the determination of total hydroxyl in cellulose esters on these characteristics. They removed the excess phenyl isocyanate by precipitation of the carbanilate, dissolved the precipitate in methylene chloride-methanol (90:10, w/w), and determined the concentration of carbanilate from the ultraviolet absorption at 280 mμ. The original concentration of total hydroxyl was then calculated.

Isocyanates have an intense infrared absorption band near 4.4 μ, a region containing potentially few interfering absorptions. McGinn and Spaunburgh (9) followed the decrease in intensity of the isocyanate band as a function of time to determine the rates of reaction of aryl diisocyanates and polyesters to form polyurethanes.

Willeboordse and Critchfield (10) based a differential reaction rate method for the analysis of binary and ternary mixtures of alcohols on the reaction of the hydroxyl groups and phenyl isocyanate. The reaction was followed by the disappearance of the isocyanate band at 4.42 μ.

METHOD FROM WILLEBOORDSE AND CRITCHFIELD (10)

Reagents and Apparatus. Dimethylformamide. Redistill, or dry with Linde Molecular Sieve Type 5A, until the water content is less than 0.01%. Store in a capped bottle.

Toluene. Redistill from stannous octoate catalyst (100 grams per gallon of toluene). The water content should be less than 0.005%. Store in a capped bottle.

Phenyl isocyanate reagent. Add 10.0 ml (10.95 grams) of phenyl isocyanate to 50 ml of dry toluene, dilute to 100 ml with additional toluene, and mix thoroughly. Store in the volumetric flask. If any crystalline precipitate forms, discard the reagent.

Catalyst. Add 0.404 gram of stannous octoate to 50 ml of dry toluene, dilute to 100 ml with additional toluene, and mix thoroughly. Prepare fresh catalyst daily.

Infrared spectrophotometer. Beckman Model IR 5A or the equivalent with 0.10-mm calcium fluoride cells.

Calibration. Prepare standard solutions of phenyl isocyanate in anhydrous toluene-dimethylformamide (90:10, v/v) in the concentration range from 1 to 5 mM per 50 ml of solution. Determine the absorbance values at the 4.42-μ peak and plot this value versus isocyanate concentration. The graph should be a straight line.

Procedure. Transfer a sample containing 5 mM of hydroxyl to a 50-ml volumetric flask with toluene and dilute to almost 30 ml with toluene. Pipet 5 ml of dimethylformamide into the flask and then add 10 ml of the catalyst solution. Note the time as soon as 5 ml of the phenyl isocyanate solution (4.60 mM) has been added. Rapidly dilute the mixture to volume with toluene. Keep the volumetric flask in a constant-temperature bath maintained at 25°C. At 5-minute intervals, withdraw small aliquots (± 1 ml) with a hypodermic syringe, inject into the cell, and measure the absorbance at 4.42 μ. Calculate the isocyanate concentration values during the course of the analysis from the absorbance data with the aid of the calibration curve.

In accordance with the conventional graphical method (11) extrapolate to zero time the line representing the slope of the less reactive alcohol to obtain the amount of more reactive alcohol.

Mixtures of primary and secondary alcohols, of primary and tertiary alcohols, and of secondary and tertiary alcohols (Table 1–8) can be determined accurately. The last two series of data presented in Table 1–8 also show that mixtures of primary alcohols can be analyzed when the amount of more reactive alcohol is relatively small. The method is useful for the determination of the primary hydroxyl contents of polypropylene glycols.

Water reacts with phenyl isocyanate faster than with primary hydroxyl groups, and corrections can be applied on the basis that 1 mole of isocyanate is consumed by 1 mole of water. Primary and secondary amines interfere in this determination because of their comparable rates of reaction with phenyl isocyanate. Tertiary amines do not interfere; neither do carbonyl compounds, carboxylic acids, or acetals. Most aromatic hydroxyl compounds do not react with phenyl isocyanate.

6. Trityl Chloride Reaction. Primary hydroxyl groups in cellulose esters have been determined by tritylation and ultraviolet measurement of the triphenylcarbinol produced (8). The reaction may be written as follows:

$$ROH + TrCl + C_5H_5N \rightarrow ROTr + C_5H_5N \cdot HCl \qquad (3)$$

where Tr represents trityl (triphenylmethyl).

The absorbance at 259 mμ varies directly with the trityl content. Cellulose acetate has an average absorbance of 0.015 under the conditions used, and the absorbance of the trityl ether was corrected by this amount. At the end of the reaction the mixture was diluted with pyridine to precipitate the triphenylcarbinol. A portion of the precipitate was dissolved in methylene chloride-methanol (90:10, w/w) to make a 0.1% solution, and the absorbance read at 259 mμ.

Hendrickson (12) based a differential kinetics method for the determination of primary hydroxyl groups in polyglycols on the trityl chloride reaction. The disappearance of the hydroxyl band in the infrared spectrum at about 3.05 μ was monitored to follow the reaction.

Saunders, Schwarz, and Stewart (13) reported that, if trityl-containing compounds are treated with acid, a vivid yellow color absorbing at 420 mμ is produced. The yellow color is due to the formation of the triphenyl cation. Samples containing the trityl ether or N-tritylamino group are soluble in organic solvents. Aliquots of the solution are placed in test tubes, and the solvent is carefully removed with a stream of nitrogen. Then 10 ml of concentrated sulfuric acid is added, taking care to wash down the sides of the tubes, and the material mixed thoroughly. The color development is measured at maximum absorption, 420 mμ, against a blank of sulfuric acid. The color can be measured at any time. Triphenylcarbinol is used to determine a calibration curve. Linearity is observed to about 35 μg of trityl group.

Table 1-8. Analysis of Binary Mixtures of Primary, Secondary, and Tertiary Alcohols

Constituents		Faster-reacting alcohol, %	
Faster-reacting alcohols	Slower-reacting alcohols	Calculated	Found
1-Propanol	2-Propanol	12.8	12.9
		25.6	25.5
		38.3	37.9
		50.7	48.8[a]
		63.1	59.6[a]
1-Butanol	2-Butanol	12.5	12.4
		25.0	24.8
		37.5	37.0
1-Propanol	2-Butanol	20.1	20.0
		40.0	39.4
1-Butanol	2-Propanol	15.7	15.4
		32.8	32.5
1-Propanol	2-Methyl-2-propanol	13.1	13.0
		26.4	26.2
		39.7	39.3
1-Butanol	2-Methyl-2-propanol	12.5	12.6
		25.0	25.0
2-Propanol	2-Methyl-2-propanol	12.1	11.8
		24.0	23.3
2-Butanol	2-Methyl-2-propanol	12.5	12.3
		25.0	24.7
		37.5	36.9
1-Propanol	2-Ethylhexanol	13.1	13.3
		26.4	26.8
		39.7	40.6
1-Butanol	2-Ethylhexanol	12.5	13.0
		25.0	25.9

[a] The amount of primary alcohol relative to secondary alcohol is too large for an accurate representation of the secondary alcohol rate curve since this portion of the curve is near the end of the reaction.

7. Infrared Determination of the Acetate Group Formed on the Acetylation of Hydroxyl Groups. Kramm, Lamonte, and Mayer (14) determined the hydroxyl content of oxidized polyethylene by an infrared analysis of the recovered polymer after quantitative acetylation with acetic anhydride. The acetyl group content was measured by the use of the 8.03-μ C—O stretching vibration band typical of acetate esters, and this was equated to the hydroxyl content. The method was standardized by the infrared spectrometric examination of polymers for which the hydroxyl content had been previously determined by acetylation with acetic anhydride-^{14}C and subsequent radiochemical analysis.

8. Determination of Adjacent Hydroxyls. Periodate readily attacks adjacent hydroxyl groups but has no effect on monohydroxy compounds or polyhydroxy compounds in which the hydroxyl groups are separated by one or more carbon atoms:

$$\text{RCHOHCHOHR}' + \text{HIO}_4 \rightarrow \text{HIO}_3 + \text{RCHO} + \text{R}'\text{CHO} + \text{H}_2\text{O} \qquad (4)$$

Malaprade (15) first reported the selective oxidizing power of periodate. Dal Nogare, Norris, and Mitchell (16) used reaction 4 as a basis for a method to determine 1,2-propylene glycol in the presence of ethylene glycol. The glycols are oxidized to the aldehydes, and the acetaldehyde that is produced by only the 1,2-propylene glycol is converted to iodoform and determined spectrometrically.

METHOD FROM DAL NOGARE, NORRIS, AND MITCHELL (16)

Procedure. Dilute a sample of the neutral mixture weighing 2 to 4 grams to 250 ml with water in a volumetric flask. Mix, transfer a 5-ml portion of the sample solution to a 100-ml distilling flask fitted with an efficient condenser, and add 45 ml of water. Add 0.5 gram of periodic acid just before distillation. Begin the distillation immediately and continue at the rate of 3 to 5 ml/minute until only about 1 ml of solution remains in the flask. Avoid evaporation to dryness. Collect the distillate in a 50-ml volumetric flask submerged in an ice bath. If necessary to minimize losses, modify the tip of the condenser to extend into the body of the receiver flask. When the distillation is complete, warm the distillate to room temperature and adjust to volume. Analyze aliquots containing 0.1 to 0.5 mg of acetaldehyde by the iodoform procedure of Dal Nogare, Norris, and Mitchell (16), described in Chapter 2, pp. 80–81. Results obtained by the original authors using this procedure are shown in Table 1–9.

The observation that acetaldehyde has a relatively strong absorption peak at 277 mμ but that formaldehyde is virtually transparent in the ultraviolet

Table 1-9.   Analysis of Samples Containing 1,2-Propylene Glycol (16)

| Composition of sample, weight % | | Acetaldehyde, mg | | Acetaldehyde recovery, % |
Propylene glycol	Ethylene glycol	Calculated	Found	
95.3[a]	0.0	0.521	0.494	94.8
		0.475	0.467	98.3
18.1	74.3	0.232	0.245	105.6
		0.269	0.261	97.1
7.9	83.7	0.103	0.097	94.2
		0.127	0.127	100.0
1.0	97.0	0.155	0.125	80.7
		0.155	0.131	84.5
0.55	97.5[a]	0.078	0.074	94.9
		0.078	0.072	92.4

[a] Remainder principally water and nonvicinal glycols.

region led Baumel (17) to develop a method incorporating the ultraviolet end determination after periodate oxidation and distillation of the resulting aldehydes. A difficulty encountered in this scheme was the evolution of iodine during the distillation of the aldehydes. Iodine and iodine-containing acids formed by iodine and water absorb strongly in the ultraviolet region and interfere with the acetaldehyde determination. Baumel postulated that the iodine probably results from an iodic acid reaction with propylene glycol to liberate hydriodic acid, which in turn reacts with iodic acid and periodic acid to yield iodine. He found that, if the reaction is run in very dilute aqueous solution, the iodic acid reaction is not favored and the aldehydes distill in a clear form.

METHOD FROM BAUMEL (17)

Apparatus.   Spectrophotometer. Beckman Model DU with a hydrogen discharge lamp and 1-cm silica absorption cells.

Distillation assembly. A 1-liter round-bottom flask with a 75° adapter connected to a water-cooled condenser, and a 105° condenser discharge adapter connected to a 10-inch length of 4-mm glass tubing, which dips to the bottom of a 50-ml Nessler tube.

Procedure. Weigh about 5 grams of the glycol mixture and dilute to 250 ml with water. Pipet 50 ml of this solution into the 1-liter round-bottom flask. Add 300 ml of water, 100 ml of periodic acid solution (45 grams/liter), and a few glass beads and immediately connect the flask to the condenser adapter. Collect the distillate in a 50-ml Nessler tube containing 10 ml of water, the Nessler tube being half-immersed in an ice bath. Lower the tip of the 4-mm distillate discharge tube below the surface of the water in the Nessler tube. Heat the flask and distill until the liquid in the Nessler tube is almost to the 50-ml mark. Raise the tip above the surface of the distillate and allow it to come to full volume. Insert a one-hole rubber stopper with a thermometer into the Nessler tube and allow the temperature of the sample to rise to the temperature of the sample cell compartment of the spectrophotometer. Read the absorbance at 277 mμ and compare with a calibration curve prepared using known amounts of pure 1,2-propylene glycol.

Table 1–10 shows results obtained by Baumel for the analysis of synthetic mixtures containing ethylene glycol, propylene glycol, and 2.5% borax.

Table 1-10. Analysis of Synthetic Mixtures Containing Ethylene Glycol, 1,2-Propylene Glycol, and Borax (17)

1,2-Propylene glycol, %	
In sample	Average found[a]
3 08	3.08 (4)
9.95	10.04 (4)
14.70	14.75 (4)
19.40	19.37 (3)

[a] Numbers in parentheses indicate numbers of determinations.

Leibman and Ortiz (18) presented a more general method involving periodate oxidation for the determination of neutral glycols. The method can be applied with appropriate modifications to a number of different dihydroxy compounds.

METHOD FROM LEIBMAN AND ORTIZ (18)

Procedure. Add to a 15-ml glass-stoppered centrifuge tube, in the order given, the following: 2 ml of the aqueous sample solution, 1 ml of 10 N

sulfuric acid, and 1 ml of 0.1 M sodium periodate. Mix thoroughly after each addition. Keep the tube unstoppered at an appropriate temperature for a definite period of time (Reaction A, Table 1–11). If the temperature is above ambient, place the tube in an ice bath for 5 minutes at the end of the reaction. Add 0.5 ml of thioacetamide solution (0.867 M—dissolve 650 mg of thioacetamide in distilled water and dilute to 10 ml; prepare fresh daily).

Table 1-11. Conditions for the Specific Application of the Method of Leibman and Ortiz (18)

Glycol	Reaction A[a], temperature	Absorbance[b]
cis-1,2-Dihydroxyindane	80°C	0.417 ± 0.011
trans-1,2-Dihydroxyindane	80°C	0.416 ± 0.014
cis-1,2-Dihydroxycyclohexane	Room	0.705 ± 0.011
trans-1,2-Dihydroxycyclohexane	Room	0.700 ± 0.011
Phenylethylene glycol	Room	0.632 ± 0.021

[a] Reaction A was run for 60 minutes and Reaction B at room temperature for 45 minutes in each case.
[b] Absorbance at 375 mμ obtained when the method was applied to 100 μM of glycol. The wavelength of maximum absorption in all these cases was 375 mμ.

After 5 to 10 minutes at room temperature, mix the contents of the tube thoroughly for 30 seconds with a vibrating mixer. Add 0.5 ml of 2,4-dinitrophenylhydrazine solution (100 mg of 2,4-dinitrophenylhydrazine in 100 ml of 2 N hydrochloric acid), mix, and keep the tube unstoppered for a measured time at a certain temperature (Reaction B). Add 5 ml of chloroform, stopper the tube, and shake vigorously several times, allowing the liquid to drain down and venting the pressure between each shaking. Centrifuge the tube at about 1000 × g for 5 minutes. Sulfur usually collects at the interface as well as at the bottom of the chloroform layer. With a capillary connected by way of a suction flask to a vacuum source, aspirate the aqueous layer and, as much as possible, the material at the interface. Introduce a 3-ml pipet into the chloroform layer; use a Propipette controller with appropriate manipulation of the E port and bulb to maintain a positive pressure in the pipet and thus prevent entry of any residual water or interface material into the pipet. After wiping the exterior of the pipet carefully with tissue, transfer a sample of the chloroform extract to a 1-cm cuvette and measure the absorbance at an appropriate wavelength.

To obtain a blank, apply the identical procedure to 2 ml of a solution similar to the experimental sample but containing no glycol. Compare the absorbance to those measured on extracts obtained in a like manner from solutions of known concentrations of a reference glycol.

The specific conditions for application of the method are shown in Table 1-11.

Dixon and Lipkin (19) demonstrated that the consumption of periodate by vicinal hydroxyl groups may be followed spectrometrically on the basis of the absorption band of metaperiodate, which has a maximum at about 223 μm. The determination was successful for 10^{-8} to 10^{-6} mole of samples of ribofuranosides.

D. Phenols

The phenolic groups and the phenolate ion in particular exhibit strong electron-donating effects that make the simple phenolic compounds readily susceptible to substitution by electrophilic agents such as the nitroso group and diazonium salts. Substitution is directed para or ortho to the hydroxyl group. Many of the substituted phenols are highly colored, and the substitution reactions can be used for their analysis.

1. Azo Dye Formation. Diazonium salt substitution to form a colored azo dye is the basis for one of the most generally applicable methods for the determination of trace quantities of phenols:

$$\langle\bigcirc\rangle - OH \quad + \quad ArN \equiv NCl \quad \longrightarrow \quad ArN = N - \langle\bigcirc\rangle - OH \quad + \quad HCl \qquad (5)$$

The two aromatic amines most commonly used for the formation of the diazonium salt are sulfanilic acid and p-nitroaniline. The coupling takes place in the para position of the phenol if this position is unoccupied; otherwise, the coupling may take place at the ortho position, although at a much slower rate. If both the ortho and para positions are occupied, coupling ordinarily does not occur. Sources of interference include aromatic amines, many of which may be nitrosated by any excess nitrous acid in the mixture. The decomposition products of diazonium compounds are generally colored; therefore, material present in the reaction solution such as metallic salts that accelerate the decomposition can cause color to form and thus interfere in the determination.

DeMeis (20) used diazotized p-nitroaniline to determine phenol in animal tissues. The color obtained was measured in 1-cm cells at 500 mμ. Beer's law was obeyed for up to 50 μg of phenol. Takeucki, Furusawa, and Takayama

(21) determined phenol in methyl methacrylate by treating an aqueous solution with diazotized *p*-nitroaniline, piperidine, and sodium hydroxide and measuring the orange color at 460 mμ. Smith and King (22) determined steam-volatile phenols present in cigarette smoke condensate by the use of diazotized *p*-nitroaniline. A wavelength of 490 mμ was used for aqueous solutions. Later the same workers (23) recommended an extraction of the acidified solution with carbon tetrachloride and measurement of the dye at 365 mμ. Rud' and Skochilova (24) presented a method for small amounts of phenols in hydrocarbons based on the color formed with diazotized *p*-nitroaniline. They extracted 100 to 200 grams of the sample three times with 10-ml portions of 1% sodium hydroxide solution, combined the extracts, and diluted them to 100 ml. A 50-ml portion of the diluted extracts was mixed with 1 ml of a fresh solution of the diazonium salt, prepared by mixing in the cold equal volumes of 0.1% *p*-nitroaniline hydrochloride and 0.05% sodium nitrite solution. Satisfactory results were claimed over the range 0.001% to 0.01%.

Fox and Gauge (25) used diazotized sulfanilic acid to determine phenols. This method, as adapted by Siggia (26), is as follows.

METHOD FROM FOX AND GAUGE (25) AND SIGGIA (26)

To diazotize sulfanilic acid, add 1 part of 1:3 sulfuric acid solution to 5 parts of sulfanilic acid solution (7.6 grams/liter), followed by 5 parts of sodium nitrite solution (3.4 grams/liter). Prepare the solution of diazotized sulfanilic acid fresh 5 minutes before use and keep it cold to minimize decomposition.

Dissolve the sample in water or in the weakest sodium hydroxide solution that is sufficient to produce solution. If the sample is a complex mixture, insoluble in water, dissolve it in a water-immiscible solvent such as benzene, petroleum ether, or carbon tetrachloride and extract the solution with aqueous caustic. Use the aqueous layer containing the phenolate for the analysis.

To the sample solution or aqueous extract, add 8% sodium hydroxide solution in the ratio of 5 ml of 8% sodium hydroxide solution per 100 ml of sample solution. Treat standards and sample in the same manner. Add enough diazotized sulfanilic acid to obtain optimum color development. Determine the amounts of reactants and the optimum reaction time for each type of sample.

The final solution must be alkaline if the coupling reaction is to proceed satisfactorily. Measure the color and compare with standards, either visually or preferably in a spectrophotometer. Less than 1 ppm cresols, phenols, and naphthols can be determined.

2. Formation of Indophenol Dyes. Gibbs (27) developed qualitative and quantitative tests for as little as 5 μg/ml of phenols, using 2,6-dichloro- and 2,6-dibromo-N-chloroquinonimines to produce the 2,6-dihaloindophenol dyes (equation 6):

$$O={\bigcirc}=NCl \;+\; {\bigcirc}-OH \xrightarrow[\text{buffer}]{\text{alkaline}} O={\bigcirc}=N-{\bigcirc}-OH \;+\; HCl \qquad (6)$$

Gibbs carried out the reaction in buffered solutions at pH 9.1 to 9.5. To secure greater sensitivity, Ettinger and Ruchhoft (28) recommended extraction of the dye with n-butanol. There is a shift in the light absorption characteristics when the dye is extracted with the alcohol, as shown in Table 1-12. Complete color development requires up to 24 hours. 1-Naphthol and thymol react rapidly, but phenol and cresols require 6 to 24 hours. p-Cresol does not react with the Gibbs reagent.

Table 1-12.   Wavelength, mμ, of Maximum Absorption (28)

Compound	Aqueous solution	n-Butanol extract
Phenol	610–630	670
o-Cresol	600	660
m-Cresol	610–620	660–670
1-Naphthol	580–600	640
o-Chlorophenol	650–670	680–700
p-Chlorophenol[a]	600–620	670

[a] Presumably yields the same indophenol as phenol. The absorption curves are almost identical.

METHOD FROM ETTINGER AND RUCHHOFT (28)

Reagents.   Borate buffer. Dilute 3.1 grams of boric acid, 3.5 grams of potassium chloride, and 32 ml of 1 N sodium hydroxide to 1 liter. Make up 5 ml of this solution to 100 ml and, if the pH is not 9.4, add more sodium hydroxide until 5 ml of the buffer solution produces a pH of 9.4 when made up to 100 ml.

Gibbs's reagent. Dissolve 0.2 gram of 2,6-dibromoquinone chlorimide in 50 ml of ethanol. Filter any residue. Although this reagent should be used immediately, if refrigerated it may be used for periods up to 1 week. Dilute

4.5 ml of the alcoholic concentrate to 100 ml and use the dilution immediately or at least within 30 minutes.

Procedure.   To a 500-ml sample add 0.7 ml of 10% phosphoric acid, distill 400 ml rapidly in an all-glass still, and then distill 50 ml more slowly. Stop the distillation, add 50 ml of water, and distill 50 ml more. Chill the distillate if its temperature is over 20°C. Use a series of aliquots, one of which should contain 50 to 100 ppb of the phenolic material or 300 ml if the phenolic material totals less than 100 ppb. If other than a 300-ml sample is used, dilute to 300 ml. Add 15 ml of borate buffer, mix well, and add 5 ml of the dilute Gibbs reagent. Mix thoroughly and allow to stand for 6 to 24 hours. Add 75 ml of *n*-butanol and shake well. Separate the alcohol in a separatory funnel and filter to clarify. Dilute the filtrate to 60 ml with *n*-butanol. Determine the absorbance of the extract at the appropriate wavelength indicated in Table 1-12. Read the sample absorbance against a blank.

Mohler and Jacob (29) compared this method with the 4-aminoantipyrine method and decided that the latter was faster, more accurate, and more precise. Bozhevol'nov (30) claimed that the Gibbs reaction was more specific than one based on diazotized sulfanilic acid and that 1- and 2-naphthol, quinol, pyrogallol, pyridine, and *p*-cresol do not interfere in the determination of phenol. Feigl, Anger, and Mittermann (31) studied the reaction for qualitative use and reported that all meta-substituted phenols give a positive reaction but that most phenols with ortho substituents containing a C=O or an N=O group (e.g., *o*-nitrophenol, salicylic acid, methyl salicylate, and *p*-hydroxybenzaldehyde) are unreactive. Salicylamine and salicylanilide, however, give positive reactions because the ortho substituent is present as the imido tautomer, in which there is no C=O groups.

Guilbault, Kramer, and Hackley (32) used *N*-(benzenesulfonyl)quinonine (equation 7) and a kinetic technique to eliminate the long reaction time delay involved in the usual Gibbs reaction:

METHOD FROM GUILBAULT, KRAMER, AND HACKLEY (32)

Procedure. Add 1 ml of 1.5 N ammonium hydroxide to 1 ml of a 0.0005 M methyl Cellosolve solution of N-(benzenesulfonyl)quinonine in a 1.0-cm spectrophotometer cell. Adjust the absorbance to read zero at the appropriate wavelength (Table 1-13). At zero time add 0.5 ml of the solution of phenol to be analyzed (1 to 100 μg) and automatically record the rate of change in absorbance of the solution with time (a Beckman Model D spectrophotometer was used). Calculate the amount of phenol present in the solution from a calibration plot of change in absorbance versus time for a standard phenol sample.

Table 1-13. Amount Determinable, Lowest Detectable Concentration, and λ_{max} of Various Phenols (32)

Compound	Amount determinable, μg/ml	Lowest detectable concentration, μg/ml	λ_{max}, mμ
Phenol	1–100	1.0	630
o-Chlorophenol	10–250	10.0	640
2,3-Dimethylphenol	3–300	3.0	610
m-Aminophenol	1–50	0.8	640
5-Amino-1-naphthol	5–100	5.0	620
1-Naphthol	1–100	0.6	650
2-Naphthol	5–100	5.0	620

This procedure was used to determine phenol, o-chlorophenol, 2,3-dimethylphenol, m-aminophenol, and 2-naphthol. For 5-amino-1-naphthol and 1-naphthol it was recommended that the absorbance be read after 3 minutes and the amount of phenol be calculated from calibration data plots of absorbance versus concentration.

The amounts of phenols which can be determined with a standard deviation of about $\pm1.3\%$ are shown in Table 1-13. The wavelengths of the maxima and the lowest detectable concentrations of the various phenols are also given.

3. Formation of Antipyrine Dyes.   The formation of antipyrine dyes as a result of the condensation of 4-aminoantipyrine with phenols in the presence of an alkaline oxidizing agent (33) (equation 8) has been used extensively as the basis for colorimetric methods to detect trace amounts of phenols (29, 34–40):

(8)

The reaction is applicable to phenolic compounds in which the para position is not blocked by an alkyl, aryl, nitro, benzoyl, nitroso, or carbonyl group. Because groups such as halogen, carboxyl, sulfonic acid, hydroxyl, and methoxyl are removed under the reaction conditions normally used, these para-substituted phenols will react. Enol-keto systems interfere.

Optimum light absorption of the aqueous solutions of the colored products has been found at 510 mμ. The sensitivity is greatly increased by extractive concentration of the reaction products with chloroform (33). A wavelength of 460 mμ is recommended for the chloroform solutions. The red aqueous reaction mixtures are not very stable and should be read within 30 minutes. The chloroform extracts, on the other hand, are stable up to 3 hours.

The dye formed with pentachlorophenol is blue in contrast to the orange or red colors formed by phenol and the lower chlorophenols (38). Only the blue color produced by pentachlorophenol and 2,3,5,6-tetrachlorophenol can be extracted quantitatively into benzene; the red phenazone dye is in- soluble, and the dyes produced by the lower chlorophenols except 2,3,5,6- tetrachlorophenol are only partly soluble. Bence (38) presented a method for the determination of pentachlorophenol in air based on these charac- teristics. A wavelength of 589 mμ was used. The method detects down to 0.15 μg of pentachlorophenol per liter of air. The antipyrine dyes formed with catechol and possibly resorcinol are sufficiently acidic so that they are not extracted from the aqueous phase with chloroform. Rosenblatt, Demek, and Epstein (39) took advantage of these properties to determine small amounts of guaiacol, a monophenol, in the presence of catechol, a poly- phenol.

The procedure of Ettinger, Ruchhoft, and Lishka (36) is fairly typical and includes the chloroform extraction.

METHOD FROM ETTINGER, RUCHHOFT, AND LISHKA (36)

Procedure. Add to a 300-ml portion of the sample sufficient concentrated ammonium hydroxide to produce a pH in the final reaction mixture of 9.8 to 10.2. Mix well, add 2 ml of 3.0% aqueous aminopyrine solution, and mix again. Then add 20 ml of 2.0% aqueous potassium ferricyanide solution. Start the extractions with chloroform after 3 minutes. Use 25 ml of chloroform for the first extraction and 15 ml for each of the two succeeding extractions. Combine the three extracts, make up to 50 ml with chloroform, and compare with a corresponding reagent blank at 460 mμ. Compare the absorbance obtained with a calibration curve prepared using known concentrations of the phenol.

4. Nitrous Acid Reaction. Stoughton (41) devised a method that depends on the treatment of the phenol with nitric acid and sulfuric acid at about 100°C to form a nitrosophenol which rearranges in the presence of excess alcoholic ammonium hydroxide to form the highly colored quinoid salt. The resulting colored solutions were compared with those from standard phenols by a visual colorimeter. Westlaufer, Van Natta, and Quattlebaum (42) applied a modified variation to determine traces of phenols in hydrocarbon solvents. The phenol was extracted first with a small portion of dilute caustic solution.

Lykken, Treseder, and Zahn (43) also used an extraction procedure; however, sodium nitrite was substituted for the nitric acid as the source of nitrous acid and the reaction was carried out at room temperature.

The reactions involved are as follows:

$$RC_6H_4OH \quad + \quad HONO \quad \xrightarrow{H_2SO_4} \quad RC_6H_3\overset{\displaystyle OH}{\underset{\displaystyle NO}{\big|}} \quad + \quad H_2O \qquad (9)$$

$$RC_6H_3\overset{\displaystyle OH}{\underset{\displaystyle NO}{\big|}} \quad \xrightarrow[\text{NH}_4\text{OH}]{\text{alcoholic}} \quad \left[RC_6H_3\overset{\displaystyle O}{\underset{\displaystyle NO}{\big\|}} \right]^- \quad + \quad H^+ \qquad (10)$$

To detect lower concentrations in water solution the phenolics were extracted from a slightly acidified sample with ether. After evaporation the phenolic residue was dissolved in the reaction solvent. The light absorption of the greenish yellow reaction product was measured at 420 mμ.

METHOD FROM LYKKEN, TRESEDER, AND ZAHN (43)

Procedure. Introduce 10 ml of 10% potassium hydroxide into a 125-ml separatory funnel and add the volumes of sample and octane indicated in Table 1-14. Shake 5 minutes, allow the phases to separate, and remove the lower layer into a 100-ml volumetric flask. Add a 5-ml portion of potassium hydroxide to the separatory funnel and repeat the extraction. Again repeat the extraction with 5 ml of water. Combine the extracts in the 100-ml volumetric flask.

Table 1-14.   Volumes of Sample and Solvent Recommended

Phenol content, mg/100 ml	Volume of sample, ml	Volume of octane, ml
0–100	50 ± 0.2	0
50–200	25 ± 0.1	25
100–500	10 ± 0.1	40
200–1000	5 ± 0.05	45

To the combined extracts add glacial acetic acid slowly with cooling of the flask in ice water to make a total volume of exactly 100 ml at room temperature. Pipet a 1- to 5-ml aliquot containing 0.05 to 0.5 mg of phenol into a 50-ml volumetric flask. If necessary, add sufficient acetic acid buffer solution (800 ml of acetic acid, 150 ml of 10% potassium hydroxide, and 50 ml of water) to make a total volume of 5 ml. With a medicine dropper add 5 drops of concentrated sulfuric acid and 2 drops of saturated sodium nitrite solution. Mix and allow to stand for 15 to 30 minutes (30 to 45 minutes for phenol); then slowly add alcoholic ammonium hydroxide solution (450 ml of anhydrous ethanol or isopropanol, 300 ml of 14 N ammonium hydroxide, and 250 ml of water), while cooling the flask in ice water, to make a total volume of exactly 50 ml at room temperature. Allow the solution to stand for 1 hour or preferably overnight and read the absorbance at 420 mμ. Read the corresponding concentration from a calibration curve prepared using known amounts of phenol.

A precision of better than $\pm 1\%$ was claimed, and the systematic error was reported as less than 2% of the actual concentration. Aniline and xylidine interfere.

5. Decoloration of Diphenylpicrylhydrazyl by Phenols.   Diphenyl-picrylhydrazyl has been recommended as a colorimetric analytical reagent for phenols (44). Diphenylpicrylhydrazyl is an intense, violet-colored, stable free radical which becomes decolored on reaction with phenols; the decrease in violet color serves as a measure of the quantity of phenol present. Hogg, Lohmann, and Russel (45) formulated the reaction as shown in equation 11:

(11)

METHOD FROM PAPARIELLO AND JANISH (44)

Procedure. Pipet 4 ml of phenol sample (2×10^{-2} to 2×10^{-5} mM/ml), 1 ml of methanol, 1 ml of acetate buffer solution (aqueous 1 N acetate adjusted to pH 5.0), and 4 ml of diphenylpicrylhydrazyl reagent (2×10^{-4} mM/ml of methanol) into a glass-stoppered test tube. In another test tube prepare a reagent blank with 5 ml of methanol, 1 ml of buffer, and 4 ml of reagent. Mix and allow the test tubes to stand at room temperature or at 60°C for 15 minutes to 1 hour. (The time and temperature of reaction depend on the rate of reactivity of the particular phenol being considered.) Measure the absorbance of the sample and of the reagent blank against methanol at 515 mμ in 1-cm cells in a suitable spectrophotometer. Read the concentration corresponding to the difference between the absorbance values of the blank and the sample on a calibration curve prepared from known concentrations of the same phenol.

Both aromatic amines and mercaptans are reactive with diphenylpicrylhydrazyl and interfere if present. Recovery results shown by Papariello and Janish for phenol, 1-naphthol, 2-naphthol, 4-hexylresorcinol, eugenol, hydroquinone, and diethylstilbestrol were all essentially 100%.

6. Halogenation of Phenols. To overcome interferences in the determination of phenols based on infrared measurements, the absorption maximum at 2.79μ due to O—H vibration can be shifted to 2.84 μ by bromination (46). Intramolecular hydrogen bonding between the hydrogen of the hydroxyl group and the bromine atom in a position adjacent to the hydroxyl group causes the shift. Therefore, the shift is constant for phenols for which substitution is ortho to the hydroxyl group. No response can be expected from phenols which are already substituted in both the 2 and the 6 positions, or in which steric hindrance prevents substitution of bromine in either of these positions. Organic acids show a small amount of infrared absorption at 2.84 μ, and it is recommended that any acids present be removed from the carbon tetrachloride extract with sodium bicarbonate solution.

METHOD FROM SIMARD ET AL. (46)

Add 100 grams of potassium bromide solution (12 grams/liter) and 80 ml of 1:3 hydrochloric acid solution for each liter of sample contained in a separatory funnel. Shake for 5 minutes. Then add 30 ml of 10% sodium thiosulfate solution for each liter of sample and also add the volume of carbon tetrachloride indicated in Table 1-15. Shake for 15 minutes. Drain off 25 ml of the carbon tetrachloride layer and shake vigorously for 5 minutes in a separatory funnel with 125 ml of a 2% solution of sodium bicarbonate. Withdraw the carbon tetrachloride layer, filter once through paper to remove suspended water, and scan for phenols from 2.6 to 2.9 μ.

Table 1-15.   Carbon Tetrachloride Volume for Sample Concentration (46)

Sample size, liters	Minimum detectable concentration, ppm		
	Carbon tetrachloride used, ml:		
	20	60	100
1	0.01	0.03	0.05
2	0.005	0.015	0.025
3	0.003	0.01	0.02

Calibrate the spectrophotometer for phenols at 2.84 μ with a series of known concentrations of 2,4,6-tribromophenol in carbon tetrachloride.

If a m-dihydroxyphenol is iodinated in the presence of an o-dihydroxyphenol, a dark-colored addition compound is formed. This precipitate dissolves on the addition of an equal volume of acetone to give a grape-blue color. The intensity of this color is proportional to either the m- or the o-dihydroxyphenol, depending on which is present in the lesser amount. The optimum concentration range is 0 to 50 ppm. Of 25 phenols tested by

Willard and Wooten (47) only the *o*- and *m*-dihydroxyphenols gave this reaction.

METHOD FROM WILLARD AND WOOTEN (47)

Buffer. Acetic acid-sodium acetate, molar in acetate ion. For the determination of resorcinol or catechol, the pH should be 5.7; for the determination of phloroglucinol, 6.0.

Procedure for Resorcinol. Take a neutral sample of no more than 15 ml containing no more than 0.75 mg of resorcinol and add 10 ml of 0.05% catechol solution and 15 ml of 0.1 N iodine solution. After 1 minute titrate the excess iodine with 0.1 N thiosulfate solution and starch. Transfer the sample to a 100-ml volumetric flask, add 50 ml of acetone to dissolve the precipitate, and dilute to volume with water. Measure the absorbance at 725 mμ. Refer to a standard curve for the resorcinol content.

Procedure for Catechol. Follow the same procedure but add 0.05% resorcinol rather than catechol. Although the product is the same in both cases, separate standardization curves are necessary.

Procedure for Phloroglucinol. Follow the resorcinol procedure but use a buffer of pH 6.0. A new standardization curve is required. At least twice the theoretical amount of catechol must be present if the reaction is to be complete in 1 minute.

Resorcinol in the presence of 50 times as much of each of the following phenols was determined by this method with an error of less than 1%: *o*-cresol, *m*-cresol, *p*-cresol, phenol, *o*-phenylphenol, *p*-phenylphenol, *o*-tert-butylphenol, *p*-tert-butylphenol, *p*-tert-amylphenol, salicylaldehyde, *p*-hydroxybenzaldehyde, *m*-hydroxybenzoic acid, *p*-hydroxybenzoic acid, salicylic acid, *p*-aminophenol, *m*-aminophenol, *o*-aminophenol, *o*-nitrophenol, and *m*-nitrophenol. Hydroquinone interferes in large excess but not when it is present in the same order of magnitude as the resorcinol.

II. GAS CHROMATOGRAPHIC METHODS*

Compounds containing hydroxyl groups tend to tail in gas chromatography and to give poor quantitative data, mainly because their adsorption by the support, which is often largely irreversible, cannot be entirely eliminated. These defects can usually be overcome (and quantitative analysis made possible) by converting the alcohol to a suitable derivative (see discus-

* Written by Morton Beroza and May N. Inscoe.

sion on derivatization in Appendix). The great variety of alcohols and the many substrates in which they are found have led to the widespread use of a large number of derivatives, each with specific conditions for providing optimum performance in a given situation.

Most derivatives of the hydroxyl group used for quantification are esters or ethers. Steric factors play a significant role in derivatization; for example, tertiary alcohols cannot be derivatized quantitatively by procedures that are satisfactory for primary and secondary alcohols.

A. Esterification for Gas Chromatography

A listing of typical reactions used to esterify hydroxyl compounds is given in Table 1-16. Anhydrides with acid and basic catalysts and acid chlorides

Table 1-16. Typical Esterification Reactions Used for Quantitative Determinations of OH Compounds by Gas Chromatography

Ester	Typical reagents	Types of compounds and references
Acetate	Acetic anhydride + pyridine or acidic catalyst	Antipyretics,[1] fatty alcohols and other hydroxy compounds,[2] hydroxy aldehydes,[3] phenols,[4] pantothenates,[5] sterols,[6,7] monosaccharides as their alditols[8]
Chloroacetate	Chloroacetyl chloride + aq. NaOH	Phenols,[9] sterols[10]
Trifluoroacetate	Trifluoroacetic anhydride; trifluoracetic anhydride + Na trifluoroacetate	Alcohols in wax,[11] OH compounds and phenols,[12] panthothenates,[5] sugars[13]
Heptafluorobutyrate	Heptafluorobutyric anhydride	Sterols,[14-19] vitamins[20]
Pentafluorobenzoate	Pentafluorobenzoyl chloride	Alcohols and phenols[21]
Hexadecafluoro-nonanoate	9-H-Hexadecafluorononanoyl chloride + pyridine	Sterol[22]
Benzoate	Benzoyl chloride + aq. NaOH	Glycerol[23]
3,5-Dinitrobenzoate	3,5-Dinitrobenzoyl chloride	C_1–C_{12} aliphatic and several aromatic alcohols[24]

[1] T. Hattori, S. Kawai, and M. Nishiumi, *Bunseki Kagaku*, **14**, 586 (1965).

with acid acceptors are preferred over the classical carboxylic acid and alcohol plus catalyst because they react more rapidly and completely and apparatus requirements are simple. Since esterification reagents usually react with enols and amino compounds, these compounds can be expected to interfere with the analysis of alcohols and phenols in most instances unless efforts are made to remove them. Their inclusion in the analysis may not be harmful, however; for example, if the derivative of a potential interferer has a retention time (t_R) different from that of the compound of interest, it can be ignored.

A number of fluorinated derivatives have been utilized to increase the volatility of OH compounds and also to improve the sensitivity of analysis through the use of electron capture detection.

B.   Ether Formation for Gas Chromatography

Typical procedures for forming ethers of OH compounds are listed in Table 1-17.

[2] A. Prevot and C. Barbati, *Rev. Franc. Corps Gras*, **15**, 157 (1968).

[3] L. M. Andronov and Yu. D. Norikov, *Zh. Analit. Khim.*, **20**, 1007 (1965).

[4] D. B. Katague and C. A. Anderson, Jr., *J. Agr. Food Chem.*, **14**, 505 (1966).

[5] A. R. Prosser and A. J. Sheppard, *J. Pharm. Sci.*, **58**, 718 (1969).

[6] D. M. Lawrence, *J. Chromatog.*, **36**, 344 (1968).

[7] J. W. Goldzieher, C. Matthijssen, C. Gual, B. A. Vela, and A. de la Pena, *Am. J. Obstet. Gynecol.*, **98**, 759 (1967).

[8] J. S. Sawardeker, J. H. Sloneker, and A. Jeanes, *Anal. Chem.*, **37**, 1602 (1965).

[9] W.-T. Chin, T. E. Cullen, and R. P. Stanovick, *J. Gas Chromatog.*, **6**, 248 (1968).

[10] R. A. Landowne and S. R. Lipsky, *Anal. Chem.*, **35**, 532 (1963).

[11] R. Kleiman, F. R. Earle, and I. A. Wolff, *J. Am. Oil Chemists' Soc.*, **46**, 505 (1969).

[12] S. Wilk, S. E. Gitlow, and D. D. Clarke, *Abstracts of Papers*, American Chemical Society, 2nd Middle Atlantic Regional Meeting, New York City, February 1967, p. 48.

[13] M. Vilkas, H.-I.-Jan, G. Boussac, and M.-C. Bonnard, *Tetrahedron Letters*, **14**, 1441 (1966).

[14] D. Exley, *Biochem. J.*, **107**, 285 (1968).

[15] I. R. Sarda, P. E. Pochi, J. S. Strauss, and H. H. Wotiz, *Steroids*, **12**, 607 (1968).

[16] K. Honda and S. Kushinsky, *Mikrochim. Acta*, **1969**, 1182.

[17] A. K. Munson, J. R. Mueller, and M. E. Yannone, *Biochem. Med.*, **3**, 187 (1969).

[18] G. L. Nicolis and J. L. Gabrilove, *J. Clin. Endocrinol. Metab.*, **29**, 1519 (1969).

[19] G. L. Kumari, W. P. Collins, and I. F. Sommerville, *J. Chromatog.*, **41**, 22 (1969).

[20] P. W. Wilson, D. E. M. Lawson, and E. Kodicek, *J. Chromatog.*, **39**, 75 (1969).

[21] A. Zlatkis and B. C. Pettit, *Chromatographia*, **2**, 484 (1969).

[22] M. A. Kirschner and J. P. Taylor, *Anal. Biochem.*, **30**, 346 (1969).

[23] G. A. R. Decroix, J. G. Gobert, and R. De Deurwaerder, *Anal. Biochem.*, **25**, 523 (1968).

[24] W. G. Galetto, R. E. Kepner, and A. D. Webb, *Anal. Chem.*, **38**, 34 (1966).

Table 1-17. Typical Procedures Using Ethers for Quantitative Determination of OH Compounds by Gas Chromatography

Ether	Typical reagents	Types of compounds and references
Methyl	Diazomethane	Phenols,[1,2,3] carboxylic and phosphorus acids[3]
	Tetramethylammonium hydroxide	Phenols[4]
	Trimethylanilinium hydroxide	Phenolic alkaloids[5]
	Sodium + methyl sulfate	Monohydric phenols[6]
Ethyl and higher alkyl	Diazoethane; diazoalkane + BF$_3$	Phenols, carboxylic and phosphorus acids,[3] phenolic acids[7]
Trimethylsilyl	Hexamethyldisilazane (HMDS) + trimethylchlorosilane (TMCS) + pyridine	Sugars and carbohydrates,[8-12] glycols,[13] fatty alcohols,[14] sterols[15]
	Bis(trimethylsilyl)acetamide (BSA)	Flavanols,[16] drugs,[17] sterols[15,18]
	HMDS + trifluoroacetic acid	Mono- to tetrasaccharides[19]
	HMDS + BSA + trimethylsilyl diethylamine	Hexachlorophene[20]
	Bis(trimethylsilyl)trifluoroacetamide	Phenols[21]
	N-Trimethylsilylimidazole	Sterols[22,23]
Halomethyldimethylsilyl	Chloro- and bromomethyldimethylchlorosilane + diethylamine	Phenols from pesticides,[24] sterols[25]
Pentafluorobenzyl	Pentafluorobenzyl bromide + K$_2$CO$_3$	Phenols[26]
2,4-Dinitrophenyl	1-Fluoro-2,4-dinitrobenzene	Phenols[27]

[1] G. Yip and S. F. Howard, *J. Assoc. Offic. Anal. Chemists,* **51**, 24 (1968).

[2] W. H. Gutenmann and D. J. Lisk, *J. Assoc. Offic. Agr. Chemists,* **48**, 1173 (1965).

[3] C. W. Stanley, *J. Agr. Food Chem.,* **14**, 321 (1966).

[4] H. G. Henkel, *J. Chromatog.,* **20**, 596 (1965).

[5] E. Brochmann-Hanssen and T. O. Oke, *J. Pharm. Sci.,* **58**, 370 (1969).

[6] A. C. Bhattacharyya, A. Bhattacharjee, O. K. Guha, and A. N. Basu, *Anal. Chem.,* **40**, 1873 (1968).

[7] M. Wilcox, *Anal. Biochem.,* **32**, 191 (1969).

[8] C. C. Sweeley, R. Bentley, M. Makita, and W. W. Wells, *J. Am. Chem. Soc.,* **85**, 2497 (1963).

[9] C. C. Sweeley and B. Walker, *Anal. Chem.,* **36**, 1461 (1964).

Diazomethane has been widely used for forming methyl ethers of phenols, enols, and carboxylic, phosphorus, and other acids. (It does not react with neutral alcohols.) The methyl ether is determined by GC:

$$R—OH + CH_2N_2 \rightarrow R—OCH_3 + N_2$$

Diazomethane is conveniently prepared immediately before use by alkaline hydrolysis of N-methyl-N-nitroso-p-toluenesulfonamide and distillation according to the directions of the manufacturer (Aldrich Chemical Co., Milwaukee, Wis.). The yellow ethereal solution of the chemical is usually added to an ether solution of the compound, and N_2 gas evolves as the diazomethane reacts; when the yellow color persists, indicating that an excesss of reagent is present, this excess is removed by partial evaporation of the solvent. Operations are carried out in a hood since diazomethane is toxic and potentially explosive. A convenient method of preparing diazomethane from N-methyl-N'-nitro-N-nitrosoguanidine (Aldrich Chemical Co.), using extraction rather than distillation, has been described by Stanley (48). Diazomethane succeeds in methylating phenols and hindered acids when other procedures fail (49).

Methyl ethers have also been formed by pyrolysis of tetramethylammonium salts of phenols in the injection port of a gas chromatograph. This method

[10] J. H. Sloneker in H. A. Szymanski, ed., *Biomedical Applications of Gas Chromatography*, Vol. 2, Plenum, New York, 1968, pp. 87–135.

[11] P. K. Davison and R. Young, *J. Chromatog.*, **41**, 12 (1969).

[12] M. Kimura, M. Tohma, Y. Okazawa, and N. Murai, *J. Chromatog.*, **41**, 110 (1969).

[13] K. C. Leibman and E. Ortiz, *J. Chromatog.*, **32**, 757 (1968).

[14] A. Prevot and C. Barbati, *Rev. Franc. Corps Gras*, **15**, 157 (1968).

[15] E. M. Chambaz and E. C. Horning, *Anal. Biochem.*, **30**, 7 (1969).

[16] A. R. Pierce, H. N. Graham, S. Glassner, H. Madlin, and J. G. Gonzalez, *Anal. Chem.*, **41**, 298 (1969).

[17] F. Fish and W. D. C. Wilson, *J. Chromatog.*, **40**, 164 (1969).

[18] G. P. Gaidano, G. Molino, L. Perrotti, F. Matta, and G. Boccuzzi, *Boll. Soc. Ital. Biol. Sper.*, **44**, 1494 (1968).

[19] K. M. Brobst and C. E. Lott, Jr., *Cereal Chem.*, **43**, 35 (1966).

[20] P. J. Porcaro and P. Shubiak, *Anal. Chem.*, **40**, 1232 (1968).

[21] T. J. Sprinkle, A. H. Porter, M. Greer, and C. M. Williams, *Clin. Chim. Acta*, **25**, 409 (1969).

[22] M. G. Horning, A. M. Moss, and E. C. Horning, *Biochim. Biophys. Acta*, **148**, 597 (1967).

[23] M. G. Horning, A. M. Moss, E. A. Boucher, and E. C. Horning, *Anal. Letters*, **1**, 311 (1968).

[24] C. A. Bache, L. E. St. John, Jr., and D. J. Lisk, *Anal. Chem.*, **40**, 1241 (1968).

[25] D. B. Gower and B. S. Thomas, *J. Chromatog.*, **36**, 338 (1968).

[26] F. K. Kawahara, *Anal. Chem.*, **40**, 1009 (1968).

[27] I. C. Cohen, J. Norcup, J. H. A. Ruzicka, and B. B. Wheals, *J. Chromatog.*, **44**, 251 (1969).

has been found to be rapid and convenient to use with aqueous solutions, which may be evaporated without loss of compound.

Probably the most successful as well as the most useful method of derivatizing the OH group to facilitate the GC of OH compounds has been the formation of silyl ethers, such as trimethylsilyl [$(CH_3)_3Si$—] and trifluoromethylsilyl [$(CF_3)_3Si$—]. A large number of silylation methods have now been described in the literature, and at least one volume has been devoted exclusively to this subject (50). Recent references cited in Table 1-17 exemplify current methodology. They illustrate the types and variety of compounds derivatized, the silylation reagents, and the procedures. Many of the methods incorporate specialized cleanup procedures for the crop, tissue, or fluid under investigation, and many utilize internal standards to facilitate quantification and to make determinations more accurate.

The method described in the following section is a typical one that is very widely used. Additional comments on silylation methods follow under "Results and Discussion."

C. Determination of Nonvolatile Hydroxy Compounds (sugars and related substances) as Trimethylsilyl Derivatives by Gas Chromatography

METHOD OF SWEELEY ET AL. (51)

Sugars in anydrous pyridine containing hexamethyldisilazane (HMDS) and trimethylchlorosilane (TMCS) form trimethylsilyl (TMS) derivatives within a few minutes at room temperature. The reaction mixture is injected into the gas chromatograph for analysis of the derivatives.

$$ROH + (CH_3)_3SiNHSi(CH_3)_3 \xrightarrow[\text{pyridine}]{\text{TMCS}} ROSi(CH_3)_3 + NH_3$$

Apparatus. A gas chromatograph with a flame ionization detector is equipped with a 6-foot by $\frac{1}{4}$-inch o.d. stainless steel column containing 3% SE-52 on 80/100 mesh acid-washed, silanized Chromosorb W (Johns Manville Corp., New York, N.Y.). Chromatographs with other detectors (e.g., argon ionization) and other columns (e.g., 10% Carbowax 1540 or 15% Apiezon M on the same support) may also be used.

Reagents. Hexamethyldisilazane and trimethylchlorosilane (considered a catalyst) are reagent-grade chemicals available from most GC supply houses. Pyridine should be anhydrous and should be kept over KOH pellets.

Procedure. Ten milligrams of sample is shaken for 30 seconds with 1 ml anhydrous pyridine, 0.2 ml HMDS, and 0.1 ml TMCS in a 1-dram (3.5-ml) plastic-stoppered vial or similar container and then allowed to stand for

5 minutes or longer at room temperature. (Upon addition of TMCS the solution becomes cloudy because of the precipitation of ammonium chloride. The precipitate need not be removed; it does not interfere with the ensuing gas chromatography.) Should the sample remain insoluble, the vial may be heated for 2–3 minutes at 75 to 85°C. When no rearrangements are likely, the sample may be dissolved in warm pyridine before adding the HMDS and TMCS. If the amount of sample is limited, proportionally less of the reagents is used. From 0.1 to 0.5 μl of the reaction mixture is injected into the gas chromatograph. The peak areas of the TMS ethers produced are determined and are related to standards similarly analyzed.

Results and Discussion.   The foregoing procedure, which is easily and rapidly carried out, has been found generally applicable to a wide variety of polyhydroxy compounds (about 100 carbohydrates and related substances were studied). The quantity of pyridine is not critical, so that the final volume can be adjusted with this solvent for quantitative analysis. The stability of the sugar derivatives in the stoppered vials is said to be good for at least several days.

The method is especially well suited for multicomponent analyses. Nonpolar or mildly polar liquid phases (e.g., SE-52 and OV-17) and temperature programming are preferred for analyses of derivatives of sugars with a wide range of volatilities (including tetroses) in a single sample; C_4 to C_7 sugars give reasonable t_R's at 140°C with SE-52. The more polar liquid phases (e.g., OV-210 and OV-225), although less useful for compounds with a wide range of volatilities, will often resolve compounds not separable on the nonpolar liquid phases. Polar phases with reactive protons are usually unsuitable for chromatographing TMS derivatives because silylation reagents (and often their derivatives) may react with labile hydrogens and change column properties, or the derivatives may be degraded and thereby upset quantification. Silicone liquid phases, especially the OV Series (OV-1, OV-101, OV-17, OV-210, and OV-225), appear to be the most satisfactory from the standpoint of inertness, resolution characteristics, and stability at high temperatures. In addition to normal conditioning of the column before use, several large injections of the silylating agent are generally recommended for further conditioning of the column for the individual analysis. Supports should be silanized before the liquid phase is applied. Low-loaded columns (3% liquid phase) are suggested for compounds having low volatilities (long t_R's). Glass columns appear to be best for chromatographing TMS compounds, although stainless steel is also used. Glass injection port liners have been especially recommended when silyl compounds are subject to breakdown in the injection port at high temperatures. This breakdown appears to be due to metal catalysis since TMS derivatives are generally stable to 300°C.

Modifications of silylation procedures are frequently necessary. Many compounds will not silylate completely in 5 minutes (e.g., because of steric hindrance of OH), and reaction times will have to be increased. The reaction mixture can be sampled at intervals, and the shortest time giving a maximum product peak is selected for use. Ketones may cause difficulty by reacting partially to give enol TMS ethers; the formation of a methoxime derivative (52) before silylation has been suggested to avoid this difficulty with steroids (53). Pyridine is widely employed for silylation reactions because it is an excellent solvent and is an HCl acceptor in reactions using the chlorosilanes. Other solvents used are dimethyl sulfoxide, dimethylformamide, tetrahydrofuran, and acetonitrile. On occasion, particularly with small samples or low concentrations, the reaction is conducted in the silylating reagent and no solvent is used. Avoidance of moisture is always desirable and in some cases necessary because many TMS derivatives are hydrolyzed readily by water; loss of derivative by hydrolysis can defeat quantification and introduce extraneous peaks on the chromatogram.

The halodimethylsilyl derivatives are used mainly to achieve greater sensitivity in analyses utilizing electron capture detection. The response of the detector to the halogen is great enough to allow nanogram and even subnanogram amounts of the derivative to be determined.

The following list of important silylation agents offered commercially includes their formulas and some suggested abbreviations (in parentheses between name and formula):

Hexamethyldisilazane   (HMDS) $(CH_3)_3SiNHSi(CH_3)_3$.
Trimethylchlorosilane   (TMCS) $(CH_3)_3SiCl$.
N,O-Bis(trimethylsilyl)acetamide   (BSA) $(CH_3)_3SiOC(CH_3)=$
    $NSi(CH_3)_3$.
N,O-Bis(trimethylsilyl)trifluoroacetamide   (BSTFA) $(CH_3)_3SiOC-$
    $(CF_3)=NSi(CH_3)_3$.
Chloromethyldimethylchlorosilane   (CMDMCS) $(CH_2Cl)(CH_3)_2SiCl$
    (also Br analog).
1,3-Bis(chloromethyl)-1,1,3,3-tetramethyldisilazane
    $[(CH_2Cl)(CH_3)_2Si]_2NH$ (also Br Analog).
Trimethylsilylimidazole   (TSIM) $(CH_3)_3SiNCH=NCH=CH$.
Trimethylsilyldiethylamine   (TMSDEA) $(CH_3)_3SiN(C_2H_5)_2$.
Dimethylchlorosilane   (DMCS) $H(CH_3)_2SiCl$.
Tetramethyldisilazane   (TMDS) $H(CH_3)_2SiNHSi(CH_3)_2H$.

N,O-Bis (trimethylsilyl)acetamide is a very reactive TMS donor. It reacts with alcohols to form TMS ethers, often quantitatively at room temperature, and also reacts with phenols, ureas, organic acids, amino groups, aromatic amides, and some enolic groups (54). An equally reactive TMS donor is BSTFA, which is used when the by-products of BSA interfere; thus the mono(trimethylsilyl)trifluoroacetamide and trifluoroacetamide have lesser t_R's than the corresponding unfluorinated analogs from BSA. Both BSA and BSTFA are sensitive to moisture; neither requires an acidic catalyst. Another silylation agent, TSIM, reacts with OH groups only to produce TMS ethers of alcohols, phenols, organic acids, hydroxy amines, steroids, flavanols, glycols, nucleotides, hydroxy acids, and barbituates. The last two on the list, DMCS and TMDS, are used to form dimethylsilyl ethers, which are more volatile than the corresponding TMS ethers. However, the proton on the silicon is reactive, and the reagent may act as a reducing agent or add to double bonds under some conditions.

An interesting development is a "silicon-specific" flame photometric detector for GC reported by Morrow et al. (55). It is based on the flame emission of silicon-containing compounds at 251.6 nm. The TMS ethers of C_1 to C_7 n-alcohols and three aromatic phenols were used to illustrate the performance of the detector. The authors indicate that the detector can be used to estimate the number of active hydrogens in OH compounds from their TMS derivatives.

D.   Other Derivatization Procedures

Other procedures for forming derivatives of OH compounds are presented in Table 1-18. The methods include oxidation of hydroxyl groups to aldehydes and ketones, bromination of phenols, a condensation reaction, and the stoichiometric formation of acetylene from calcium carbide by methanol. Compounds (e.g., those containing CH_3CHOH—) that form iodoform on treatment with sodium hypoiodite can be determined with excellent sensitivity by GC with electron capture detection (56).

E.   GC Determination of α-Hydroxycarboxylic Acids and Other Compounds with Vicinal Oxygen Functions

Lactic acid can be determined by injecting a solution of it with a large excess of periodic acid into the GC injection port held at 100°C (57). The peak height of the acetaldehyde formed in the reaction is related linearly to the lactic acid in the sample:

$$CH_3CHOHCOOH + HIO_4 \rightarrow CH_3CHO + CO_2 + H_2O + HIO_3$$

The reaction has been shown to be a general one for α-hydroxy acids, and the acid concentration is proportional to the peak height of the aldehyde or

Table 1-18.   Other Derivatives Used for Quantitative Determination of OH Compounds by Gas Chromatography

Derivative	Reagent	Compounds and references
Aldehyde	Periodic acid	Alkyl α-glycerol ethers,[1] mandelic acid[2]
Ketone	Chromic acid	Methylated bile acids[3]
	$NaBiO_3$ in acetic acid	17-Hydroxysteroids[4]
Quinone	$FeCl_3$	α-Tocopherol[5]
Bromide	Bromine	Phenols from carbamates[6,7]
Methyl haloacetone ketals	(1) Haloacetone; (2) diazomethane	C_{18}, C_{19}, C_{21} steroidal alcohols[8]
Acetylene	Calcium carbide at 300°C	Methanol in hydrocarbons[9]

[1] R. A. Gelman and J. R. Gilbertson, *Anal. Biochem.*, **31**, 463 (1969).

[2] J. D. Nicholson, *Analyst*, **94**, 413 (1969).

[3] E. Evrard and G. Janssen, *J. Lipid, Res.*, **9**, 226 (1968).

[4] E. P. Schulz, *Rev. Soc. Quim. Mex.*, **12**(5), 214A (1968).

[5] J. G. Bieri, R. K. H. Poukka, and E. L. Prival, *J. Lipid Res.*, **11**, 118 (1970).

[6] C. H. Van Middelem, T. L. Norwood, and R. E. Waites, *J. Gas Chromatog.*, **3**, 310 (1965).

[7] W. H. Gutenmann and D. J. Lisk, *J. Agr. Food Chem.*, **13**, 48 (1965).

[8] G. A. Sarfaty and H. M. Fales, *Anal. Chem.*, **42**, 288 (1970).

[9] Y. Watanabe and K. Isomura, *Bunseki Kagaku*, **15**, 1215 (1966).

ketone formed by the HIO_4 when a Carbowax 20M on firebrick column is used (58). A 4:1 minimum molar ratio of periodic acid to hydroxy acid is required. The method was illustrated with α-methyllactic, α-methyl-α-hydroxybutyric, α-hydroxyvaleric, and mandelic acids. The injection port temperature was 200°C except for the mandelic acid analysis, when it was 238°C. The analysis should be generally useful for compounds having the following structures:

$$\begin{matrix} OH & OH \\ | & | \\ R-CH-CH-R', \end{matrix} \qquad \begin{matrix} O & O \\ \| & \| \\ R-C-C-R', \end{matrix} \qquad \begin{matrix} O & OH \\ \| & | \\ R-C-CH-R' \end{matrix}$$

and possibly for some epoxides as well. The presence of compounds with these groupings will interfere if their product happens to be identical to, or to have the same t_R as, the one being produced by the hydroxy acid under analysis.

For example,
$$\begin{matrix} O & O \\ \| & \| \\ CH_3-C-C-CH_3 \end{matrix}$$
will interfere in analyses of lactic acid; both produce acetaldehyde.

F. Subtraction of OH Compounds for GC Analysis

Alcohols, especially the lower ones, and monohydric phenols can be determined directly by gas chromatography. The Porapaks have been especially useful in these applications (59, 60, 61). In one GC technique, compounds with certain functional groups are removed from the carrier gas stream by including in the system a chemical that reacts with the group. A typical reaction loop is shown in Figure 1-2. It can be included in the system before or after the GC column, as convenient. Thus, in most cases primary and secondary (but not tertiary) alcohols can be completely "subtracted" from a mixture injected into a gas chromatograph by placing a boric acid loop in the GC pathway (62, 63); compounds with other functional groups are not subtracted. The boric acid forms borates of low volatility with most alcohols, and although the borates are retained in the loop, they gradually bleed off as extremely broad peaks. It was also found that certain allylic unsaturated secondary alcohols are not subtracted, probably because they are dehydrated to conjugated dienes and thereby lose their hydroxyl groups (64).

Chromatograms made with and without the subtraction agent in the system can show which are the alcohol peaks; by difference the amount of each component that is present can be determined. In one typical arrangement, the alcohols are subtracted with a 6-inch loop containing 1 part of powdered boric acid mixed with 20 parts by weight of the Carbowax 20M packing used to chromatograph the alcohols. Compounds with aromatic hydroxyl groups (phenols) are not subtracted by the boric acid loop, but their peaks are generally broadened and delayed, usually less than twofold. Regnier and Huang (65) have used a somewhat similar approach. Placing the reaction loop after the GC column will avoid contaminating the column, for example from the bleed of reactor chemical or products.

In another procedure (66), monohydric alcohols in solution treated with trichloroacetyl isocyanate react quantitatively and their peaks no longer

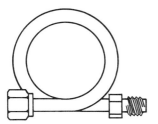

Figure 1–2. Subtraction or reaction loop.

appear on chromatograms. Of fifteen alcohols tested, some dehydration of tertiary alcohols occurred, one tertiary ester isomerized, and one epoxide rearranged; the trichloroacetyl isocyanate esters could be extracted with dilute alkali, and with some exceptions the alcohols could be regenerated essentially quantitatively by refluxing with strong alkali.

A peak-shift technique has been used to shift the peaks of compounds containing alcoholic or phenolic groups from their original positions on a chromatogram to other locations. In this procedure the injection of the sample is followed by another of the derivatizing agent, for example, acetic or propionic anhydride (67). The original peak of the OH compound disappears, and a new one due to the ester appears. In many instances the conversion to acetates or propionates is essentially complete, allowing quantitative analysis to be made with much less effort than is required for derivative formation before injection. The technique is especially helpful for high-molecular-weight hydroxy compounds which migrate little or not at all in the GC column until derivatized.

G.   Active Hydrogen Analysis

Gas chromatographic procedures have been devised to determine OH groups in a molecule or in resins by active hydrogen analysis. See Chapter 8, Section II, on this topic.

III.   ELECTROANALYTICAL METHODS*

A.   Conductometric Titration

There are a variety of electrochemical methods for the determination of the hydroxyl group.

A general route for the analysis of this group is based on a conductometric titration with the sodium or lithium salt of methylsulfinyl carbanion, "dimsyl ion" (68–71). Dimsylsodium in dimethyl sulfoxide is a very strong base and will react with extremely weak acids, including alcohols and weak carboxylic acids. Water, another hydroxy compound, also reacts with dimylsodium in DMSO; in many cases, water and alcohol give separate breaks in the titration curve and thus may be distinguished from each other.

Table 1-19 lists the results obtained for a variety of alkyl and aromatic alcohols and glycols. Secondary and tertiary alcohols could be distinguished from water. Water was more acidic than these alcohols, and the first part of

* Written by Alan F. Krivis.

the titration was due to the reaction of water. The volume of titrant from the first break to the second break correlated with the alcohol content. The diols tested showed an interesting response. The first hydroxyl group was more acidic than water and the second less acidic; the overall titration sequence was as follows: first hydroxyl—water—second hydroxyl. Aromatic alcohols were more acidic than water.

METHOD OF L. K. HILLER, JR. (68)*

Solvents and Reagents.   Matheson, Coleman, and Bell Spectrograde methyl sulfoxide (dimethyl sulfoxide) was used throughout these experiments and was stored in 1-liter bottles over activated Linde Type 4A Molecular Sieves. The conductivity of the solvent at ambient room temperature (about 25°C) was $3.5 \pm 0.2 \times 10^{-7}$ ohm^{-1} cm^{-1}. The residual water content or, rather, the total acidic impurities in the solvent, measured by titration with dimsylsodium, was typically about 2.2 mM, or 40 ppm of water. Sodium hydride was obtained as the 50% mineral oil dispersion (Metal Hydrides, Inc.). 2-Naphthol (MCB reagent grade), used in standardizing the dimsylsodium reagent, gave the following analysis:

Calculated for $C_{10}H_8O$: C, 83.3; H, 5.59; O, 11.1.

Found: C, 83.2; H, 5.6; O, 11.1.

National Cylinder Gas dry nitrogen was used in an all-copper-and-glass tube line, connection being made with the shortest possible lengths of Tygon.

Preparation of Titrant.   The dimsylsodium reagent was prepared by adding 200 ml of dimethyl sulfoxide to 0.6 to 1.0 gram of 50% NaH-mineral oil which had previously been washed with two 10- to 15-ml portions of petroleum ether to remove the mineral oil. The reaction was allowed to proceed under nitrogen at 50 to 60°C with stirring and required 3 to 4 hours for completion; it was judged to be complete with the cessation of hydrogen gas evolution. The reagent solution was clear and pale green. The reagent could be stored under nitrogen in the reaction flask until needed for use. As an extra precaution enough mineral oil was added to the flask to form a layer 1 cm thick over the reagent solution. The reagent can be stored at room temperature for at least a week with only a 1 to 3% loss in activity. Approximately fifty batches of reagent were prepared without difficulty. [*Caution:* Pressure explosions have been reported as a result of adding NaH (3.27 moles) to dimethyl sulfoxide (19.5 moles) and heating to 50° with mechanical stirring (72, 73).]

Apparatus.   Conductance measurements may be carried out with a conductivity bridge capable of ±1% precision. A dip-type cell with a cell constant of 0.1 is adequate.

* Reprinted in part by permission of the copyright owner, the American Chemical Society.

Table 1-19. Conductometric Titrations with Dimsylsodium in Dimethyl Sulfoxide Solvent

Compound No.	Compound	(Moles reagent)/ (Moles sample)[a]	Comments[b]
1	Water	0.97 ± 0.00 (2)	A
2	Methanol	1.04 ± 0.01 (2)	A
3	*tert*-Butanol	1.00 ± 0.02 (4)	A,B,C
4	1-Decanol	1.04	A,D
5	6-Undecanol	1.02 ± 0.02 (3)	A,B,C
6	1-Dodecanol	1.02 ± 0.01 (2)	A,B,C
7	2-Dodecanol	0.93 ± 0.04 (2)	A,B,C
8	1-Tetradecanol	1.03 ± 0.01 (2)	A,D
9	2-Tetradecanol	1.02 ± 0.00 (2)	A,D
10	2-Hexadecanol	0.91 ± 0.01 (2)	A,D
11	Cyclohexanol	1.12 ± 0.07 (2)	A,B,C
12	1,2-Propanediol	1.91 ± 0.01 (2)	F,G
13	1,3-Propanediol	2.01 ± 0.03 (3)	F
14	Benzhydrol	0.99 ± 0.02 (3)	A,J
15	Triphenylmethanol	1.07	A
16	Hydroquinone	1.01	B,J
17	Pyrogallol	0.94	B,G
18	*o,o'*-Biphenol	1.05[c] 2.17	B,G H,L
19	8-Amino-2-naphthol	0.91 ± 0.00 (2)	B,J,K
20	2-Aminoethanol	1.08	A,J
21	2-*tert*-Butylaminoethanol	1.02	A,J
22	2-Ethylaminoethanol	1.04	A,J
23	1-Amino-2-propanol	1.10	A,J
24	2,2',2''-Nitrilotriethanol	3.02	F
25	2-(2-Aminoethylamino)-ethanol	1.09	A,J
26	2-Amino-2-(hydroxymethyl)-1,3-propanediol	1.01 2.13	B,G F

[a] Average value and average deviation are given. Number of trials is given in parentheses if more than one.

[b] Comments refer to conductometric titrations.
  A. Method I.
  B. Method II.
  C. Less acidic than water by Method II.
  D. Cannot distinguish from water by Method II.
  E. Incompletely soluble.

Titration Procedure. An adapter tube (reducing bushing) was fitted to one of the necks of the reaction flask to receive a male ground-glass tube to which a serum cap was secured. The reagent was transferred from the reaction flask to the storage bulb of a 5-ml analytical buret with a 50-ml syringe fitted with a stopcock and a 6-inch 18-gauge hypodermic needle. The buret was modified by making a glass tubing T-connection between the storage bulb and the graduated buret above the solution level to allow the reagent to be stored under nitrogen. The titration cell was a 180–ml tall form beaker fitted with a No. 10 rubber stopper which had holes drilled to accommodate a gas inlet tube (for nitrogen), a Teflon (Du Pont) catheter (which delivered the reagent from the buret), a conductance cell, and a hole to allow for solvent and sample delivery. The nitrogen gas was passed through a solution of 80 ml of dimethyl sulfoxide, 0.05 M in dimsylsodium, to remove any acidic impurities. Atmospheric water, carbon dioxide, and oxygen will react with the reagent and therefore must be excluded. All titrations were performed at ambient temperatures (about 25°C).

Conductometric titrations were carried out in one of two ways, depending on whether or not the conductometric behavior of the residual water in the solvent could be distinguished from that of the sample. In Method I, the residual water was titrated first to a conductometric end point and then the sample was added and titrated. Method II involved simply titrating the sample and the residual water together. This method was used when a definite portion of the conductometric titration curve could be assigned to the reaction of water. Fifty-milliliter aliquots of dimethyl sulfoxide were used in all conductometric titrations; only one sample could be titrated per solvent aliquot to achieve meaningful results. Approximately 15 minutes was required for each titration. All conductance measurements were corrected for changes in the solution volume during the titrations. End points were determined by drawing the best straight lines through the conductivity data points and were measured as the points of intersection. Above each titration curve figure a solid line interrupted by arrows indicates the theoretical end point for each acid function being titrated.

F. Total by Method I.

G. First acidic proton more acidic than water.

H. Total by Method II.

I. Second acidic proton less acidic than water by Method II.

J. More acidic than water.

K. Highly colored solution.

L. Second acidic proton more acidic than water by Method II.

ᵉ Bracket indicates data from the same trial.

B. Polarographic Determination of Diols

The Malaprade reaction can from the basis for the analysis of vicinal diols in several different ways. For example, if 1,2-diol compounds are involved, oxidation with periodic acid will produce formaldehyde. Formaldehyde may be distilled from the reaction mixture and measured polarographically (74, 75), or the excess periodate in the reaction mixture can be destroyed with arsenite and the formaldehyde determined directly in the reaction mixture (76).

The limitation to the methods cited concerns the formation of formaldehyde and the necessity, therefore, for a 1,2-glycol structure in the sample. However, if the periodate is measured polarographically, the method becomes general for any diol, polyol, sugar, or aminoalcohol (76). Furthermore, because the rate of reaction of periodate with diols varies from compound to compound, it is possible also to differentiate some mixtures of diols.

The method, as described below, measures the decrease in the periodate concentration by monitoring the polarographic wave at around -0.4 V versus a mercurous sulfate reference electrode.

EXPERIMENTAL METHOD [BASED ON THE STUDY BY ZUMAN AND KRUPICKA (76)]

Apparatus.   Any of the commercial polarographs, manual or recording, can serve for this analysis. An H-cell (77) with a mercurous sulfate reference electrode should be used. To avoid photolytic effects, the reaction flask and cell should be coated with glyptal or black paint. A constant-temperature bath (25°C) thermostats the reaction flask and cell and their contents.

Solutions.   A strongly buffered background electrolyte solution is needed. A solution containing 0.1 M sodium acetate and 0.1 M acetic acid at pH 4.7 has been found to be useful.

The periodate stock solution is 0.005 M potassium metaperiodate (reagent grade) in water; this should be standardized by an iodometric titration.

Stock solutions of the sample should be prepared so that the molarity of the unknown is approximately 0.005 M.

Procedure.   A blank and a sample should be run simultaneously. Transfer 160 ml of buffer solution to each of two separate painted flasks immersed in the constant-temperature bath. Purge both with nitrogen for 10 minutes, add 6 ml of periodate stock solution to each, and purge for another 5 minutes. To only one flask, add 0.5 ml of diol stock solution, record the time of addition, purge with nitrogen, and divert the nitrogen to blanket the solution. After the reaction is completed, transfer an aliquot of the deoxygenated reaction mixture to the polarographic cell and record the diffusion current at -0.4 V. (The length of reaction time may vary and, in some cases, may be as long as 24 hours. This aspect should be checked with standard samples before analyzing unknowns.) Run an aliquot of the flask containing only periodate immediately after finishing the sample run and determine the diffusion current. Compare both diffusion currents to a previously prepared calibration curve to determine the concentration of periodate before and after reaction. Calculate the diol concentration from the decrease in periodate concentration. The reaction is given by equation 1:

$$
\underset{\underset{\text{OH}}{\overset{\displaystyle |}{\underset{}{}}}\underset{\underset{\text{OH}}{\overset{\displaystyle |}{\underset{}{}}}}{R \cdot CH - CH - R_1} + HIO_4 \rightarrow R - \overset{\overset{\displaystyle O}{\parallel}}{C}H + R_1 - \overset{\overset{\displaystyle O}{\parallel}}{C}H + HIO_3 + H_2O \quad (1)
$$

C. Coulometric Titration of Phenols

An excellent method for the analysis of phenols is based on a titrimetric bromination of the aromatic ring, as indicated in equation 2:

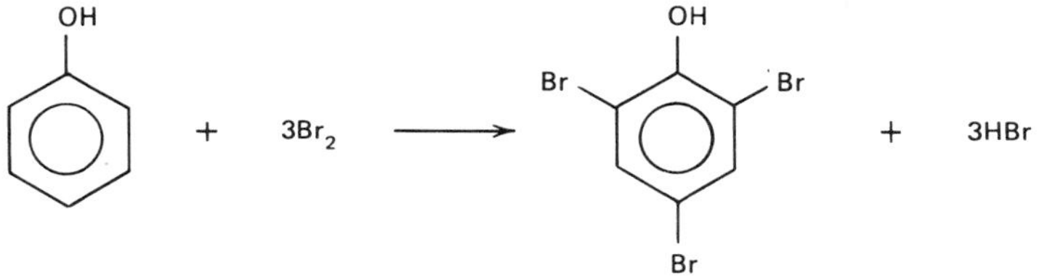

The reaction is rapid and quantitative, and several simple but effective end-point detection systems are available. Volumetric titrations with standard bromine solutions, however, leave a great deal to be desired, including the availability of a usable standard bromine solution. Bromine solutions are so unstable that their preparation and use can be classed as highly undesirable.

One route around the instability obstacle employs a standard bromate solution which reacts with bromide in acid solution to give bromine (equation 3):

$$
BrO_3^- + 5Br^- + 6H^+ \rightarrow 3Br_2 + 3H_2O \quad (3)
$$

The reaction to form bromine is essentially instantaneous so that in effect the titrant is "bromine." This approach can be quite useful, particularly when combined with the biamperometric end-point system to be described shortly. However, there are two major disadvantages of the bromate-bromide volumetric titration: the medium required is water, and high concentrations of acid are needed. Both of these may be detrimental to certain organic analyses.

The electrolytic generation of bromine offers another way to carry out bromine titrations. This technique eliminates the various problems discussed above. Bromide ion can be readily oxidized to bromine by passing a constant current between two platinum electrodes dipped into a bromide solution; this forms the basis for a coulometric titration.

In a coulometric titration the titrant is electrolytically generated *in situ* by impressing a constant current on the electrodes and measuring the length of time needed to reach the end point of the titration. From the known current, i, and the measured time, t, the coulombs, Q consumed in the titration may be calculated (equation 4):

$$Q = i \times t \tag{4}$$

From these data, the unknown concentration can be calculated by use of Faraday's laws (equation 5):

$$\frac{W}{MW} = \frac{Q}{nF} \tag{5}$$

where W is the sample weight, MW the molecular weight, n the number of electrons involved in the reaction, and F the faraday (96487).

The advantages of a coulometric titration are very real. Foremost among them is the elimination of standard solutions; the real reagent is the electron, dispensed and measured directly into the beaker. Standard samples also are dispensed with because the analyses are based on an absolute standard (the Faraday) and relative calibrations are often unnecessary. Because the titrant is generated at the time of its use and does not remain stored, unstable reagents can be exploited; many less stable than the bromine under discussion here have been utilized. Finally, because electricity is the reagent and time the measurement, automation of a coulometric titration is much easier than that of a volumetric titration. Further discussion of this topic will be found in Chapter 19 on automated methods of analysis.

Bromine may be generated in a variety of solvents, aqueous and non-aqueous, and at different pH levels. A few of these solvents will be described under "Experimental Method," p. 51, so that a range of unknowns can be handled.

The end point of the titration is detected most easily by a very simple current-measuring circuit. Two platinum wires immersed in the solution and having a small voltage impressed on them from a flashlight or doorbell battery will show an appreciable current flow in the presence of even minute quantities of free bromine. Before the end point no excess bromine is present and little current flows; after the end point a current flows in the detection circuit. Since the current is linearly related to bromine concentration, the

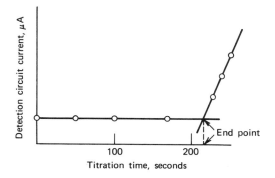

Figure 1-3. Typical coulometric titration curve.

increasing current after the end point has been reached may be extrapolated back to locate the end point of the titration, as shown in Figure 1-3.

A more rapid end-point location scheme using the same equipment is also available. In this case, a fixed current level after the end point is chosen. Let us say that it is 5 μA, as shown by the X on Figure 1-3. At this point, we have an excess of bromine. However, by running a blank titration until this detecting current level is reached, we can find out exactly how much of an excess is involved, in terms of titration time. This blank titration time should be subtracted from all sample titration times to give the titration time for the unknown itself.

Titration to a fixed current level has another useful aspect in addition to rapidity: it is possible to use an inexpensive meter-relay to monitor the current and to shut off the power supply and timer when the end point is reached. In this way, a semiautomatic titration can be set up in any laboratory.

EXPERIMENTAL METHOD [ADAPTED FROM KRIVIS ET AL. (78), AS DISCUSSED IN CHAPTER 12]*

Titration Cell and Electrodes. The titration cell consists of a convenient-sized beaker (50 to 150 ml), in which the solution and the electrodes are placed. Stirring by magnetic stirrer and bar is the most convenient method.

The electrodes are best supported from above the beaker. Leeds & Northrup sells a very useful electrode holder for coulometric titrations. This holder is a plastic disk which has a rod attached to it for clamping the entire unit to a ring stand. The disk itself has a variety of holes molded into it to support different types of electrodes and glass tubes. Electrodes of almost any

---

* Reprinted in part by permission of the copyright owner, the American Chemical Society.

kind may be fabricated to fit the holder, or they may be purchased from Leeds & Northrup or other laboratory supply houses.

Figure 1-4 shows an outline drawing of the electrode-cell arrangement. The anode is a coil of platinum wire (16 gauge) wrapped around the isolation chamber for the cathode. The isolation chamber is a glass tube with a very fine frit sealed in its bottom, and the cathode consists of a coil of platinum inserted into the isolation chamber. Solvent, minus the sample, can be used inside the cathode chamber.

The detecting electrodes may be small pieces of platinum wire sealed in the bottom of glass tubes. The types sold by most laboratory supply houses

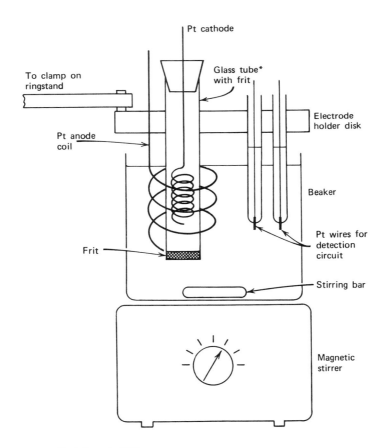

*Isolation chamber

Figure 1–4. Coulometric apparatus used for brominations.

as contacts for electrochemical cells are inexpensive and work very well (e.g., Sargent-Welch S-29414). The glass tubing may be partially filled with mercury, and a wire dipped into the mercury pool to make electrical contact with the platinum electrodes. In some cases, a wire has been welded to the platinum and extended up through the tubing so that an alligator clip may be clamped directly to this extension wire.

Circuits. Figure 1-5 shows the electrical hookup of the apparatus. The detecting circuit includes a battery, variable resistor, and microammeter (10-0-10 scale is useful). The variable resistor should be adjusted so that 0.2 V is impressed across the detecting electrodes. The value of the variable resistor is not critical but should be high enough so that the battery will not be drained too rapidly.

Brominating Solutions. A number of different solutions can be used. Three of the more popular ones are as follows:

1. Acetic acid solution. This is composed of 60% glacial acetic acid, 26% methanol, and 14% 1 M KBr (aqueous).
2. Aqueous solution. A solution 0.3 M HCl and 0.1 M KBr may be used.
3. Methanolic solution. A solution which is 0.3 M HCl and 0.1 M KBr can be prepared in 85% methanol.

If these solvents are too acidic, the pH may be raised to 5 to 6. Successful titrations have been carried out at even higher pH levels.

Procedure. Place a magnetic stirring bar into a beaker (100 ml) and transfer an accurately weighed sample containing about 0.01 mM of phenol. Add 50 ml of the chosen brominating solution and dip the electrodes into the solution. Make sure that sufficient brominating solution is in the cathode isolation chamber so that the level there is above that of the test solution. Start the stirrer and then titrate at a generating current of about 20 mA. (The sample size and generating current should be adjusted for a particular type of compound or sample to give a titration time of about 200 seconds.) The end point is reached when a stable 5-μA current is shown on the detecting circuit meter. Run a blank titration and subtract the blank from the sample titration. Calculate the phenol* in the sample.

Calculations.

$$\text{Per cent of phenol} = \frac{(i)(t)(94.11)(100)}{(96487)(2n)(\text{Sple wt})}$$

---

* Substituted phenols may differ in the stoichiometry of their reactions with bromine. Hence the stoichiometry should be established before any analyses are carried out.

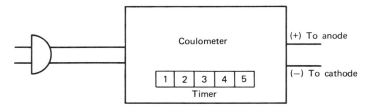

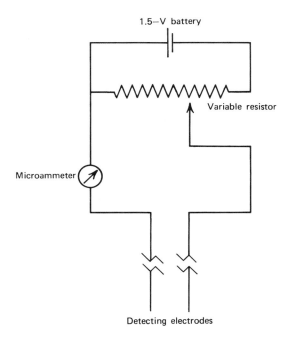

Figure 1–5.   Coulometric circuit used for brominations.

IV.   NUCLEAR MAGNETIC RESONANCE METHODS*

A.   Hydroxyl Groups

The determination of the hydroxyl group by the use of nuclear magnetic resonance (nmr) spectrometry can be accomplished by derivatization,

* Written by Harry Agahigian.

through changes in the hydroxyl chemical shift by dilution, or by means of deuterium exchange. The choice of the approach depends on the nature of the problem to be studied. For example, the simpler alcohols can be determined directly by noting the chemical shifts and obtaining an integral. If there is any doubt, an exchange with deuterium will establish the hydroxyl function unequivocally. A deuterium exchange may not be necessary, however, as dilution of the sample will cause a change in the hydroxyl chemical shift because the latter is concentration dependent.

The technique which is more applicable is derivatization. This involves preparing the derivatives of the various alcohols and noting the chemical shifts and concentrations of the esters that can be formed.

The method of S. L. Manatt (79) involves converting the alcohols to the trifluoroacetic acid esters and noting the ^{19}F chemical shifts of the trifluoromethyl function to determine whether the alcohols are primary, secondary, or tertiary. The trifluoroacetate ester of the alcohols exhibits the shielding effect, wherein the primary is less than the secondary. In turn, the secondary is less than the tertiary; that is, in relation to the trichlorofluoromethane standard, the primary alcohol derivatives appear at lowest field:

$$
(CF_3CO)_2O + ROH \rightarrow CF_3\overset{\overset{\displaystyle O}{\displaystyle \|}}{C}\!-\!OR + CF_3\overset{\overset{\displaystyle O}{\displaystyle \|}}{C}\!-\!OH
$$

A noteworthy feature is the quantitative data that are obtained with the integral. The advantage of the nmr technique is that the conversion of the hydroxyl to the ester can be followed by scanning the proton spectrum and noting the disappearance of the hydroxyl function.

The conversion to the trifluoroacetic acid ester is accomplished by the addition of trifluoroacetic anhydride to the alcohols or alcohol solutions (using an inert solvent) with the anhydride in excess. The anhydride is removed by means of vacuum or extraction with dilute sodium bicarbonate. The completion of the reaction can be determined by noting the ^{1}H nmr at 60 MHz.

The advantage of this method lies in the large differences among the ^{19}F chemical shifts of the various derivatives. The alcohols reported were methanol, ethanol, n-propanol, the butanols, propargyl, benzyl, pentanols, cyclopentanol, cyclohexanol, cycloheptanol, cholesterol, lanosterol, and mixtures of these. This method may be applied also to amines, phenols, and thio derivatives. The utilization of ^{19}F nmr, however, is considered a limitation as ^{1}H is the most commonly used nmr instrumentation. A feature which cannot be overlooked is the equivalence of the fluorines of the trifluoromethyl function, which is advantageous because the ^{19}F signal will be threefold more intense. Interferences in the strict sense are not critical in that the purity can be determined in the original scan.

The utilization of dichloroacetic anhydride as a derivatizing agent of alcohols was reported by Babiac, Barrante, and Vickers (80):

$$
(CHCl_2CO)_2O + ROH \rightarrow -CHCl_2\overset{\overset{\displaystyle O}{\|}}{C}-OR
$$
$$
+
$$
$$
CHCl_2COOH
$$

the preparation of the derivative is essentially identical to that reported by Manatt. The advantage of this method is that it is accomplished by means of ^1H nmr. The authors demonstrate the effect of solvent on the chemical shift of the proton of the dichloromethyl function of the ester. The resonance frequency of this proton is such that generally there will not be overlapping of peaks. The data include mixtures of benzyl alcohol, 1,3- and 1,2-propylene glycols, methanol, ethanol, isopropanol, and tertiary butanol. Although the quantitative data were not given, derivatization is an advantage of this method over the other approaches. Dramatic solvent effects are described, comparing spectra obtained in deuteriochloroform and deuteriated dimethyl sulfoxide. This method was reported as applied to ethanolamines at the 1970 Pittsburgh Analytical Chemistry Conference (J. Babiac and H. Agahigian, "NMR Analysis of Mono-, Di-, and Triethanolamines, Utilizing Derivatization Techniques").

Another derivatization method utilizes the reaction of the trichloroacetyl isocyanate with alcohols, whereupon the protons alpha to the oxygen are measured (81):

$$
CCl_3\overset{\overset{\displaystyle O}{\|}}{C}-NCO + ROH \rightarrow CCl_3\overset{\overset{\displaystyle O}{\|}}{C}-\overset{\overset{\displaystyle H}{|}}{N}-\overset{\overset{\displaystyle O}{\|}}{C}-OCH_2R
$$

A disadvantage of this method is the relative availability of the isocyanate. The tertiary alcohols cannot be studied in this manner, although the appearance of the N—H and the disappearance of the hydroxyl group can be followed. This method is not as direct or generally applicable as the two preceding methods.

Reported more recently (82) is the use of hexafluoroacetone to characterize alcohols by means of the following reaction:

$$
(CF_3)_2C{=}O + ROH \rightleftharpoons CF_3\overset{\overset{\displaystyle OR}{|}}{\underset{\underset{\displaystyle OH}{|}}{C}}CF_3
$$

The advantages of this method are that the derivatives can be prepared *in situ* and that the number of fluorines per active hydrogen is 6:1. Reagent-grade ethyl acetate is used as a solvent, and 1 to 5 mg of sample is required in 0.5 ml of solution. The hexafluoroacetone solution is prepared by passing the gas into ethyl acetate at 0°C until an 0.5 M solution is obtained. The stock solution is placed in the nmr tube, and then 5 mg of sample is added. The ^{19}F resonances are recorded in relation to external trifluoroacetic acid.

The alcohols studied were methanol, ethanol, *n*-propanol, *n*-butanol, 2-methylpropanol, 1-pentanol, 2-mercaptopropanol, 2-mercaptoethanol, 2-methyl-2-nitropropanol, 2-phenylethanol, allyl alcohol, 2-propyne-1-ol, benzyl alcohol, glycolic acid, and trifluoroethanol. Secondary and tertiary alcohols as well as phenols have been reported. The glycols studied by this method were ethylene glycol, ethylene glycol monomethyl ether, tetra-hydroxyfurfuryl alcohol, 1,3-propanediol, 1,4-butanediol, 1,4- butenediol, 2-methyl-2-nitro-1,3 propanediol, 1,2-propanediol (differentiates primary and secondary hydroxyls), glycerol (differentiates primary and secondary hydroxyls), and ethanolamines. The fact that this method requires the use of ^{19}F nmr is its only disadvantage, that is, the ^{19}F nmr unit is not always available.

Ingham et al. (83) used another method which applied the same approach to characterize polyalkylene oxide terminal groups. Their study demonstrated not only that the primary to secondary hydroxyl ratios might be obtained but also that subtle differences in structure had effects on the ^{19}F chemical shifts. Both the acetate and trifluoroacetate derivatives were studied. However, the trifluoroacetates yielded more information. The di-trifluoro-acetate of 1,2-ethanediol, 1,2-propanediol, commercial dipropylene oxide,[*] and polypropylene oxides[*] of molecular weights 425 and 2000 were prepared and studied. The authors found, on the basis of the ^{19}F chemical shifts, that the hydroxyl groups of the PPG 425 and PPG 2000 are all secondary. The presence of two types of secondary hydroxyl may indicate a stereochemical difference. In addition to the polypropylene oxides, polypropylene oxide-polyethylene oxide copolymers were also reported and the primary to secondary hydroxyl ratios determined.

The diol esters are prepared by dissolving the diol (10% by weight) in 100 ml of benzene. To this, redistilled trifluoroacetic acid is added dropwise. After standing for 2 hours at room temperature, 100 ml of water is added and the mixture stirred. The benzene phase is washed with 5% sodium carbonate and water and dried over sodium sulfate. The benzene is then removed. Once the extent of reaction is determined, the sample is ready for ^{19}F nmr analysis.

_____

[*] These are also known as dipropylene glycol and polypropylene glycols, respectively.

Related to these data is the report by Babiac and Agahigian at the Pittsburgh Conference, March 1–7, 1970, of a series of polyol dichloroacetate esters, in which the trifluoroacetate method and the dichloroacetate derivatization methods were compared. These data were obtained at 60 and 100 MHz.

The use of tris(dipivalomethanate)europium for obtaining simple first-order spectra or more readily analyzable spectra was reported by Saunders and Williams (84). The dramatic effects on benzyl alcohol and *n*-hexanol are described. Other functional groups such as ketones, esters, ethers, and amines can be studied in the same manner. Although this technique is related primarily to structure elucidation, the application to functional group analysis is obvious. The europium derivative associates with the functional group, and the effect is decreased as the distance of the protons from the functional group increases.

B. Phenols

The characterization of phenols by nmr spectrometry has been reported, as well as determinations of the components of phenol mixtures (85). The analysis of phenols is somewhat more direct than that of alcohols because the hydroxyl peaks are not concentration dependent over a range of 0.05 to 1.0 *M* and usually appear in a range from 8 to 13 ppm in relation to tetramethylsilane. The spectra are obtained at 60 MHz in hexamethylphosphoramide (HMPA) solution. The choice of the HMPA is dependent on its capacity to effect the largest shifts of the phenolic hydroxylic peak to lowest field. Equimolar solutions of phenol, 2-*tert*-butylphenol, 2,6-dimethylphenol, 2-methyl-6-*tert*-butylphenol, and 2,6-di-*tert*-butylphenol were prepared in carbon tetrachloride and HMPA, and this chemical shift change due to hydrogen bonding was shown with dramatic results. The phosphoramide must be treated with 250 mg of orthoboric acid, which is added to 100 ml of freshly distilled HMPA to suppress association and react with impurities. The spectra were obtained on samples at 0.6 *M* concentration.

The monosubstituted phenols studied were 2-acetylphenol, 4-aminophenol, 2-, 3-, and 4-bromophenols, 2-, 3-, and 4-methoxyphenols, 2-, 3-, and 4-phenoxy and phenylphenols. A series of polysubstituted phenols has also been reported. Although there may be some overlapping, generally ortho-substituted phenols appear at lower field than the meta or para. Phenolic hydroxyls of monosubstituted phenols appear in a range from 9.3 to 10.3 ppm, and polyalkylphenols at from 8.9 to 9.2 ppm. A linear plot was obtained when the phenolic chemical shift for meta- and para-substituted derivatives was plotted against the Hammett constant, and also when ortho-substituted phenols were plotted against Taft's substituent constant. In this method the interferences would come from aldehydes and contaminants such as water and acidic materials, which would lead to exchange with the phenolic hydroxyl altering the chemical shift.

A related study was reported earlier by Lindeman and Nicksic (86). These authors acetylated the hydroxyl by means of acetyl chloride and measured the methyl of the acetate function. Although the integrals were obtained in the region of these resonances, some overlapping occurred. The number of protons is three times that of the hydroxyls; however, there is some limitation on conversion to the acetate because of steric hindrance. The ortho/para ratio of phenolic compounds was determined. The *o-*, *m-*, and *p*-cresols, the *o-*, *m-*, *p*-ethylphenols, and *o-* and *p-sec*-butylphenols were studied. The other alkylphenols reported were the heptyl, octyl, nonyl, decyl, C_{15} to C_{20}, and polypropylene.

Another derivatization method that is applicable but requires ^{19}F nmr analysis has been reported by Leader (87). The phenols studied were hydroquinone, resorcinol, *o-* and *p*-cresol, 2-chlorophenol, 4-chlorophenol, and 4-chloro-3-methylphenol. The ^{19}F chemical shift for the phenols studied were 2.65 to 2.73 ppm from hexafluoroacetone H_2O adduct.

Ballantine and Pillinger (88) utilized the effect of substituents on the chemical shifts of the aromatic protons to determine functional group orientation. The diamagnetic and paramagnetic effects make it possible to analyze unknown polysubstituted phenols.

V. RADIOCHEMICAL METHODS*

Radiochemical methods for determining aliphatic hydroxyl groups are based almost exclusively on esterification of these groups with labeled acid anhydrides and acid chlorides. Techniques that use two isotopes are valuable for the estimation of micro and submicro amounts of hydroxylated compounds, particularly steroids and sterols in biological fluids. Chromatographic procedures are widely employed in these radioisotope methods for isolation of derivatives. Macro quantities of many compounds may be determined by combining a labeled reagent with direct isotope dilution. Diazomethane-^{14}C and radioactive halogens are useful reagents for phenols.

A. Conversion to Acetates

As a radioactive reagent (radioreagent) for primary and secondary hydroxyl groups, acetic anhydride has many advantages. In the presence of pyridine, the reaction of excess anhydride with these groups proceeds readily

---

* Written by D. Campbell.

and often quantitatively at room temperature; essentially complete esterification can be achieved at elevated temperatures. Excess anhydride is removable by hydrolysis and subsequent extraction with aqueous alkali, as well as by chromatographic separation. The compound can be labeled with either ^{14}C or tritium and therefore is especially useful in approaches that require two isotopes. Available specific activities of greater than 2 curies per millimole (Ci/mM) and of up to 0.1 Ci/mM for the ^3H- and the ^{14}C-labeled reagent, respectively, make possible highly sensitive methods that employ either isotope. In addition, the compound is easily separated from nonvolatile impurities by distillation. Labeled acetic anhydride is valuable for estimating steroids and sterols in micro and submicro amounts, as well as for determining macro quantities of many other hydroxylated substances.

Reagents of relatively high specific activity are used extensively for determining steroids and sterols in extracts from biological fluids by acetylation of their hydroxyl groups. The concentration of these substances in such fluids is extremely low, and the specimen available for analysis may contain less than 1 μg of the compound of interest. Primary, secondary, and tertiary hydroxyl groups are present, and certain steroids (e.g., hydrocortisone) may contain all three types. In addition to the expected differences in reactivity associated with the three classes of hydroxyl groups, significant differences in rate of acetylation also exist among the secondary hydroxyl groups; the rate of reaction of the secondary group depends on the location of the group in the molecule (89). Since only micro and submicro amounts of these hydroxylated compounds are usually present in the samples to be analyzed, the use of two isotopes is advantageous: one, the reference isotope, serves to establish the amount of derivative isolated chromatographically; and the other, the indicator isotope, monitors recovery, extent of reaction, and purity. The reference isotope is always in the anhydride used to treat the sample. The indicator isotope may be either in the compound to be determined or in the acetyl moiety of the product.

Although introduction of the indicator radioisotope into the anhydride is the more practicable approach from the viewpoint of synthesis, the steroid or sterol itself may be made radioactive if a suitable labeling method is available. The potential application of labeled substrates for determining the extent of reaction was recognized in early reports that concerned the use of radioisotope techniques for amino acids (90, 91). Steroids and sterols are difficult to extract quantitatively from biological fluids, and addition of the radioactive substrates to such fluids as indicators is a very convenient means of monitoring recovery. When labeled substrate is added to the fluid before extraction, the proportion of indicator activity ultimately recovered permits the accumulative loss throughout the analysis to be assessed accurately, including that attributable to incomplete acetylation. Tritiated steroids of high specific activity have been prepared for this purpose (92) by the Wilzbach technique, and a large number of steroids labeled with ^{14}C are now commercially available.

The indicator-labeled steroid or sterol added to the fluid or dried extract should preferably have a specific activity sufficiently high to make the weight added negligible in comparison with the amount of endogenous substance in the specimen taken. If the compound used possesses less than this minimum desirable specific activity, the weight as well as the amount of activity added must be known in order to obtain by subtraction the correction for the hydroxyl groups that are introduced concomitantly. An important advantage inherent in the use of labeled substrate, especially in view of the differences in reactivity of the hydroxyl groups with structure and position, is that derivatization need not be quantitative. The extent of reaction, however, may be determined in a separate experiment by adding a known amount of substrate activity to the dried extract before acetylation. The fraction of added activity recovered then corresponds to the extent of reaction, provided that there are no handling losses after the addition.

If acetylation is known to be quantitative, the corresponding acetate, which may be prepared from either the indicator-labeled anhydride or substrate, may be added after treatment of the dried extract with the reference-labeled anhydride to monitor recovery and purification. This approach, when used in conjunction with the chromatographic separation methods required in steroid analysis, is advantageous in that the weight of labeled derivative employed need not be known exactly; only a knowledge of total indicator activity added is necessary. Moreover, since the introduction of additional reactive groups is not a consideration, the compound employed as indicator is useful at a molar specific activity lower than that required when the labeled substrate itself is used. Possible isotope effects that arise in chromatography (93) with the modification of the indicator technique are evaluated with a mixture of the pure correspondingly labeled derivatives.

Chromatographic separation of the radioactive derivative of interest from other components of the reaction mixture has been performed both with and without the addition of nonradioactive derivative (carrier). If carrier is used, the amount added must not exceed that which can be accommodated by the procedure utilized to isolate the product. When an indicator isotope is not employed, the treated sample must be handled quantitatively throughout the series of operations used to remove excess reagent. An aliquot of the final sample solution is subjected to chromatographic analysis to separate the acetate, and the activity in the latter is determined, commonly by liquid scintillation counting, after elution. The method may be standardized by quantitatively acetylating a known weight of the substrate of interest with the same lot of anhydride and isolating and assaying a definite fraction of the product. The number of millimoles (M_s) of steroid or sterol in the sample of fluid or extract used for analysis is given by

$$M_s = \frac{A}{S_a F} \qquad (1)$$

where A = activity in the portion of the derivative isolated (μCi),

   S_a = specific activity of the pure acetate derivative (μCi/mM),

   F = fraction of the final sample solution taken for separation.

Alternatively, the specific activity of the acetic anhydride in microcuries per milliequivalent is determined and used to calculate the number of millimoles of the compound of interest in the sample:

$$M_s = \frac{A}{S_r N F} \qquad (2)$$

where S_r = specific activity of the anhydride (μCi/meq) and

   N = number of reactive hydroxyl groups in the molecule.

The latter approach is necessary when the reactive hydroxyl groups of the substrate cannot be esterified completely without significant degradation. If an indicator isotope was used to assess losses due to nonquantitative extraction, acetylation, or handling during analysis of the sample, the appropriate corrections are made to M_s as calculated from equations 1 and 2. The liquid scintillation counting assay method is extremely valuable in this application, since it permits convenient resolution of the ^{14}C/^{3}H pair used in the reference-indicator techniques.

Because of the importance of quantitative derivatization, the acetylation of hydroxyl groups in steroids and sterols has been studied extensively to establish conditions favoring complete esterification (89). This investigation revealed that a 1:5 (v/v) acetic anhydride/pyridine ratio was most suitable for obtaining quantitative acetylation and a single product in 24 hours at room temperature. Steroid concentrations were 0.6 to 100 μg/0.1 ml of acetylating reagent. With the exception of the hindered tertiary 17α and secondary 11β–OH groups, all hydroxyl groups normally found in steroids and sterols were acetylatable at room temperature; significant differences, however, were observed in reactivity. The relative order of reactivity was as follows: 3-phenol > 21 OH > 3β–OH > 6β–OH > 20α–OH > 20β–OH > 16α–OH > secondary 17β–OH > secondary 17α–OH. These differences in rates of reaction have been used to identify steroids by selective acetylation (94, 95).

After the early reports of the use of labeled acetic anhydride in analyses for aldosterone (96–98), the radioreagent was employed for determining hydroxyl groups in steroids and sterols by procedures that involve only a single isotope (99–104). Although these techniques are predominantly simple radioreagent methods, the determination of stigmasterol in soy sterols (99) combines acetic-^{14}C anhydride with direct isotope dilution. The labeled acetate is added to the sterol mixture, which is then acetylated. The radioactive substance isolated and assayed is stigmasterol acetate tetrabromide, which is formed by treating the acetate with elemental bromine. Isotope dilution with the substrate as radioreagent has been applied to the determination of stilboestrol (105).

Another proposed approach to determining hydroxyl groups in steroids (and in other compounds) that requires only ^{14}C-labeled acetic anhydride has been termed "derivative ratio analysis" (106). A known amount of an hydroxylated compound is added to the specimen, which is then treated with the anhydride. The reaction is performed under conditions that will result in quantitative esterification of all hydroxylated substances of interest, including the compound added. A definite quantity of the mixture of derivatives is then resolved by gas chromatography, and the activity in each separated derivative is determined by a suitable assay method. From the measured molar specific activity of acetylated internal standard and the amount of activity in labeled compounds derived from other hydroxylated components the respective amounts of the latter may be calculated. When only extremely low concentrations of the substance to be determined are present, carrier derivative may be added to assist in the separation. Removal of all derivatives from the column must be complete. In the only application given, testosterone was determined with methyl ricinoleate as internal standard.

In general, the great versatility of the reference-indicator method for determining steroids and sterols by acetylation makes it more useful than the single-isotope technique. It has been widely applied for determining aldosterone and other steroids in biological fluids (97, 107–114). Chemical transformations of the derivatives are frequently employed to facilitate the removal of labeled contaminants. Biochemical applications of both one- and two-isotope methods have been discussed extensively (92, 115).

Acetylation with acetic-1-^{14}C anhydride in pyridine, a technique that was suggested originally on the macro scale for estimating acetyl groups in the reaction products (116), is applicable also to materials that contain primary and secondary hydroxyl groups. Stable substances have been esterified by heating 1 mM of the compound in 1 ml of anhydrous pyridine in the presence of 2 mM of anhydride for each expected hydroxyl group for 4 hours at 125°C. Sodium acetate is useful as a catalyst for the solvent-free esterification at

125°C of compounds that undergo decomposition when treated in pyridine. The reaction conditions used, however, should not be more rigorous than are necessary for quantitative acetylation. Conditions as mild as possible are desirable to suppress the equimolar condensation of acetic anhydride with pyridine (117). The extent of this side reaction depends significantly on time and temperature, but decreases markedly with increasing acetic acid content of the solution. Unless the labeled acetates are isolated chromatographically, the presence of the condensation product, which is not completely hydrolyzed by alkali, has an adverse effect on accuracy and sensitivity. Mono- and diglycerides undergo slow ester interchange when treated with acetic anhydride in pyridine (118), and this exchange must be considered in applying the method to all samples that contain esters.

Pyridine and labeled acetic acid are removed by extracting an ether solution of the products successively with dilute solutions of hydrochloric acid and sodium hydroxide. If the total hydroxyl content of the sample is to be determined, this extraction method for purifying the acetylated product to constant specific activity may be employed if it does not result in losses of either labeled or unlabeled components of the original mixture. Mechanical losses are permissible provided that these also do not alter the relative composition of the product. This requirement for isolation of a representative specimen of gross acetylated product differentiates the technique for total hydroxyl content from that for number of acetyl groups, in which purification of only a specific derivative is required. The milliequivalents of acetyl group per gram of gross, quantitatively acetylated product, E, is given by

$$E = \frac{S_p}{S_r} \tag{3}$$

where S_p = specific activity of product (μCi/gram), and S_r = specific activity of the anhydride (μCi/meq).

The total hydroxyl content, C, in milliequivalents per gram, is obtained from

$$C = \frac{E}{1 - 0.0420E} \tag{4}$$

provided that the only difference in composition between sample and product is that due to introduction of the acetyl group (cf. equation 2 of Section III, Chapter 14). This condition thus obviates the need for either knowledge of the weight acetylated or quantitative handling, from the viewpoint of mechanical losses, during removal of anhydride. The procedure is particularly useful for determining low concentrations of hydroxyl groups in polymeric materials. Purification of the acetylated product must be performed carefully to avoid volatilizing low-molecular-weight derivatives that may be present. The specific activity of the anhydride is determined either directly (116) or after its conversion to a suitable derivative, such as *N*-acetyl-*p*-toluidine (119) or acetanilide. Although liquid scintillation counting is the preferred assay method, flow gas counting also permits recovery of the specimen (116).

Acetic-^{14}C anhydride was used to derivatize hydroxyl groups in oxidized polyethylene, and the acetylated polymers were employed as standards in the calibration of an infrared procedure (119). The reaction was performed in a 200:15 (v/v) xylene–pyridine mixture to accommodate the solubility characteristics of polyethylene. The hydroxyl content of the original polymer can be obtained from equation 4.

Determinations of (3-methoxy-4-hydroxyphenyl)hydroxyacetic acid in serum have been made with (Ac-1-^{14}C)$_2$O (120) and (AC-^{3}H)$_2$O (121) that had specific activities of 0.1 mCi/ml and 5 mCi/ml, respectively. A linear calibration plot was obtained with the ^{3}H-labeled reagent for the acid in the range of 1 to 40 ng.

The attractiveness of the anhydride as a radioreagent is enhanced by the relative ease with which preparations of high specific activity may be purified by distillation under vacuum. Impurities that arise from self-radiation, that is, by degradation of other radioactive anhydride molecules by beta particles produced in radioactive decay, consist essentially of nonvolatile products. Distillation of benzene solutions of acetic-^{3}H anhydride of high specific activity before use and at intervals of 1 to 2 weeks during use has been recommended (122).

B. Conversion to Esters of Substituted Benzoic Acids

Many hydroxylated compounds are quantitatively esterified when refluxed with acid chlorides in pyridine, and others often react completely under less rigorous conditions. Excess chloride may be removed by extraction with dilute alkali after hydrolysis, or by precipitation of the derivative in water after the addition of ethanol.

A method that combines the use of 3-chloro-4-methoxybenzoyl-^{36}Cl chloride with direct isotope dilution has been proposed for determining macro amounts of reactive hydroxyl compounds (123). The reagent is prepared in a two-step synthesis by treating *p*-methoxybenzoic acid with elemental ^{36}Cl and then converting the product to the acid chloride with thionyl chloride. The initial step is preparation of the pure labeled ester of the compound to be determined. One milliequivalent of hydroxyl compound is then dissolved in 3 ml of pyridine, and 0.8 gram of nonradioactive 3-chloro-4-methoxybenzoyl chloride is added. In the analysis of methanol,

3 meq is used with 1.2 gram of the reagent. After a 20-minute reflux period, 3 ml of ethanol is added to the cooled solution. The labeled derivative is then quickly introduced as a weighed aliquot of a dioxane solution, and the product is immediately precipitated in cold water. The specified method of purification requires recrystallizing the compound until the melting point is within 0.1°C of that of the pure substance; chemical purity is then determined by melting point depression. Recrystallization until constant specific activity is attained [or, more rigorously, until the specific activity of the substance in the filtrate agrees with that of the fraction collected (124)] may also be applied if sufficient derivative is available. The use of more than one solvent for the crystallization is also desirable. An important advantage of this procedure is that the only radioactivity added to the sample is in the form of the derivative of the compound to be determined. Radiochemical purity thus corresponds to chemical purity.

In the radioassay method used originally with this isotope dilution technique, a definite quantity of pure labeled derivative is added to a known amount of pure inactive derivative approximately equal to that obtained with the specimen to furnish a reference standard. Although the specific activities of sample and standard preparations have been measured by end-window G-M counting, liquid scintillation counting is a more convenient and sensitive radioassay technique. The weight in grams, W, of derivative formed from the specimen taken is obtained from the relationship shown in equation 5, which does not require a knowledge of the specific activity of the pure labeled derivative added:

$$W = W_1 \left(\frac{W_3}{r W_2} + \frac{1}{r} - 1 \right) \tag{5}$$

where W_1 = weight of labeled derivative added to the sample (grams),
$\quad\quad\ W_2$ = weight of labeled derivative added to the standard (grams),
$\quad\quad\ W_3$ = weight of pure nonradioactive derivative taken for the standard (grams),
$\quad\quad\ \ r$ = ratio of the specific activity of the derivative isolated from the sample to that of the standard.

This approach to radioassay is advantageous with end-window G-M counting, since it permits the specific activity of the pure radioactive ester to be brought within the range of that of the derivative prepared from the sample. Alternatively, the specific activity of the radioactive derivative added to the sample is determined, and equation 6 for conventional direct isotope dilution is applied:

$$W = W_1 \left(\frac{S_0}{S_f} - 1 \right) \tag{6}$$

Here, S_0 is the specific activity of the pure labeled derivative added (μCi/gram), and S_f is the specific activity of the derivative isolated (μCi/gram). The modification that makes use of equation 6 is particularly advantageous when liquid scintillation counting is available, since this counting method is conveniently applicable over a very wide range of specific activities. When equation 6 is used, weights are selected so that $0.1 < W_1/W < 0.2$; the optimum value of W_1/W is close to 0.125 (125). The value obtained for W from either equation 5 or equation 6 is then corrected for chemical purity as determined by melting point depression constants, if insufficient derivative is available for purification to constant specific activity. The accuracy of the method was very good with reagent-grade methanol and distilled technical ethylene glycol, as shown by the results given in Table 1-20. In these analyses, end-window G-M counting, equation 5, and melting point depression corrections were used.

Table 1-20. Determination of Methanol and Ethylene Glycol by Isotope Dilution with 3-Chloro-4-methoxybenzoyl-^{36}Cl Chloride

Compound determined	Weight taken, mg	Weight recovered, mg	Per cent recovered
Methanol	97.2	96.5	99.3
	95.6	97.3	101.8
	95.6	95.6	100.0
	96.7	96.5	99.8
	96.6	97.2	100.6
Ethylene glycol	39.8	39.3	98.5
	39.0	38.3	98.2
	40.6	40.4	99.5
	39.2	38.7	98.7

Reprinted from *Anal. Chem.*, **27**, 390 (March 1955). Copyright 1955, by the American Chemical Society. Reprinted by permission of the copyright owner.

Although the melting point of the diester of ethylene glycol was within 0.2°C of that of the pure derivative, purification of derivatives to this extent by recrystallization in the presence of substantial amounts of homologs may be difficult. In the analysis of methanol, for example, ethyl 3-chloro-4-methoxybenzoate, which arose in the destruction of excess reagent with ethanol, was present to the extent of 2% in the assayed product. The melting point of the latter differed by 1°C from that of the pure methyl ester after two crystallizations from methanol and one from petroleum ether. Although the method is capable of high accuracy, the need to correct for nonradioactive impurities, either by melting point determinations or by purification to constant specific activity, and the necessity of preparing the pure labeled derivative of each substance to be determined are disadvantageous.

A radioreagent method for hydroxyl compounds that requires neither corrections for nonradioactive impurities nor preparation of the pure labeled derivative of each substance of interest is based on *p*-iodobenzoyl-^{131}I chloride (126, 127). High sensitivity, moreover, makes the technique applicable in principle to any weighable amount of sample. Compounds to be determined, however, must yield derivatives quantitatively, and the *p*-iodobenzoyl esters must also be amenable to isolation or separation on a liquid chromatographic column. After transfer of an aliquot of a solution of the ester, or mixture of esters, to the column, development is performed with an appropriate solvent in the usual manner. Use is then made of the gamma emission of ^{131}I in monitoring the column from top to bottom with an electrically driven scintillation detector connected to a recording count-rate meter to obtain a graphical representation of the distribution of activity. A known amount of a suitable ester prepared with the same lot of radioreagent is isolated on an identical reference column, and the latter is also monitored. The peak areas on both chromatograms are then integrated. The number of millimoles, M_s, of each hydroxyl compound in the sample used for derivatization is obtained from

$$M_s = \frac{A}{S_r N F} \tag{7}$$

where A = activity in the zone of the respective ester (μCi),
 S_r = specific activity of the reagent (μCi/mM),
 N = number of reactive hydroxyl groups in the compound,
 F = fraction of sample solution applied to the column.
The 8-day half-life of ^{131}I requires monitoring of the reference column whenever sample columns are assayed. Important advantages of the method are that individual hydroxyl compounds in a mixture can be determined if the derivatives are separable and that the column can be monitored repeatedly during development. Disadvantages are the requirements for specialized column-monitoring equipment and substantially more shielding than is necessary with pure beta emitters.

C. Conversion to Esters of Substituted Benzenesulfonic Acids

p-Iodobenzenesulfonyl chloride (pipsyl chloride) and *p*-iodobenzenesulfonic acid anhydride (pipsan) have been used as radioreagents to determine micro and submicro quantities of steroids by techniques that are similar to those employed with acetic anhydride. Each reagent can be labeled conveniently with either ^{35}S or one of the isotopes of iodine (^{131}I, ^{125}I) for use in methods that depend on quantitative derivative formation. Both the anhydride (128–130) and the acid chloride (130–133) have been applied to the determination of corticosteroids and estrogens by esterification of one or more of their hydroxyl groups. If the compound to be determined is available in the ^{14}C- or ^{3}H-labeled form, the resolvable pairs—(^{131}I, ^{125}I)/^{14}C, (^{131}I, ^{125}I)/^{3}H, and ^{35}S/^{3}H—may be used to develop methods that do not require quantitative derivatization. A technique that involves the use of tritiated substrate, pipsyl-^{35}S chloride for derivative formation, and the pipsyl-^{131}I derivative as a second indicator has been described for quantitatively estimating submicro amounts of estradiol (134). Labeled steroids, however, have not been used extensively with pipsyl chloride and pipsan.

Present methodology favors ^{35}S in the reagent used to form the derivative of the steroid, and ^{125}I in the indicator derivative (133). The weaker gamma radiation of ^{125}I, in comparison with ^{131}I, requires substantially less shielding and also reduces self-decomposition in the reagent and derivatives. The longer half-life (ca. 60 days) of ^{125}I is also advantageous, since reagents and indicator derivatives do not have to be prepared so often. Moreover, flow gas counting permits the determination of ^{35}S in the presence of ^{125}I with only slight interference, and solid scintillation counting may be used to assay ^{125}I with even less interference from the sulfur isotope (133). Both isotopes of iodine, as well as ^{35}S may be determined by liquid scintillation counting (135).

The anhydride of p-toluenesulfonic acid ("tosan"), labeled with ^{35}S at a specific activity of 150 mCi/mM (136), was suggested recently for determining steroids (137). The compound is significantly more soluble in pyridine than is pipsan.

Although the sensitivity of methods that employ ^{35}S, ^{131}I, and ^{125}I is high, the relatively short half-lives of these isotopes necessitate frequent preparation of reagents and indicator derivatives. This disadvantage has tended to limit the use of pipsyl chloride and the sulfonic acid anhydrides for hydroxyl groups in comparison with acetic anhydride, which may be labeled with longer-lived radioisotopes. An advantage of sulfonic acid esters of aliphatic hydroxyl groups over carboxylic acid derivatives, however, is their greater resistance to hydrolysis.

Radioisotope methods for determining hydroxyl end groups in polyesters prepared by the condensation polymerization of dicarboxylic acids with diols are desirable, since the concentration of such groups in the final products is often very low. If excess diol is used to ensure that the terminal functional groups are essentially all hydroxyl, an accurate and sensitive technique for determining these groups also provides an absolute method for estimating the number average molecular weight of the polymer. Treatment of such a polyester with excess p-toluenesulfonyl-^{35}S chloride in pyridine has been used to quantitatively convert the polymer to a labeled derivative (138). The concentration of radioactive moiety in the product is determined by equation 3, and from this result the number average molecular weight of the polymer is calculated.

D. Conversion to Esters after Addition of Labeled Hydroxyl Compound

Methanol and ethanol have been determined in an aqueous solution by a modification of the radioreagent-direct isotope dilution method in which the substance to be determined is made the radioreagent and a quantitative derivatizing reaction is unnecessary (139). Definite amounts of ^{14}C-labeled alcohols of known specific activity were added to the sample, and the latter was treated with 3,5-dinitrobenzoyl chloride. The esters were then isolated by liquid column chromatography. The weight of each alcohol in the sample is obtained from equation 6, in which weights are expressed in gram-moles and specific activities are in radioactivity units per mole. The unlabeled reactant need not be used in excess, provided that addition of less than an equivalent amount will yield a useful quantity of derivative. In principle, all reactive hydroxylated compounds in the sample may be determined with a single specimen if a separation method is available that will furnish enough of each pure derivative for a determination of specific activity. The method has high potential sensitivity, since the weights of the separated 3,5-dinitro-benzoates may be determined by absorption spectrophotometry. It is limited, however, to compounds that can be obtained in labeled form at a useful specific activity.

E. Conversion of Phenols to Esters

Certain of the radioreagents that have been used for determining aliphatic hydroxyl groups are applicable as well to phenols in which steric hindrance does not prevent quantitative esterification. Results for macro amounts of phenol and pyrocatechol (as the diester) with 3-chloro-4-methoxybenzoyl-^{36}Cl chloride and direct isotope dilution (123) are given in Table 1-21. An important advantage of the method is its applicability in the presence of significant amounts of water.

The phenolic group of estrogens reacts readily with p-iodobenzenesulfonyl chloride in a 0.75% solution of borax in 75% acetone (131). A high yield of ester was obtained when cholesterol was treated with p-iodobenzoyl-^{131}I chloride under mild conditions (126).

Table 1-21.   Determination of Phenol and Pyrocatechol by Isotope Dilution with 3-Chloro-4-methoxybenzoyl-^{36}Cl Chloride

Compound determined	Weight taken, mg	Weight recovered, mg	Per cent recovered
Phenol	103.8	103.5	99.7
	103.5	103.5	100.0
	103.8	103.9	100.1
	100.8	101.8	101.0
	101.2[a]	101.9	100.7
	103.2[a]	103.3	100.1
	103.5[a]	104.0	100.5
	104.6[b]	105.3	100.7
	104.6[c]	104.7	100.1
	104.4[c]	103.7	99.3
	104.0[d]	98.1	94.3
Pyrocatechol	64.9	64.0	98.6
	63.5	63.2	99.5
	64.3	63.9	99.4
	65.3	64.4	98.6

Reprinted from *Anal. Chem.*, **27**, 390 (March 1955). Copyright 1955, by the American Chemical Society. Reprinted by permission of the copyright owner.

[a] 5 mg of *o*-, 5 mg of *m*-, and 5 mg of *p*-cresol added to sample.
[b] 10 mg of water added to sample.
[c] 25 mg of water added to sample.
[d] 50 mg of water added to sample.

F.   Conversion of Phenols to Halogenated Derivatives

Phenols that are unsubstituted in the ortho or para position react rapidly with elemental bromine to yield brominated derivatives. Use has been made of this high reactivity to determine estrogens by treating the sample with elemental ^{82}Br in anhydrous acetic acid (140). After addition of the 2,4-dibromo derivatives of estrone, 17β-estradiol, and estriol as carriers, the derivatives were isolated by liquid column chromatography and purified to constant specific activity. Although the method is sensitive to 1 ng, the 35-hour half-life of ^{82}Br is disadvantageous for routine applications.

Salicylic acid has been determined in a manner similar to that used for the 3,5-diiodo derivative by treatment with excess ^{131}ICl in glacial acetic acid. The radioreagent-reverse isotope dilution technique (cf. Chapter 11), developed for application to 20 to 200 mg of the compound (141), was later refined to permit the determination of 1 μg (142).

G. Other Methods for Phenols

Phenol groups are sufficiently acidic to react readily with diazomethane to form methyl ethers under conditions comparable to those employed for derivatization of carboxylic acids (cf. Chapter 3). Use of this ^{14}C-labeled reagent is particularly advantageous in that its gaseous nature facilitates removal of excess from the reaction mixture. Diazomethane-^{14}C, which was generated by decomposition of N-methyl-^{14}C-N-nitroso-p-toluenesulfonamide with alkali, was first used to convert a phenolic glucoside to the methyl ether (143). It has since served to detetmine phenolic hydroxyl groups on the surface of carbon black (144).

Dimethyl-^{3}H sulfate has been employed as a radioreagent for the qualitative identification of estradiol and estrone by methylation of their phenolic groups (145). The labeled compound appears to merit further study as a quantitative reagent under carefully controlled conditions.

Complexation with ^{60}Co is the basis of a procedure for determining quercetin and rutin (146). After deposition of the substances on chromatographic paper, the spots were exposed to ammonia, treated with a reagent solution that contained ^{60}Co(NO$_3$)$_2$, and dried. Excess cobalt was removed by ascending chromatography, and the spots of the cobalt complexes ($R_f = 0$) were cut out and assayed. When ^{60}Co that had a specific activity of 0.1 μCi/gram was used, the detection limit of the phenolic compounds was 1 nM.

REFERENCES

Absorption Spectrophotometric

1. L. J. Schmauch and H. M. Grubb, *Anal. Chem.*, **26**, 308 (1954).
2. D. P. Johnson and F. E. Critchfield, *Anal. Chem.*, **32**, 865 (1960).
3. M. W. Scoggins, *Anal. Chem.*, **36**, 1152 (1964).
4. F. E. Critchfield and J. A. Hutchinson, *Anal. Chem.*, **32**, 862 (1960).
5. M. W. Scoggins and J. W. Miller, *Anal. Chem.*, **38**, 612 (1966).
6. G. Gutnikov and G. H. Schenk, *Anal. Chem.*, **34**, 1317 (1962).
7. K. H. Reuther and E. Bayer, *Berichte*, **89**, 2541 (1956).
8. C. J. Malm, L. J. Tanghe, B. C. Laird, and G. D. Smith, *Anal. Chem.*, **26**, 188 (1954).
9. C. E. McGinn and R. G. Spaunburgh, *Dyestuffs*, **42**, 224 (1958).

10. F. Willeboordse and F. E. Critchfield, *Anal. Chem.*, **36,** 2270 (1964).

11. S. Siggia and J. G. Hanna, *Anal. Chem.*, **33,** 896 (1961).

12. J. G. Hendrickson, *Anal. Chem.*, **36,** 127 (1964).

13. R. M. Saunders, H. P. Schwarz, and J. C. Stewart, *Anal. Chem.*, **39,** 550 (1967).

14. D. E. Kramm, J. N. Lamonte, and J. D. Mayer, *Anal. Chem.*, **36,** 2170 (1964).

15. L. Malaprade, *Compt. Rend.*, **186,** 382 (1928).

16. S. Dal Nogare, T. O. Norris, and J. Mitchell, Jr., *Anal. Chem.*, **23,** 1473 (1951).

17. I. M. Baumel, *Anal. Chem.*, **26,** 930 (1954).

18. K. C. Leibman and E. Ortiz, *Anal. Chem.*, **40,** 251 (1968).

19. J. S. Dixon and D. Lipkin, *Anal. Chem.*, **26,** 1092 (1954).

20. R. H. DeMeis, *Science*, **108,** 391 (1948).

21. T. Takeuchi, M. Furusawa, and Y. Takayama, *Japan Analyst*, **4,** 568 (1955).

22. G. A. L. Smith and D. A. King, *Analyst*, **89,** 305 (1964).

23. G. A. L. Smith and D. A. King, *Analyst*, **90,** 55 (1965).

24. E. K. Rud' and S. Y. Skochilova, *Zavodsk. Lab.*, **22,** 919 (1956).

25. J. J. Fox and J. H. Gauge, *J. Chem. Ind.*, **39,** 260T (1920).

26. S. Siggia, *Quantitative Organic Analysis via Functional Groups*, 3rd ed., John Wiley, New York, 1963, p. 70.

27. H. D. Gibbs, *J. Biol. Chem.*, **71,** 445 (1927).

28. M. B. Ettinger and C. C. Ruchhoft, *Anal. Chem.*, **20,** 1191 (1948).

29. E. F. Mohler, Jr., and L. N. Jacob, *Anal. Chem.*, **29,** 1369 (1957).

30. E. A. Bozhevol'nov, *Tr. Vses. Nauchn.-Issle, Inst. Khim. Reaktivov*, 38 (1956).

31. F. Feigl, V. Anger, and H. Mittermann, *Talanta*, **11,** 662 (1964).

32. G. G. Guilbault, D. N. Kramer, and E. Hackley, *Anal. Chem.*, **38,** 1897 (1966).

33. E. Emerson, *J. Org. Chem.*, **8,** 417 (1943).

34. S. Gottlieb and P. B. Marsh, *Ind. Eng. Chem., Anal. Ed.*, **18,** 16 (1946).

35. R. W. Martin, *Anal. Chem.*, **21,** 1419 (1949).

36. M. B. Ettinger, C. C. Ruchhoft, and R. J. Lisha, *Anal. Chem.*, **23,** 1783 (1951).

37. P. S. Jones, D. Thigpen, J. L. Morrison, and A. P. Richardson, *J. Am. Pharm. Assoc.*, **45,** 268 (1956).

38. K. Bence, *Analyst*, **88,** 622 (1963).

39. D. H. Rosenblatt, M. M. Demek, and J. Epstein, *Anal. Chem.*, **26,** 1655 (1954).

40. R. J. Lacoste, S. H. Venable, and J. C. Stone, *Anal. Chem.*, **31,** 1246 (1959).

41. R. W. Stoughton, *J. Biol. Chem.*, **115,** 293 (1936).

42. A. Westlaufer, F. J. Van Natta, and H. B. Quattlebaum, *Ind. Eng. Chem., Anal. Ed.*, **11,** 438 (1939).

43. L. Lykken, R. S. Treseder, and V. Zahn, *Ind. Eng. Chem., Anal. Ed.*, **18,** 103 (1946).

44. G. J. Papariello and M. A. M. Janish, *Anal. Chem.*, **38,** 211 (1966).

45. J. S. Hogg, D. H. Lohmann, and K. E. Russel, *Can. J. Chem.*, **39,** 1588 (1961).

46. R. G. Simard, I. Hasegawa, W. Bandaruk, and C. E. Headington, *Anal. Chem.*, **23,** 1384 (1951).

47. H. H. Willard and A. L. Wooten, *Anal. Chem.*, **22,** 670 (1950).

Gas Chromatographic

48. C. W. Stanley, *J. Agr. Food Chem.*, **14**, 321 (1966).

49. C. M. Williams, *Anal. Biochem.*, **11**, 224, (1965).

50. A. E. Pierce, *Silylation of Organic Compounds*, Pierce Chemical Co., Rockford, Ill., 1968.

51. C. C. Sweeley, R. Bentley, M. Makita, and W. W. Wells, *J. Am. Chem. Soc.*, **85**, 2497 (1963).

52. H. M. Fales and T. Luukkainen, *Anal. Chem.*, **37**, 955 (1965).

53. G. P. Gaidano, G. Molino, L. Perrotti, F. Matta, and G. Boccuzzi, *Boll. Soc. Ital. Biol. Sper.*, **44**, 1494 (1968).

54. J. F. Klebe, H. Finkbeiner, and D. M. White, *J. Am. Chem. Soc.*, **88**, 3390 (1966).

55. R. W. Morrow, J. A. Dean, W. D. Shults, and M. R. Guerin, *J. Chromatog. Sci.*, **7**, 572 (1969).

56. R. D. Stevens and C. H. Van Middelem, *J. Agr. Food. Chem.*, **14**, 149 (1966).

57. N. E. Hoffman, J. J. Barboriak, and H. F. Hardman, *Anal. Biochem.*, **9**, 175 (1964).

58. N. E. Hoffman and P. J. Conigliaro, *Develop. Appl. Spectry.*, **4**, 299 (1965).

59. O. L. Hollis, *Anal. Chem.*, **38**, 309 (1966).

60. M. E. Le Pera, *J. Gas Chromatog.*, **6**, 335 (1968).

61. R. N. Baker, A. L. Alenty, and J. F. Zack, Jr., *J. Chromatog. Sci.*, **7**, 312 (1969).

62. R. M. Ikeda, D. E. Simmons, and J. D. Grossman, *Anal. Chem.*, **36**, 2188 (1964)

63. F. W. Hefendehl, *Naturwissenschaften*, **51**, 138 (1964).

64. B. A. Bierl, M. Beroza, and W. T. Ashton, *Mikrochim. Acta*, **1969**, 637.

65. F. E. Regnier and J. C. Huang, *J. Chromatog. Sci.*, **8**, 267 (1970).

66. P. A. Hedin, R. C. Gueldner, and A. C. Thompson, *Anal. Chem.*, **42**, 403 (1970).

67. M. W. Anders and G. J. Mannering, *Anal. Chem.*, **34**, 730 (1962).

Electroanalytical

68. L. K. Hiller, Jr., *Anal. Chem.*, **42**, 30 (1970).

69. G. G. Price and M. C. Whiting, *Chem. Ind. (London)*, **1963**, 775.

70. E. C. Steiner and J. M. Gilbert, *J. Am. Chem. Soc.*, **85**, 3034 (1963).

71. C. D. Ritchie and R. E. Uschold, *J. Am. Chem. Soc.*, **90**, 2821 (1968).

72. *Chem. Eng. News*, **44**, 48 (Apr. 11, 1966).

73. G. L. Olsen, *Chem. Eng. News*, **44**, 7 (June 13, 1966).

74. B. Warshowsky and P. J. Elving, *Ind. Eng. Chem., Anal. Ed.*, **18**, 253 (1946).

75. P. J. Elving, B. Warshowsky, E. Shoemaker, and J. Margolit, *Anal. Chem.*, **20**, 25 (1948).

76. P. Zuman and J. Krupicka, *Collection Czech. Chem. Commun.*, **23**, 598 (1958).

77. I. M. Kolthoff and J. J. Lingane, *Polarography*, 2nd Ed., Interscience, New York, 1952; L. Meites, *Polarographic Techniques*, 2nd Ed., Interscience, New York, 1965.

78. A. F. Krivis, E. S. Gazda, G. R. Supp, and P. Kippur, *Anal. Chem.*, **35**, 1955 (1963).

Nuclear Magnetic Resonance

79. S. L. Manatt, *J. Am. Chem. Soc.*, **88**, 1323 (1966).

80. J. S. Babiac, J. R. Barrante, and G. C. Vickers, *Anal. Chem.*, **40**, 610 (1968).

81. V. W. Goodlett, *Anal. Chem.*, **37**, 431 (1965).

82. G. R. Leader, *Anal. Chem.*, **42**, 16 (1970).

83. J. D. Ingham, D. D. Lawson, S. L. Manatt, N. S. Rapp, and J. P. Hardy, *J. Macromol. Chem.*, **1**, 75 (1965).

84. J. K. M. Saunders and D. H. Williams, *Chem. Commun.*, 422 (1970).

85. M. W. Dietrich, J. S. Nash, and Robert Keller, *Anal. Chem.*, **38**, 1479 (1966).

86. L. P. Lindeman and S. W. Nicksic, *Anal. Chem.*, **36**, 2415 (1964).

87. G. R. Leader, *Anal. Chem.*, **42**, 16 (1970).

88. J. A. Ballantine and C. T. Pillinger, *Tetrahedron*, **23**, 1691 (1967).

Radiochemical

89. O. V. Dominguez, J. R. Seely, and J. Gorski, *Anal. Chem.*, **35**, 1243 (1963).

90. A. S. Keston, S. Udenfriend, and R. K. Cannan, *J. Am. Chem. Soc.*, **71**, 249 (1949).

91. A. S. Keston, S. Udenfriend, and M. Levy, *J. Am. Chem. Soc.*, **72**, 748 (1950).

92. R. E. Peterson, *in* S. Rothchild, ed., *Advances in Tracer Methodology*, Vol. 1, Plenum, New York, 1962, pp. 265–273.

93. P. D. Klein, *in* S. Rothchild, ed., *Advances in Tracer Methodology*, Vol. 2, Plenum, New York, 1965, pp. 145–154.

94. C. D. West, B. Damast, and O. H. Pearson, *J. Clin. Endocrinol. Metab.*, **18**, 15 (1958).

95. W. G. Wiest, *J. Biol. Chem.*, **238**, 34 (1963).

96. S. A. Simpson and J. F. Tait, *Mem. Soc. Endocrinol.*, **2**, 9 (1953).

97. P. Avivi et al., *Proc. Radioisotope Conf.*, *2nd, Oxford, Engl.*, **1**, 313 (1954).

98. S. A. Simpson et al., *Helv. Chim. Acta*, **37**, 1163 (1954).

99. R. A. Donia, A. C. Ott, and N. Drake, *Anal. Chem.*, **29**, 464 (1957).

100. D. L. Berliner, *Proc. Soc. Exptl. Biol. Med.*, **94**, 126 (1957).

101. V. P. Hollander and J. Vinecour, *Anal. Chem.*, **30**, 1429 (1958).

102. E. Demey and W. G. Verly, *Arch. Intern. Physiol. Biochim.*, **66**, 62 (1958).

103. E. Demey-Ponpart, *Bull. Soc. Chim. Biol.*, **41**, 795 (1959).

104. M. Kuroda, H. Werbin, and I. L. Chaikoff, *Anal. Biochem.*, **9**, 75 (1964).

105. R. Fleming, *J. Pharm. Pharmacol.*, **12** (Suppl.), 217T–219T (1960).

106. A. Karman, I. McCaffrey, and B. Kliman, *Anal. Biochem.*, **6**, 31 (1963).

107. B. Kliman and R. E. Peterson, *J. Biol. Chem.*, **235**, 1639 (1960).

108. R. Koedding, W. Lamprecht, H. P. Wolff, J. Karl, and K. R. Koczorek, *Z. Anal. Chem.*, **181**, 574 (1961).

109. R. Koedding, H. P. Wolff, J. Karl, and M. Torbica, *Symp. Deut. Ges. Endokrinol.*, **8**, 321 (1962); *Chem. Abstr.*, **65**:15749b.

110. T. J. Benraad and P. W. C. Kloppenborg, *Clin. Chim. Acta*, **12**, 565 (1965).

111. A. K. A. Wahid and B. Singer, *5th Arab. Sci. Congr.*, Baghdad, Part 3, 509 (1966); *Chem. Abstr.*, **69**:74415k.

112. W. Nowaczynski, J. Silah, and J. Genest, *Can. J. Biochem.*, **45**, 1919 (1967).

113. C. W. Bardin and M. B. Lipsett, *Steroids*, **9**, 71 (1967).

114. W. G. Wiest, *Steroids*, **10**, 257 (1967).

115. J. K. Whitehead and H. K. Dean, *Methods Biochem. Anal.*, **16**, 1 (1968).

116. R. H. Benson and R. B. Turner, *Anal. Chem.*, **33**, 344 (1961).

117. I. Fleming and J. B. Mason, *J. Chem. Soc.*, (C), **1969**, 2510.

118. H. Mangold, *J. Am. Oil Chemists' Soc.*, **38**, 708 (1961).

119. D. E. Kramm, J. N. Lomonte, and J. D. Moyer, *Anal. Chem.*, **36**, 2170 (1964).

120. L. P. O'Gorman, *Clin. Chim. Acta*, **19**, 485 (1968).

121. L. P. O'Gorman, *Clin. Chim. Acta*, **23**, 247 (1969).

122. H. H. Henderson, F. Crowley, and L. E. Gaudette, *in* S. Rothchild., ed., *Advances in Tracer Methodology*, Vol. 2, Plenum, New York, 1965, pp. 83–86.

123. P. Sorensen, *Anal. Chem.*, **27**, 388 (1955).

124. M. M. Rapport and B. Lerner, *J. Biol. Chem.*, **232**, 63 (1958).

125. V. I. Shamaev, *Zh. Anal. Khim.*, **22**, 988 (1967).

126. W. M. Stokes, F. C. Hickey, and W. A. Fish, *J. Am. Chem. Soc.*, **76**, 5174 (1954).

127. W. M. Stokes, W. A. Fish, and F. C. Hickey, *Anal. Chem.*, **27**, 1895 (1955).

128. E. Bojesen, *Scand. J. Clin. Lab. Invest.*, **8**, 55 (1956).

129. E. Bojesen and H. Degn, *Acta Endocrinol.*, **37**, 541 (1961).

130. E. Bojesen, A. Keston, and M. Karsiotis, *Abstr. Comm. XIX, Intern. Physiol. Congr.*, Montreal, 1953, p. 220.

131. D. C. Leegwater, *Nature*, **178**, 916 (1956).

132. R. Svendsen, *Acta Endocrinol.*, **35**, 161 (1960).

133. H. A. Andersen, E. Bojesen, P. K. Jensen, and B. Sorensen, *Acta Endocrinol.*, **48**, 114 (1965).

134. J. E. O'Grady, *Biochem. J.*, **106**, 77 (1968).

135. B. A. Rhodes, *Anal. Chem.*, **37**, 995 (1965).

136. L. Thuneberg, *Intern. J. Appl. Radiation Isotopes*, **16**, 413 (1965).

137. E. Bojesen, O. Buus, R. Svendsen, and L. Thuneberg, *in* H. Carstensen, ed., *Qualtitative and Quantitative Analysis of Steroid Hormones*, Vol. I, Marcel Dekker, New York, 1967, pp. 1–53.

138. E. G. Hoffman and H. Hoberg, *Z. Elektrochem.*, **58**, 646 (1954).

139. V. Ya Efremov, M. B. Neiman, and V. N. Panfilov, *Tr. Komiss. po Analit. Khim., Akad. Nauk SSSR*, **9**, 361 (1958).

140. W. R. Slaunwhite, Sr., and L. Neely, *Anal. Biochem.*, **5**, 133 (1963).

141. H. A. Swartz and J. E. Christian, *J. Am. Pharm. Assoc., Sci. Ed.*, **47**, 701 (1958).

142. C. E. Breckinridge, Jr., and J. E. Christian, *J. Am. Pharm. Assoc., Sci. Ed.*, **49**, 330 (1960).

143. A. Stoll, J. Ruschmann, A. von Wartburg, and J. Reut, *Helv. Chim. Acta*, **41**, 993 (1956).

144. E. Papirer and J.-B. Donnet, *Bull. Soc. Chim. France*, **1966**, 2033.

145. J. E. O'Grady and P. J. Heald, *Nature*, **205**, 390 (1965).

146. V. Pereira Crespo, J. S. Veiga, and F. P. Coelho, *Proc. 15th Internat. Congr. Pure Appl. Chem. (Anal. Chem.)*, Lisbon, 1956, Vol. II, pp. 737–740.

2 Carbonyl groups

I. ABSORPTION SPECTROPHOTOMETRIC METHODS*

A. Direct Infrared Measurement

Absorptions produced by vibrations of the C=O group in organic compounds occur in the infrared spectra from 5.45 to 6.5 μ. However, the absorptions in this region are not specific for this group in aldehydes and ketones only but include the C=O group in carboxylic acids and anhydrides, esters, and amides. Although each type of compound has a characteristic band position, there is much overlapping.

B. Direct Ultraviolet Measurement

Absorptions of the carbonyl group in the ultraviolet region as a result of the presence of the unsaturated C-O bond are of relatively low intensities. Although this region has been used to a limited extent in special cases (1–5), it is not of general usefulness for the determination of carbonyl compounds.

C. Measurement after Chemical Reaction

1. 2,4-Dinitrophenylhydrazone Formation. Carbonyl compounds condense with 2,4-dinitrophenylhydrazine to form 2,4-dinitrophenylhydrazones. This reaction (equation 1) has been extensively studied and used as a basis for the colorimetric determination of small amounts of aldehydes and ketones.

$$(1)$$

*Written by J. Gordon Hanna.

75

Mathewson (6) was apparently the first to describe a colorimetric method based on this reaction for the determination of low concentrations of acetone. Toren and Heinrich (7) suggested a general procedure for carbonyl compounds based on extraction of the 2,4-dinitrophenylhydrazones with isooctane to eliminate the interference from excess reagent. Lohman (8) proposed extraction of the 2,4-dinitrophenylhydrazones with hexane. Lappin and Clark (9) shifted the absorbance of the 2,4-dinitrophenylhydrazones from about 340 mμ by adding alkali and measured the wine-red color at 480 mμ. Mendelowitz and Riley (10) observed that the maximum absorbance occurred near 430 mμ instead of 480 mμ and that the wine-red complex was more unstable than had been reported. These last workers also reported difficulty from inorganic chloride precipitation upon the addition of potassium hydroxide to the reaction mixture.

Jordan and Veatch (11) studied the previous methods and found that the addition of water to ensure solution of inorganic chloride and substitution of a mixed hydrocarbon-alcohol solvent for the methanol solvent resulted in the accurate and rapid determination of aliphatic and simple aromatic carbonyl compounds. Within the prescribed time interval, results are accurate even though the wine-red color of the hydrazones decreases with time in the presence of potassium hydroxide.

METHOD FROM JORDAN AND VEATCH (11)

Apparatus.   Beckman DB spectrophotometer with 1-cm matched silica cells.

Reagents.   Carbonyl-free Formula 30 alcohol (95% ethanol, 5% methanol). Add to 5 liters of alcohol excess 2,4-dinitrophenylhydrazine and a few drops of concentrated hydrochloric acid, and reflux the mixture for 1 hour. Distill the alcohol from an all-glass distillation apparatus.

n-Hexane. Purify as for the Formula 30 alcohol but reflux overnight.

2,4-Dinitrophenylhydrazine solution. Prepare a saturated solution in Formula 30 alcohol.

Procedure.   Weigh a sample of appropriate size into a 25-ml volumetric flask. Add 5 ml of 3:7 *n*-hexane-Formula 30 alcohol, 2 ml of the saturated 2,4- dinitrophenylhydrazine solution, and 0.1 ml of concentrated hydrochloric acid to each flask. Prepare a reference in the same manner with 5 ml of 3:7 *n*-hexane-alcohol. Heat at 55 $\pm$ 1°C for 30 minutes. Cool rapidly to room temperature and dilute to volume with a solution containing 59 grams of potassium hydroxide and 180 ml of water diluted to 1 liter with Formula 30 alcohol. After mixing well, read the absorbance of each solution at 480 mμ in matched silica cells between 8 and 15 minutes after dilution to volume.

Calibration Curve. Prepare standard solutions containing 0.002 to 0.01 mg of carbonyl per liter (as C=O) in 3:7 *n*-hexane-Formula 30 alcohol. Pipet a 5-ml aliquot of each standard into a 25-ml volumetric flask. Add 0.1 ml of concentrated hydrochloric acid and 2 ml of the saturated 2,4-dinitrophenyl-hydrazine solution to each and treat further as described in the procedure above. Plot carbonyl concentration versus absorbance.

Johnson and Scholes (12) prepared the hydrazones in perchloric acid solution, extracted them with carbon tetrachloride, treated the extracts with ethanolic sodium hydroxide, and measured the red color produced at 420 mμ. The advantages claimed were that the hydrazine is much more soluble in this system and that less of the unchanged reagent is extracted. Basson (13) studied the Johnson and Scholes procedure and recommended, among other things, the use of aqueous potassium hydroxide solution in place of the ethanolic sodium hydroxide solution and an increase in acid concentration to improve the color stability.

2. *Chromotropic Acid Reaction for Formaldehyde.* Chromotropic acid (1,8-dihydroxynaphthalene-3,6-disulfonic acid) is an essentially specific reagent for formaldehyde and is the basis for the determination of formaldehyde in the presence of large excess amounts of other components. A purple color results when formaldehyde is warmed with the reagent in concentrated sulfuric acid. The reaction chemistry is not known with certainty. Feigl (14) proposes that the reaction may be similar to that between phenols and formaldehyde to produce hydroxyphenylmethanes and that this is followed by oxidation to a quinoid compound of the following type:

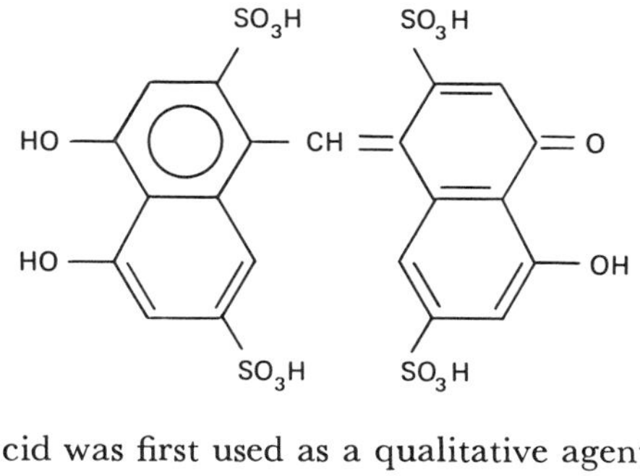

Chromotropic acid was first used as a qualitative agent for formaldehyde by Eegriwe (15). Quantitative use of the reagent was then made for biological systems by Boyd and Logan (16), who visually compared the colors formed with those of standards.

Bricker and Johnson (17) and McFayden (18) subsequently proposed spectrophotometric measurements of the color at 570 mμ. The two procedures outlined by these authors are quite similar.

METHOD FROM BRICKER AND JOHNSON (17)

Reagent. Dissolve 2.5 grams of chromotropic acid in 25 ml of water and filter any insoluble material.

Procedure. Weigh in a test tube a sample containing less than 100 μg of formaldehyde and having a volume in the range of 0.4 to 0.9 ml. Add 0.5 ml of the chromotropic acid solution. Slowly add 5 ml of concentrated sulfuric acid and shake to mix. Place in a boiling water bath for 30 minutes. Cool and dilute to 50 ml in a volumetric flask. Read the absorbance of the solution at 570 mμ. Then read the formaldehyde concentration from a calibration curve relating absorbance to concentration prepared with standard formaldehyde solutions.

Bricker and his coworkers (17, 19) investigated the possible interferences from a number of different types of compounds. They reported that a reaction takes place with other aldehydes but that the purple color is specific for formaldehyde. Acetic acid, formic acid, oxalic acid, acetone, glycerol, glucose, mannose, pyridine, and benzene do not give serious interferences. Bricker et al. also concluded that methanol and ethanol do not interfere but that higher alcohols, diacetone alcohol, and methyl ethyl ketone reduce the formaldehyde color intensity.

Later, Wilson and Lynch (20) disputed the previous claims of noninterference from ethanol and recommended that ethanol interference be corrected for with an appropriate ethanol blank or by a preliminary removal of the ethanol by evaporation. Bricker and Vail (19), although they did not note interference from ethanol, removed interfering compounds by evaporation in the presence of chromotropic acid, which retains the formaldehyde. They then added sulfuric acid to develop the purple color of the formaldehyde reaction product. The evaporation was performed at 170°C in an oil bath so that compounds readily volatile at or below this temperature were eliminated.

3. Schiff's Reagent for Aldehydes. Schiff's reagent is prepared by treatment of the red dye, basic fuchsin (rosaniline hydrochloride), with aqueous sulfur dioxide to form the colorless addition product. Reaction of this addition product and an aldehyde results in the formation of a red quinoid dye:

(2)

Basic fuchsin (colored) →[H₂SO₃] Schiff's reagent (colorless) →[RCHO] Aldehyde addition product (colored)

The intensity of the color of the reaction product is dependent on the alde-hyde concentration in the reaction mixture. The instability of the color of the addition product and interference from color formed by the Schiff's reagent itself on standing present difficulties in the use of reaction 2 for quantitative purposes. The temperature and the time of reaction affect the rate of color development, and Beer's law is not followed. To obtain satis-factory results, precise control of reaction conditions must be maintained and standards must be run at the same time as the unknowns. At best, methods based on the color obtained by the use of Schiff's reagent can be considered only semiquantitative. No general procedure can be given to cover many specific situations. Readers who wish to consult some of the analytical literature on this reaction will find refs. 21–30 of interest.

4. Iodoform Reaction.   Acetaldehyde and acetone produce iodoform when acted upon by hypoiodite:

$$\underset{CH_3}{\overset{R}{\diagdown}} C{=}O \; + \; 3I_2 \; + \; 4NaOH \longrightarrow CHI_3 \; + RCOONa + 3NaI + 3H_2O \qquad (3)$$

Dal Nogare, Norris, and Mitchell (31) found that iodoform absorbs in the ultraviolet region from 400 to 260 mμ, giving three well-defined peaks at 347, 307, and 274 mμ. The absorption peak at 347 mμ was selected as being the most sensitive to change in iodoform concentration and as showing the best adherence to Beer's law for the determination of iodoform from 0 to 3 mg. Other compounds that show a high conversion to iodoform and to which the procedure probably can be applied are methyl isopropyl ketone and mesityl oxide; other methyl ketones are possible candidates.

METHOD FROM DAL NOGARE, NORRIS, AND MITCHELL (31)

Apparatus.   Beckman Model DU spectrophotometer. Use either a hydrogen or a tungsten lamp to make measurements at 347 mμ.

Reagent.   Iodine solution, about 20%. Dissolve 400 grams of potassium iodide in 800 ml of water, add 200 grams of iodine, and mix until dissolved.

Procedure.   Pipet 10 ml of 20% iodine solution into a 125-ml separatory funnel and add 3.3 ml of 20% sodium hydroxide. Mix by swirling. If the resulting solution is not distinctly orange-yellow, adjust it to this color by the dropwise addition of iodine solution. To this hypoiodite solution, add from a pipet 1 to 5 ml of the sample containing no more than 0.4 mg of acetone or acetaldehyde, and mix immediately. Stopper the funnel and allow the mix-ture to stand for 5 minutes at room temperature. If the orange-yellow color is gradually discharged during the 5-minute reaction time, add iodine solution dropwise until the color is restored. After the reaction, discharge the oidine color with a few drops of 5% sodium thiosulfate. Add 22 to 24 ml of chloroform from a graduated cylinder and extract the iodoform by shaking. Allow the phases to separate and transfer the chloroform layer to a second 125-ml separatory funnel containing approximately an equal volume of water. Shake vigorously. Transfer the washed chloroform layer to a 25-ml volumetric flask, passing it first through a funnel containing a thin bed of anhydrous sodium sulfate supported on a glass wool plug. This operation is designed to remove water droplets from the extract. Make the extract to volume by adding chloroform from a dropper, passing it first through the sodium sulfate to wash down the retained extract.

Measure the absorbance of the chloroform extract at 347 mμ against chloroform in 2.5-cm silica cells. Determine the absorbance of a blank and subtract this value from all the sample readings.

Calculations.

$$\text{Per cent of acetaldehyde} = \frac{A \times 0.421 \times 100}{\text{milligrams of sample}}$$

$$\text{Per cent of acetone} = \frac{A \times 0.284 \times 100}{\text{milligrams of sample}}$$

where A = absorbance corrected for blank at 347 mμ,
   0.421 = reciprocal of slope of acetaldehyde calibration curve,
   0.284 = reciprocal of slope of acetone calibration curve.

Interference will be encountered from compounds that give iodoform by reaction with hypoiodite. Analytical data obtained by Dal Nagore et al. for the determination of acetone in cyclohexanol are given in Table 2-1.

Table 2-1. Analytical Data for Acetone in Cyclohexanol (31)

Acetone in cyclohexanol, weight %	Absorbance	Acetone, mg	
		Calculated	Found
0.480	1.545	0.453	0.458
0.384	1.260	0.368	0.374
0.288	0.919	0.277	0.273
0.192	0.637	0.185	0.189
0.096	0.317	0.092	0.094
0.038	0.135	0.037	0.040

5. Photometric Titration of Aldehydes with Sodium Borohydride.
Cochran and Reynolds (32) proposed a direct photometric titration based
on the reduction of aldehydes by sodium borohydride:

$$4R_2C{=}O + NaBH_4 + 2H_2O \rightarrow 4R_2CHOH + NaBO_2 \qquad (4)$$

The sample is dissolved in an isopropyl alcohol-water mixture and titrated
with a standard solution of sodium borohydride in dimethylformamide. The
absorption of the aldehyde in the ultraviolet region of the spectrum is measured as a function of the volume of standard borohydride reagent. Although
the method is probably useful in some well-defined situations, it has serious
limitations for general use. It is not recommended for aldehydes in general,
and its specificity is not sufficient for most situations. The authors proposed
the method for aliphatic aldehydes and aromatic aldehydes that are unsubstituted with electron-withdrawing groups. Peroxides, acidic hydrogen
atoms, and some ketones will cause interference. Formaldehyde, acetals, and
other compounds that do not show the characteristic carbonyl absorption
peaks cannot be determined by this method.

6. Near-Infrared Measurement of Oximes.   Aldehydes and ketones
condense with hydroxylamine to form aldoximes and ketoximes, respectively (equation 5):

$$\begin{array}{c} R \\ \diagdown \\ \diagup \\ R' \end{array} C{=}O \;\; + \;\; NH_2OH \longrightarrow \begin{array}{c} R \\ \diagdown \\ \diagup \\ R' \end{array} C{=}NOH \;\; + \;\; H_2O \qquad (5)$$

Goddu (33) suggested that oximes should be determinable with both good
sensitivity and selectivity by use of the near-infrared region.

Oximes have an extremely intense fundamental hydroxyl band at 2.78 μ,
the molar absorptivities of which are 3 to 4 times those of most alcohols and
hydroperoxides and are roughly equivalent to those of phenols. Phenols,
which absorb in the same region, therefore are expected to interfere with the
determination of oximes, but alcohols, which absorb at 2.74 to 2.76 μ, and
hydroperoxides, which absorb at 2.81 to 2.84 μ, should not interfere (33).

II. GAS CHROMATOGRAPHIC METHODS*

Carbonyl compounds are readily determined directly by conventional gas
chromatography without tailing or serious adsorption effects that impede

* Written by Morton Beroza and May N. Inscoe.

quantification, provided polar groups, such as OH or COOH, are absent. When compounds contain polar groups, they are usually derivatized. (See section on appropriate functional group.) As detailed below, carbonyl compounds and their derivatives have also been subjected to chemical reactions to facilitate their quantitative estimation.

A.   Subtraction of Carbonyl Compounds for Analysis by Gas Chromatography

In a mixture of compounds, those with a carbonyl group may often be quantitatively removed before GC injection by derivatization, or they may be "subtracted" by including in the system a reaction loop containing a chemical that will react with and retain carbonyl compounds. (See Chapter 1, Section II.F.) From chromatograms made with and without the reaction loop, peaks due to carbonyl compounds are identified and their amounts determined by difference.

Aldehydes may be quantitatively subtracted with a 6-inch length of packing consisting of 5% *o*-dianisidine on 70/80 mesh Anakrom ABS (Analabs, North Haven, Conn.), the last $\frac{1}{2}$ inch of the packing being uncoated support to prevent bleeding of the reactant (34). The only ketone retained on the column was reported to be cyclohexanone, about 20 to 50% of it being retained in passage. Some epoxides, especially the heavier ones (C_{12} and higher), were partially or completely retained. The loop was reactive over a temperature range of 50 to 175°C.

Carbonyl compounds (aldehydes and ketones) may be subtracted with a 6-inch-long reaction loop containing 20% benzidine on 60/80 mesh acid-washed Chromosorb P at 100 to 175°C, the last $\frac{1}{2}$ inch again being uncoated support to minimize bleeding of the reactant. The loop efficiently subtracts aldehydes, most ketones, and epoxides (34). Sterically hindered ketones are partially subtracted. Since benzidine is highly reactive, increasing the t_R's of other types of compound and retaining them in amounts up to 40%, quantification with this loop is unreliable. Because some of the reactant bleeds from both loops, they are generally placed after the chromatographic column to avoid contaminating it.

Use of the *o*-dianisidine and benzidine loops will usually allow differentiation between an aldehyde and a ketone. A phosphoric acid loop will "subtract" epoxides but not carbonyl compounds. (See Chapter 5, Section II). Hence this loop can serve to rule out the presence of epoxides. Hydroxylamine has also been used to subtract carbonyl compounds (35).

Carbonyl compounds are completely subtracted with $LiBH_4$ and $LiAlH_4$ in the GC stream (36). However, alcohols, esters, and epoxides are also subtracted. Although NaBH$_4$ has been used to distinguish aldehydes from ketones, the results were generally not quantitative.

The quantitativeness of each analysis utilizing subtractive loops should be checked with model compounds, since there are many instances in which these loops will not be satisfactory for quantification.

B.   Reduction of Carbonyl Compounds for Determination by Gas Chromatography

Carbonyl compounds may be reduced to alcohols by borohydrides. This reaction has been used to determine as little as 10 ppm of formaldehyde in acetaldehyde by attaching to the end of the chromatographic column a Staybrite tube, 3 $\times$ $\frac{1}{4}$ inch i.d., packed with "evenly ground" potassium borohydride (37). The methanol formed from formaldehyde is detected by a flame ionization detector much more easily than is formaldehyde. Rotenone, which gives broad multiple peaks in GC, gives a single sharp peak after treatment with sodium borohydride; the ketone group in rotenone is reduced to a hydroxyl, which is then lost by dehydration upon injection into the gas chromatograph (38).

Hexosamines have been reduced by sodium borohydride at room temperature; the hexitols were then acetylated (39) or the trifluoroacetyl derivatives formed (40) for analysis by GC.

C.   Oxidation of Aldehydes for Determination by Gas Chromatography

Higher aliphatic aldehydes have been oxidized with silver oxide and the carboxylic acid products esterified with diazomethane for GC determinations (41).

D.   Acetal Formation for Gas Chromatography

It has been found convenient on occasion to form derivatives of carbonyl compounds in a mixture without preliminary isolation and then to analyze the derivatives by GC. Aldehydes and aldehydogenic moieties in plasmalogens have been converted to stable cyclic acetals that chromatograph well (42). 2,4-Dinitrophenylhydrazones of aldehydes (C$_2$ to C$_{16}$) and ketones (C$_3$ to C$_{19}$) have been converted to dithioacetals and dithioketals for analysis by GC (43). 2-Chloroethyl acetals of aldehydes have been prepared for analysis by a gas chromatograph equipped with a halogen detector (44).

E. Regeneration of Carbonyl Compounds from Derivatives for Analysis by Gas Chromatography

A variety of derivatives have been made to aid in the separation of carbonyl compounds from mixtures. Since many of these derivatives are too polar or too involatile to be estimated well by GC, methods have been devised to regenerate the original compound from its derivative practically quantitatively so that it may be amenable to analysis by GC. One method, called flash exchange (45), is especially useful for low-boiling carbonyl compounds. They have been determined quantitatively in many instances by preparing their 2,4-dinitrophenylhydrazones and placing a mixture of the product and α-ketoglutaric acid (1:3) in a capillary tube closed at one end with 1 mg of sodium bicarbonate on the bottom. The open end is inserted through the injection port septum of a gas chromatograph and the tube is rapidly heated, causing the regenerated carbonyl compound to enter the gas chromatograph for determination in the usual manner. In some instances extraneous peaks appear. A method in which a mixture of oxalic acid and p-(dimethylamino)benzaldehyde is used in place of α-ketoglutaric acid and sodium bicarbonate is said to yield better results (46). A ten-fold excess of phthalic acid is used in another procedure to give quantitative data with C_2 to C_6 carbonyl compounds (47).

Other procedures have been devised for regenerating carbonyl compounds from their derivatives. Semicarbazones taken to dryness may be regenerated to carbonyls (C_2 to C_{14} monoaldehydes and ketones, aromatic aldehydes and ketones, some diketones) by treating them with a dilute phosphoric acid solution (48). Girard-T derivatives (up to C_{11} aldehydes) have been regenerated with methylolphthalimide (49), and those from citrus oils (citral and lower saturated aliphatic aldehydes) by formaldehyde (50).

F. Other GC Procedures for Analysis of Carbonyl Compounds

Lower saturated aldehydes (C_2 to C_5) have been converted to 2,4-dinitrophenylhydrazones and then oxidized with ozone to the corresponding carboxylic acids, which were directly determined quantitatively by GC on a phosphoric acid-polyester column (51). Since ozone reacts with double bonds, this procedure cannot be used for the determination of carbonyl compounds with double bonds. In a modification of this analysis, several low-molecular-weight carbonyl compounds were also determined (52). An ozone generator may be easily constructed for the ozonization (53).

Compounds that undergo the iodoform reaction, such as methyl ketones and CH_3CHOH- compounds, can be determined quantitatively with great sensitivity. The compound(s) is treated with sodium hypoiodite to form iodoform, which is determined by GC with electron capture detection (54). An analytical curve of concentration versus response is plotted, using pure iodoform (a yellow solid melting at 130°C) to determine the amount of compound in an analysis; iodoform is prepared by treating acetone with excess sodium hypoiodite.

Methoxyamine(O-methylhydroxylamine) forms stable derivatives of carbonyl compounds suitable for GC analysis (55, 56). It has been also used to block ketones when the carbonyl group interferes in derivatization of OH groups (57). Glyoxal, methylglyoxal, and biacetyl were converted to quinoxalines for separation by GC (58).

Isotopic exchange on GC columns has been suggested for determining different types of compounds (59). Enolizable hydrogen atoms of ketones, for example, have been exchanged quantitatively in a single pass through a GC column pretreated with deuterium (60).

III.   ELECTROANALYTICAL METHODS*

A.   Amperometric Titration of Aldehydes

The reaction of 2,4-dinitrophenylhydrazine (2,4-DNP) with carbonyl compounds can serve as the basis for an amperometric titration of these compounds. The conditions of the titration make the analysis very useful for aldehydes in particular, although some ketones also can be determined.

Early studies (61) of this titration disclosed some areas of difficulty in the analysis. The precipitate which forms can be troublesome along two lines: adsorption of the titrant on the precipitate particle surface, and electrochemical reduction of the fine suspended particles. If no special precautions are taken, the precision of the analysis can deteriorate to about ±10%. Subsequently, the entire system was studied in great detail (62), and many of the difficulties in the analysis were overcome. For example, by titrating at about −0.6 V, rather than the −1.2 V previously recommended, the nonreproducible reduction of the precipitate is eliminated. Unfortunately, however, at these potentials the current is not linearly dependent on 2,4-DNP. This is a surface effect which can be eliminated by the addition of thymol, which does not allow adsorption of titrant on the precipitate particles and thus solves the problem. Another possible solution involves the use of ethanol. Samples which are water insoluble may be made up in ethanol; if the final alcohol concentration is at least 33%, no further additives are needed.

---

* Written by Alan F. Krivis.

Apparatus. Any polarograph, recording or manual, may be used for this analysis, as outlined in Chapter 1, Section III. A very convenient type of cell (63) is sold by Sargent-Welch Scientific Co. (S-29408). However, the type of cell is not at all critical, and the writer has utilized ordinary beakers for many kinds of amperometric titrations.

Titrant. The titrant (0.1 M 2,4-DNP) is prepared from the best grade of material available. Dissolve 1 gram of 2,4-dinitrophenylhydrazine in 500 ml of 2 M HCl and allow the mixture to equilibrate for a few days. Filter the solution and standardize by a coulometric bromination (*cf.* Chapters 1 and 12).

Procedure. Mix 7.5 ml of concentrated HCl, 20 ml of 0.05% thymol solution, and 0.05 mM of aldehyde sample in the cell. Dilute the solution to 50 ml and purge with nitrogen for 10 minutes. Insert the reference and the indicating electrodes in the cell. The current is to be measured at a fixed -0.6 V. Titrate with the standard 2,4-DNP solution by adding small increments. After each addition purge for 2 minutes before making a measurement. (The Sargent-Welch cell mentioned previously will permit a more rapid titration because constant purging is permissible with this cell. In cells which do not shield the indicating electrode, the nitrogen must be shut off during the actual measurement.) Plot the current versus the volume of titrant and obtain a curve similar to that shown in Figure 2-1. The end point is the intersection of the two lines.

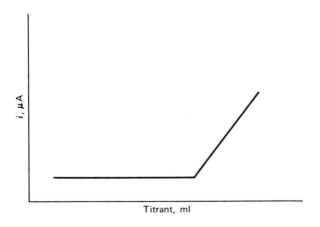

Figure 2–1. Amperometric titration curve.

If the sample is not soluble in water, dissolve it in ethanol and add to the mixture specified above. If the alcohol content is 15 to 20 ml, the thymol may be omitted.

Calculations.

Per cent of aldehyde

$$= \frac{\text{(ml titrant) (molarity titrant) (mol. wt. unknown) (100)}}{\text{(sample weight in gms) (1000)}}$$

B. Polarographic Determination of Aliphatic Aldehydes

Both of the two types of carbonyl compounds, aldehydes and ketones, can be reduced polarographically. However, it is only under unusual circumstances that aliphatic ketones are measured; nonaqueous media and tetraalkylammonium salts are necessary to reach the negative potentials where reduction of these compounds occurs. Aliphatic aldehydes are more readily reduced and therefore are detected in the usual polarographic solvent-electrolyte systems.

As a general rule, carbonyl groups (aldehydes *and* ketones) which are conjugated to unsaturated groups and/or aryl groups are more easily reduced. For this reason, it is possible to determine some ketones of commercial interest by the simple procedure outlined below. However, if difficulties are encountered with a particular sample, one of the indirect procedures described in Section C may be employed.

In neutral to basic media, many of the aliphatic aldehydes are reduced at approximately the same half-wave potential (64). Formaldehyde is the exception, in that it is reduced at a more positive potential, but it too may be determined in the same way as the other aldehydes. The procedure described below (65) will permit the analysis of most aliphatic aldehydes and many aromatic aldehydes. Specific compounds should be studied before running unknowns.

EXPERIMENTAL METHOD [ADAPTED FROM THE STUDY OF ELVING AND RUTNER (65)]*

Apparatus. Any polarograph, recording or manual, may be used in conjunction with a thermostated (25°C) H-cell containing a saturated calomel reference electrode (64). A simpler cell with a mercury pool anode also can serve for the analysis (65).

---

* Reprinted in part with the permission of the copyright owner, the American Chemical Society.

Solutions. Two different solutions (pH 6.8 and 12.7) may be used, depending on the particular unknown species and the concentration. Since some aldehydes condense very rapidly at high pH, the neutral solution may be preferred.

An aqueous stock solution of 0.12 M lithium hydroxide-lithium chloride should be prepared by dissolving 5.1 grams of LiOH and 2.4 grams of LiCl per liter of solution. The stock solution is diluted 1:1 to give a pH of 12.7.

A solution of 0.05 M lithium acetate-0.05 M lithium chloride, with enough acetic acid to adjust the pH to 6.8, should be used as a background electrolyte without further dilution. It may be prepared by dissolving 5.1 grams of lithium acetate dihydrate and 2.1 grams of lithium chloride per liter. Acetic acid should be added to bring the pH to 6.8.

Procedures. (a) Transfer 5 ml of the lithium hydroxide-lithium chloride stock solution to a 10-ml volumetric flask. Add a weighed sample of aldehyde (or an aliquot of a sample solution) of about 0.01 mM, dissolve, and make up to volume. Transfer the solution to a polarographic cell, deoxygenate with nitrogen for 10 minutes, and obtain the polarogram. Calculate the diffusion current for the polarographic wave of the unknown and compare it to a calibration curve in order to ascertain the aldehyde concentration.

Most aliphatic aldehydes have half-wave potentials in the vicinity of − 1.9 V. However, the exact value for a particular compound should be determined before running unknowns. This may be done while obtaining the data for the calibration curve.

(b) Transfer a weighed sample containing about 0.01 mM to a 10-ml volumetric flask and dissolve in a few milliliters of lithium acetate-lithium chloride solution. Make up to volume with the solution and transfer the mixed sample solution to the polarographic cell. Follow the remainder of the procedure outlined in part (a) above.

C. Indirect Polarographic Determination of Carbonyls

The reduction of simple ketones occurs at very negative potentials, in comparison to aldehydes, as mentioned in suction B. Therefore, unless unusual conditions are employed, ketones are not normally measured directly. However, many carbonyl compounds react with a variety of amines, hydrazines, hydroxylamines, etc., to form imines and oximes which are easily reducible. Two compounds, in particular, have been found to be very useful for the determination of a wide variety of aldehydes and ketones: butylamine (66) and semicarbazide (67).

There are advantages and disadvantages to the use of either reagent. In some instances, butylamine forms imines which are reduced after the second polarographic oxygen wave so that nitrogen purging of the solutions may not be necessary. However, butylamine does not have a wholly delightful odor and one might not wish to use it unless forced to do so. Semicarbazide hydrochloride almost forms its own background electrolyte, and the preparation and handling of the reagent solution are far more pleasant; semicarbazone formation, however, is quite sensitive to the reaction medium and the specific carbonyl compound. The latter sensitivity can, in fact, be analytically very valuable. The rate of reaction of even similar compounds is different, and it is possible to analyze mixtures of carbonyl compounds by a differential rate technique (67).

EXPERIMENTAL METHOD* [ADAPTED FROM VAN ATTA AND JAMIESON (66) AND KRIVIS AND SUPP (67)]

Apparatus. Any polarograph, recording or manual, may be used with an H-cell containing a saturated calomel reference electrode (*cf.* Chapter 1, Section III). The cell should be thermostated in a constant-temperature bath (25°C) to maintain a high level of precision.

Solutions. All solutions should be prepared with reagent-grade or other high-quality materials.

Ammonium chloride stock solution (5 M) may be prepared by dissolving 267.5 grams of ammonium chloride per liter of water.

An aqueous semicarbazide-buffer stock solution can be prepared by dissolving 22.3 grams of semicarbazide hydrochloride, 8.2 grams of sodium acetate, and 6.0 grams of acetic acid per liter. This stock solution is 0.2 M semicarbazide-0.1 M acetic acid-0.1 M sodium acetate and is diluted 1:1 with methanol in use.

Procedure for Butylamine Reagent. Transfer 20 ml of 5 M ammonium chloride, 17.0 ml of *n*-butylamine, and 25 ml of water to a 100-ml volumetric flask. Add a sample containing about 1 to 10 mM of carbonyl compound and make up to volume with water. Transfer an aliquot to the polarographic cell and obtain the diffusion current for the polarographic wave for the imino compound. (If the imine is reduced beyond about -1 V, it is not necessary to remove dissolved oxygen before running the polarogram. At other potentials, the solution should be purged with nitrogen for 5 to 10 minutes.) Compare the diffusion current to a previously prepared calibration curve for the compound of interest.

Procedure for Semicarbazide Reagent. Transfer 25 ml of semicarbazide solution and 0.2 ml of 1% gelatin solution to a 50-ml volumetric flask. Add a sample

---

* Portions are reprinted from *Analytical Chemistry* with the permission of the copyright owner, the American Chemical Society.

containing about 0.05 mM of carbonyl compound (dissolved in methanol) to the flask and make up to volume with methanol. Mix thoroughly, allow time for the reaction to take place, and then transfer an aliquot to the polarographic cell. Mount the electrodes and purge with nitrogen for 10 minutes. Obtain the polarogram and calculate the diffusion current for the semicarbazone reduction wave. Compare the diffusion current for the sample to a previously prepared calibration curve to determine the amount of carbonyl present.

Procedure for Semicarbazide Reagent for Analysis of Mixtures.   All solutions should be deoxygenated with nitrogen and equilibrated at 25°C before mixing. The proportions of solutions and reagents are the same as those listed above. The actual volumes will be dependent on the physical size of the particular cell in use. Basically, the solution at final dilution should contain 1 *mM* sample, 0.1 *M* semicarbazide, and 0.05 *M* acetate-acetic acid, in 50% methanol.

Transfer the required volume of buffer solution and methanol to the cell and deoxygenate for 10 minutes. Then inject the requisite volume of de-oxygenated carbonyl sample, in methanol, into the solution in the cell and mix thoroughly. The current at the potential for the semicarbazone reduction should be measured as a function of time.

A separate calibration curve for the desired semicarbazone should be prepared. The calibration data are obtained by reacting known amounts of carbonyl compound with the reagent solution and measuring the current after the reaction is complete. A plot of current versus carbonyl concentration can be made from the data.

The analysis of mixtures of carbonyls is performed by plotting log *i* versus time (Figure 2-2) and extrapolating the line for the slower-reacting species back to time zero; the extrapolated current is the total current due to the faster-reacting material. The current then is related to concentration via the calibration curve described in the preceding paragraph.

IV.   NUCLEAR MAGNETIC RESONANCE METHODS*

The hydrogen of the aldehyde function is extremely convenient in determining aldehydes, both aromatic and aliphatic. The data of Klinck and Stothers (68) on a series of aromatic aldehydes show that the formyl hydrogen chemical shift is 9.65 to 10.44 ppm from tetramethylsilane. Generally, electron-donating groups increase the shielding of the formyl hydrogen, and

---

* Written by Harry Agahigian.

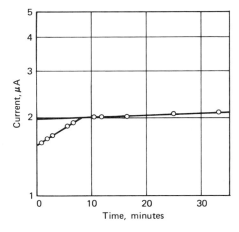

Figure 2-2. Kinetic plot for analysis of mixtures of carbonyl compounds.

electron-withdrawing groups decrease the shielding. Some of the aldehydes studied were benzaldehyde, *m*-tolualdehyde, *p*-tolualdehyde, *p*-anisaldehyde, *m*-fluorobenzaldehyde, *o*-nitrobenzaldehyde, and *o*-chlorobenzaldehyde. The ortho-substituted benzaldehydes exhibit a decrease in the shielding of the formyl hydrogen, possibly because of steric effects. In a hypothetical case in which the material is an aldehyde or a ketone, the nmr scan will establish this without resorting to chemical analysis.

A. Aldehydes

The further work of Klinck and Stothers (69) on aldehydes covers the aliphatic series as well as the α,β-unsaturated aldehydes. The factors causing change in the chemical shift of the formyl proton are the deshielding of the carbonyl function, the electronic distribution through the conjugated system, and effects due to other bond anisotropy. The following are some typical chemical shifts:

	Chemical Shift Formyl Proton
Acetaldehyde	9.73
Propionaldehyde	9.73
Octanal	9.69
2-Ethylbutanal	9.51
Acrolein	9.54
Crotonaldehyde	9.43
Furfural	9.67

These are measured in parts per million in relation to tetramethylsilane.

A total of 36 aliphatic aldehydes were determined, and the effects of conjugation are discussed. The sample preparation was carried out in the usual manner with a concentration of 5% (w/v) in carbon tetrachloride.

B. Ketones

These compounds are not as readily treated by ^{1}H nmr as some other functional groups because of the lack of a proton. However, the electronegativity of the carbonyl function and hence its effect on hydrogens on the alpha and beta carbon atoms can be used as the criterion for establishing the presence of the ketone. Acetone, the simplest ketone, has a single ^{1}H resonance line at 2.17 ppm from TMS. Methyl ethyl ketone may be easily identified because of the methyl and ethyl resonances; however, other properties may also have to be utilized. As the organic ketone structure becomes more complex, the ease of identification diminishes.

One group in which the effect of the ketone function will lead to unequivocal assignment is the aromatics. The electronegativity of this group produces a deshielding on the ortho protons; thus the presence of a resonance at 7.91 ppm strongly suggests a carboxyl or carbonyl. The aldehyde and acid, as well as the ester, function can be eliminated by viewing the rest of the nmr scan. α,β-Unsaturated ketones can be treated in the same manner. A compound such as 2,2-disubstituted indanone-1 does not contain alpha hydrogens but does contain a beta proton on the aromatic ring:

The effect of carbonyl on the aromatic hydrogen causes a deshielding similar to that in substituted benzenes.

Boykin, Turner, and Lutz (70) have studied carbonyl derivatives of cyclopropanes, using various solvents for identification through changes in chemical shifts of the substituents and the ring protons.

The derivatization of ketones and aldehydes by means of hydrazines is a method of identification that is in general use, and the publication of Karabatsos et al. (71) treats this subject in detail. The ^{1}H nmr spectra were obtained at 60 MHz. The compounds studied have the formula

$$R_1R_2C{=}NZ$$

where R_1 and R_2 = alkyl or H and Z = hydrazone substituent, and the stereochemistry is assigned relative to the functional group on the nitrogen.

The authors state that the derivatives are all in the imine form and that there is no spectral evidence for the azo or enamine form. The relative geometry with respect to the imine bond is established. For example, the alpha hydrogen of acetaldehyde 2,4-dinitrophenylhydrazone appears at lower field when cis to the dinitro-containing ring and at higher field when trans. There is no consistent trend, however, and no real correlations can be made.

Hydrazones were prepared from the following: hydrazine, *o*-nitrophenylhydrazine, *m*-nitrophenylhydrazine, *p*-nitrophenylhydrazine, 2,4-dinitrophenylhydrazine, *p*-methylphenylhydrazine, *p*-chlorophenylhydrazine, 2-carboxyphenylhydrazine, 4-carboxyphenylhydrazine, semicarbazide, and thiosemicarbazine. The carbonyl compounds used were acetaldehyde, acetone, methyl ethyl ketone, benzyl methyl ketone, diethyl ketone, methyl isopropyl ketone, diisopropyl ketone, formaldehyde, isobutylaldehyde, and other common carbonyl compounds.

The highlights of the article by Karabatsos et al. are the details related to solvent effects and the data used in determining the syn and anti configurations. Benzene, acetone, dimethyl sulfoxide, and dibromomethane were used as solvents. The authors report the chemical shifts of the alpha hydrogens, the beta hydrogens, and wherever applicable the cis and trans isomers, as well as the =NH resonances.

The nature of the nmr experiment is such that some conclusions can be reached concerning the type of carbonyl compounds present. The classification of a carbonyl compound as an aldehyde or a ketone is a trivial exercise, and often the identification of simple ketones is also relatively simple. However, in the case of mixtures or complex systems identification may be difficult; then, other spectral data will establish the presence of carbonyl or other functional groups. The ^{1}H nmr is essentially "seeing" the effect of the functional group on an adjacent proton. Carbon-13 measurements are more direct but require nonroutine instrumentation; also, there is a sensitivity problem.

V.   RADIOCHEMICAL METHODS*

The radioreagents most useful for determining the carbonyl group are, in general, the radioactive counterparts of reagents widely employed in conventional methods. Thus substituted phenylhydrazines, semicarbazide, and thiosemicarbazide, appropriately labeled, are particularly valuable in radiochemical methods. Other useful reagents are radioactive silver in the presence

* Written by D. Campbell.

of alkali; ammoniacal silver nitrate followed by elemental radioiodine; sodium cyanide, with or without hydrolysis of the cyanohydrin; and radio-silver derivatives of inactive thiosemicarbazones.

A. Conversion to Substituted Phenylhydrazones

Of the several radiochemical methods described for determining the carbonyl group, perhaps the one of greatest potential for the analysis of weighable amounts of low-molecular-weight compounds is that based on treatment of the sample with p-iodophenylhydrazine-^{131}I and separation of the substituted phenylhydrazones by liquid column chromatography (72) (cf Chapter 1, Section V). Although the only separation reported is that of the labeled derivatives of o- and m-nitrobenzaldehyde, the method is applicable in principle to any carbonyl compound quantitatively yielding a p-iodophenylhydrazone which can be isolated on a liquid chromatographic column. The analysis of mixtures is also feasible if the derivatives can be adequately separated.

The column is monitored during and after development from top to bottom with an electrically driven scintillation detector, connected to a recording count-rate meter to obtain the distribution of activity. The radiochromatogram of the separated derivatives of the isomeric aromatic aldehydes is illustrated in Figure 2-3, in which radioactivity in arbitrary units is plotted against distance from the top of the column.

The number of millimoles of a derivative on the column is obtained by integrating the area of the sample peak and comparing this area with that

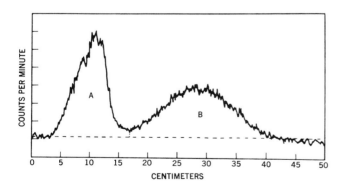

Figure 2–3.   Recorder trace of developed chromatographic column. Zone A, o-nitro-benzaldelyde p-iodophenylhydrazone-[131]. Zone B, m-nitrobenzaldehyde p-iodophenyl-hydrazone-[131]. Reprinted from *Anal. Chem.*, **27**, 1898 (December 1955). Copyright 1955, by the American Chemical Society. Reprinted by permission of the copyright owner.

obtained using a known amount of a suitable derivative prepared from the same lot of radioreagent and separated on an identical reference column. The latter is surveyed whenever a sample column is assayed to take into account the decay associated with the 8-day half-life of [131]I.

Important advantages of the technique are that individual compounds in a mixture may be determined if the derivatives are separable and that the column may be monitored during development. The method as reported, however, does not include procedures for quantitative derivatization of the carbonyl group or for separation of labeled derivatives from excess reagent.

The principle may be employed also with paper and thin-layer chromatography, using [3]H- or, preferably, ^{14}C-labeled reagent.

Carbonyl groups formed on the surfaces of thin polyethylene sheets exposed to an oxidative medium have been determined by treating exposed specimens with a 0.01 M solution of 2,4-dinitrophenylhydrazine-^{14}C in 2 N sulfuric acid (73). Excess radioreagent was removed by washing the polymer sheets with water, and the residual activity on the surfaces was assayed using a thin-window flow gas detector. The specific activity of the reagent was determined by similarly assaying a source prepared by depositing a small amount on a planchet. The count rate at the specimen surface is proportional to the surface concentration of the carbonyl group for sheets having thicknesses greater than the range of the ^{14}C beta particle in polyethylene. Assays of sheets having thicknesses less than this range require corrections for beta particles passing through the specimen.

The 2,4-dinitrophenylhydrazone of benzaldehyde-^{14}C has been employed for determining this aromatic aldehyde, using a modification of the radioreagent-direct isotope dilution technique wherein the substance to be determined is made the radioreagent (74). When liquid scintillation counting is used, the severe quenching effect of the nitro group makes nitro-substituted phenylhydrazines somewhat disadvantageous reagents for the radiochemical determination of carbonyl compounds. However, electrolytic reduction of the nitro groups of the derivatives has been shown to be an effective means of circumventing this interference (75).

B.   Conversion to p-Bromophenylhydrazones Followed by Neutron Activation

p-Bromophenylhydrazones that are separable by paper chromatography may be determined on the chromatograms by subjecting the isolated spots of the derivatives, together with a similar spot of a standard bromine compound, to a flux of thermal neutrons. The two nonradioactive isotopes of bromine are converted to radioisotopes of mass numbers 80 and 82:

$$^{79}_{35}Br + ^{1}_{0}n \rightarrow ^{80}_{35}Br + \gamma$$

$$^{81}_{35}Br + ^{1}_{0}n \rightarrow ^{82}_{35}Br + \gamma$$

The marked tendency of the stable bromine isotopes to absorb or "capture" neutrons makes bromine a sensitive indicator in this analytical technique, which has been termed "derivative activation chromatography" (76). Quantitative analysis is limited by background arising from the presence of activatable impurities in the paper; a sensitivity of approximately 0.002 μg of bromine was observed for gamma scintillation counting of ^{82}Br with a neutron flux of 10^{12} neutrons/cm^2/sec, using the arbitrary criterion of equal sample and background activities. Liquid scintillation counting after elution of the spots may also be employed, although at a sacrifice in sensitivity. The full potential of the method can be realized, however, only if a high-flux source of neutrons, such as a nuclear reactor, is available for activation.

C. Conversion to Semicarbazones, Thiosemicarbazones, and Hydrazones

Semicarbazide-^{14}C is reportedly useful for determining low concentrations of the carbonyl group in oxidized cellulose (77), in which it may occur in both the aldehyde and the ketone form in the presence of larger amounts of the carboxyl group.

Semicarbazide-^{35}S has been shown to be a valuable radioreagent for determining certain steroids by derivatization of their ketone groups (78, 79). Thiosemicarbazones of these substances are formed in very good yield, and the high specific activity at which the reagent may be prepared offers excellent sensitivity. The tendency of derivatives to adhere to paper and to silica gel requires relatively large amounts of carrier (for steroid analysis) in separation of the products by paper and thin-layer chromatography. The polarity of the derivatives is enhanced, although at a considerable sacrifice of material, by acetylating the thiosemicarbazones to the 2,4-diacetyl compounds. However, the acetylated substances are less strongly adsorbed onto glass, and losses due to such adsorption are reduced by acetylation. The need for completeness of the derivatizing reactions and for quantitative handling of solutions is obviated by adding a definite amount of activity of the tritiated steroid to be determined to the original specimen of biological fluid as an indicator of overall recovery. Specific activity of the steroid should preferably be high enough to permit the weight added to be negligible in comparison with the weight of steroid in the specimen (cf. Chapters 1 and 2 concerning the use of a second radioisotope as an indicator). Liquid scintillation counting of the $^{35}S/^{3}H$ pair may be performed using the same instrument settings employed with the $^{14}C/^3H$ pair. A modification of this assay method has been presented for determining ^3H and ^{35}S simultaneously under conditions of variable quenching (80).

Acethydrazide-3-iodopyridinium-^{131}I bromide has been proposed as a selective reagent for carbonyl groups in steroids (81). The reagent forms hydrazone derivatives with ketosteroids having carbonyl groups at position 3, if these are also α,β-unsaturated, and at positions 17 and 20, but not with those having ketone groups at position 11. Also, R_f values of the derivatives of steroids having two reactive C=O groups are much lower than those of steroids having only one such group.

D.   Oxidation with Alkaline Silver Nitrate Solution

Aldehydes are oxidized rapidly by a solution of silver nitrate in excess ammonium hydroxide (Tollen's reagent), or by silver nitrate in the presence of alkali, yielding a deposit of metallic silver:

$$R - C{\overset{\displaystyle O}{\underset{\displaystyle H}{\big\Vert}}} \quad + \quad 2Ag^+ \quad + \quad 2OH^- \quad \longrightarrow \quad R - C{\overset{\displaystyle O}{\underset{\displaystyle OH}{\big\Vert}}} \quad + \quad 2Ag \quad + \quad H_2O$$

The reaction is not specific for aldehydes, however, and a variety of other easily oxidized compounds also yield the free metal. Treatment of a paper chromatogram containing isolated reducing agents with silver nitrate in an ammoniacal or alkaline environment results in the formation of metallic silver at the spots corresponding to these compounds. Measurement by radioisotopic means of the amount of silver produced in comparison with that liberated by known quantities of the substance makes possible quantitative analyses in the micro and submicro ranges.

The method initially reported (82) employs ^{110m}Ag, which is a gamma emitter having a half-life of 253 days. Chromatograms containing the isolated substances of interest are dipped in a solution of $^{110m}AgNO_3$, dried, and sprayed with a 0.5 N solution of sodium hydroxide in aqueous ethanol to provide the alkaline silver oxidizing environment. Care is necessary in handling the treated chromatograms to avoid reduction of silver ion by the paper. After a suitable reaction or "development" time, the strips are dipped in dilute ammonia to solubilize the silver oxide and then washed to remove the silver-ammonia complex. Segments containing the elemental silver are cut out for radioassay by G-M counting. The method, when calibrated by chromatographing aliquots of a standard solution of the compound to be determined, proved useful for quantitative estimation of 0.017 to 0.22 μM

of aldehyde group in glucose and galactose and of ketone group in fructose and sorbose. Certain other reducing agents were determined in a similar manner.

Tollen's reagent, which is used extensively as a spray for the qualitative chromatographic detection of various reducing substances, is prepared to avoid large excesses of ammonia, since high concentrations of free base decrease sensitivity. When employed for the quantitative radiochemical determination of aldoses, ketoses, and other oxidizable compounds on paper chromatograms, the reagent must contain a much higher silver concentration than is required in qualitative reagents. Solutions suitable for use in a radioreagent method (83) are prepared by dissolving 9 to 17% by weight of $AgNO_3$ in concentrated ammonium hydroxide. In addition to the high Ag^+ concentration, a small amount of NaOH is necessary for reproducible spot development. After spraying, the chromatograms are heated at 105°C for an optimum time determined by the composition of the reagent. Excess Tollen's reagent is removed by washing the paper successively with 0.02% aqueous ammonia and water. The dried strips are then immersed in an acidified solution of $K^{131}I$ and KIO_3, whereupon the metallic silver is converted immediately to $Ag^{131}I$ by the elemental iodine. Uncombined ^{131}I is removed by again washing the strips with water.

The chromatograms are assayed by end-window G-M counting, using a slotted absorber to determine the activity in small, defined areas. The region of proportionality between integrated spot activity and amount of reducing agent taken must be determined for each substance examined. However, the linear relationship observed between maximum spot activity and the logarithm of spot content is more useful for rapid, routine analyses. The average error for amounts of glucose and fructose greater than 0.1 μM was $\pm6.7\%$ and $\pm3.5\%$, respectively. Nonuniformity of deposition of the sprayed reagent adversely affects both the accuracy and the reproducibility of this technique and has been recognized as a probable source of serious errors (84).

After an extensive study of the oxidative properties of acetone-silver nitrate-ammonia solutions containing an equimolar or less than equimolar proportion of ammonia (84), a modified Tollen's reagent was developed for radiochemically determining submicro amounts of monosaccharides (85). Acetone solutions of $AgNO_3$ containing less than equimolar amounts of ammonia offer higher sensitivity than those containing an equivalent quantity or an excess of base; the most effective oxidant is a reagent in which the ratio of $AgNO_3$ to ammoniacal $AgNO_3$ is 1:2. These more sensitive reagents are applied by quickly dipping the chromatograms several times into the solution, drying them each time, and, after the final immersion, developing the spots at 50°C in a humid atmosphere without drying. Excess reagent is removed by a brief treatment with 10% $Na_2S_2O_3$ and several water washings. The excellent uniformity of silver deposition afforded by this method of application, together with the use of a more efficacious developing reagent, has significantly improved accuracy and sensitivity. With these modifications, glucose can be determined in amounts of 0.008 to 0.06 μM. Average errors for the compounds studied, in the region of proportionality, ranged from ± 1.1 to 4.2%. Sensitivity is dependent to a large extent on the care with which the chromatographic solvents are removed.

E.   Conversion to Cyanohydrins or Carboxylic Acids

Aldehydes, as well as certain low-molecular-weight ketones, react with sodium cyanide under a variety of conditions to yield cyanohydrins. If it is advantageous to do so, the latter may then be hydrolyzed to carboxylic acids containing one more carbon atom than the precursor. Incorporation of ^{14}C into the reagent was first proposed for determining small concentrations of reducing end groups in polysaccharides (86):

$$R - \overset{\overset{\text{O}}{\|}}{\underset{\underset{\text{R}}{|}}{C}} + Na^{14}CN + H_2O \longrightarrow R - \overset{\overset{\text{OH}}{|}}{\underset{\underset{\text{R}'}{|}}{C}} - {}^{14}CN + NaOH$$

$$R - \overset{\overset{\text{OH}}{|}}{\underset{\underset{\text{R}'}{|}}{C}} - {}^{14}CN + NaOH + H_2O \longrightarrow R - \overset{\overset{\text{OH}}{|}}{\underset{\underset{\text{R}'}{|}}{C}} - {}^{14}\overset{\overset{\text{O}}{/\!/}}{C} - ONa + NH_3$$

Reducing sugars readily form the cyanohydrin quantitatively. Early work in this series of investigations, for example, showed that the reaction between stoichiometric amounts of NaCN and D-arabinose is essentially complete within 48 hours at 20°C in a solution 0.4 M in each, buffered with $NaHCO_3$ to prevent degradation (87).

The specific activity of the $Na^{14}CN$ may be determined by treating a definite quantity of pure glucose with excess reagent and measuring the residual activity after volatilizing unconsumed cyanide with formic acid (88). The reaction is performed in buffered solution to prevent decomposition of the glucose by NaOH normally present in solutions of NaCN. In practice, 20 λ of 0.05 M glucose solution, containing sufficient NH_4Cl to provide an excess of 30% over the sodium ion, is mixed in a sealed tube with 0.003 to

0.010 mM of Na^{14}CN dissolved in 50 λ. The tube is then heated for 24 hours at 55°C. Excess cyanide is removed as HCN by repeated evaporations to dryness with 10% formic acid solution, and the activity in the residue is measured. Nonvolatile radioactive impurities are determined by omitting the glucose.

A procedure similar to that used for standardizing the sodium cyanide, that is, without hydrolysis of the cyanohydrin, has been applied to the determination of reducing sugars and reducing end groups in polysaccharides (89) Conditions were modified somewhat, however, to permit the convenient determination of 0.0001 to 0.001 mM of sugar with a two- to threefold excess of cyanide. To 20 λ of sugar solution (containing <0.001 mM) in a tube is added 10λ of 0.6 M NH$_4$Cl solution and 10 λ of Na^{14}CN solution containing NaOH. The radioreagent solution is 0.26 M in cyanide and 0.5 M in total sodium ion. The tube is sealed immediately after addition of the reagents and then heated for 24 hours at 50 to 55°C. Excess cyanide is removed by adding 10% formic acid, evaporating to dryness at 60°C, and repeating the addition and evaporation twice with water. The specific activity of the cyanide is determined concurrently by treating a known amount of glucose in the same manner.

Triplicate analyses of the sample and glucose, as well as the blank, were recommended; the blank corrects for nonvolatile labeled impurities. Aldose monosaccharides and certain disaccharides consume 1 mole of cyanide per mole of sugar. When sugars and polysaccharides combining with more than 1 equivalent of cyanide are analyzed, controls containing known amounts of these substances must be included. Twenty-four determinations of glucose showed the method to be capable of a coefficient of variation of ±1.4%. Application of the technique (using a NaHCO$_3$ buffer at room temperature) to clinical dextrans containing reducing end groups gave results in reasonable agreement with those obtained by established chemical methods.

The usefulness of the approach has also been demonstrated in estimating 1,3′ and 1,4′ linkages in dextrans after periodate oxidation and hydrolysis (90). D-Glucose arising from 1,3′ linkages and D-erythrose produced from 1,4′ structures were transformed into labeled cyanohydrins, which were then hydrolyzed to yield the alkali metal salts of D-glycero-D-guloheptonic acid and D-arabonic acid, respectively. The hydrolysis was performed in the presence of excess cyanide without isolating the intermediates. Alkali metal salts of the acids were added as carriers before isolation and repeated crystallization of the derivatives. The method was standardized by treating known amounts of D-glucose and D-erythrose with the same lot of labeled cyanide, adding definite quantities of carrier, and comparing the specific activities of the salts isolated from the samples with those of salts prepared from standards.

F. Conversion to Radiosilver Derivatives of Inactive Thiosemicarbazones

Low concentrations of the carbonyl group in oxidized cellulose have reportedly been determined by first treating the sample with nonradioactive thiosemicarbazide, then removing excess reagent, and finally adding $^{110m}AgNO_3$, which becomes bound to the thiosemicarbazone (77):

$$R-CH{=}N-NH-\overset{\displaystyle S}{\overset{\|}{C}}-NH_2+{}^{110m}Ag^+ \rightarrow R-CH{=}N-N{=}\overset{\displaystyle S-{}^{110m}Ag}{\overset{|}{C}}-NH_2 \quad +H^+$$

However, the conditions used for forming the silver derivative must be selected to avoid interference from the carboxyl group present in the product. Copper and mercury also yield insoluble metal compounds with thiosemicarbazide and thiosemicarbazones.

REFERENCES

Absorption Spectrophotometric

1. G. L. Barthauer, F. V. Jones, and A. V. Metler, *Ind. Eng. Chem., Anal. Ed.*, **18**, 354 (1946).
2. H. L. Rees and D. H. Anderson, *Anal. Chem.*, **21**, 989 (1949).
3. D. T. Englis and D. J. Hanahan, *Ind. Eng. Chem., Anal. Ed.*, **16**, 505 (1944).
4. D. T. Englis and M. Manchester, *Anal. Chem.*, **21**, 591 (1949).
5. J. J. Pepe, I. Kniel, and M. Czuha, Jr., *Anal. Chem.*, **27**, 755 (1955).
6. W. Mathewson, *J. Am. Chem. Soc.*, **42**, 1277 (1920).
7. P. E. Toren and B. J. Heinrich, *Anal. Chem.*, **27**, 1986 (1955).
8. F. H. Lohman, *Anal. Chem.*, **30**, 972 (1958).
9. G. R. Lappin and L. C. Clark, *Anal. Chem.*, **23**, 541 (1951).
10. A. Mendelowitz and J. P. Riley, *Analyst*, **78**, 704 (1953).
11. D. E. Jordan and F. C. Veatch, *Anal. Chem.*, **36**, 120 (1964).
12. G. R. A. Johnson and G. Scholes, *Analyst*, **79**, 217 (1954).
13. R. A. Basson, *Anal. Chem.*, **38**, 637 (1966).
14. F. Feigl, *Spot Tests in Organic Analysis*, 7th Ed., Elsevier, Amsterdam, 1966, p. 434.
15. E. Eegriwe, *Z. Anal. Chem.*, **110**, 22 (1937).
16. M. J. Boyd and M. A. Logan, *J. Biol. Chem.*, **146**, 279 (1942).
17. C. E. Bricker and H. R. Johnson, *Ind. Eng. Chem., Anal Chem.*, **17**, 400 (1945).
18. D. A. McFayden, *J. Biol. Chem.*, **158**, 107 (1945).
19. C. E. Bricker and W. A. Vail, *Anal. Chem.*, **22**, 720 (1950).
20. R. H. Still, K. Wilson, and B. W. Lynch, *Analyst*, **93**, 805 (1968).

21. W. C. Tobie, *Ind. Eng. Chem.*, *Anal Ed.*, **14**, 405 (1942); *Food Res.*, **6**, 15 (1941).
22. L. Segal, *Anal. Chem.*, **23**, 1499 (1951).
23. F. C. Scott, *Analyst*, **70**, 374 (1945).
24. D. E. Kramm and C. L. Kolb, *Anal. Chem.*, **27**, 1076 (1955).
25. C. L. Hoffpauir and R. T. O'Connor, *Anal. Chem.*, **21**, 420 (1949).
26. K. Fishbeck and L. Neundeubel, *Z. Anal. Chem.*, **104**, 81 (1936).
27. G. Deniges, *Compt. Rend.*, **150**, 529 (1910).
28. W. J. Blaedel and F. E. Blacet, *Ind. Eng. Chem.*, *Anal. Ed.*, **13**, 449 (1941).
29. C. L. Hoffpauir, G. W. Buckaloo, and J. D. Guthrie, *Ind. Eng. Chem.*, *Anal. Ed.*, **15**, 605 (1943).
30. A. C. Rayner and C. M. Jephcott, *Anal. Chem.*, **33**, 627 (1961).
31. S. Dal Nogare, T. O. Norris, and J. Mitchell, Jr., *Anal. Chem.*, **23**, 1473 (1951).
32. E. Cochran and C. A. Reynolds, *Anal. Chem.*, **33**, 1893 (1961).
33. R. F. Goddu, *Anal. Chem.*, **30**, 1707 (1958).

Gas Chromatographic

34. B. A. Bierl, M. Beroza, and W. T. Ashton, *Mikrochim. Acta*, **1969**, 637.
35. Y. G. Osokin, V. S. Feldblum, and S. I. Kryukov, *Neftekhimiya*, **6**, 333 (1966).
36. F. E. Regnier and J. C. Huang, *J. Chromatog. Sci.*, **8**, 267 (1970).
37. S. Harrison, *Analyst*, **92**, 773 (1967).
38. D. G. Carlson and W. H. Tallent, *J. Chromatog. Sci.*, **8**, 276 (1970).
39. M. B. Perry and A. C. Webb, *Can. J. Biochem.*, **46**, 1163 (1968).
40. Z. Tamura, T. Imanari, and Y. Arakawa, *Chem. Pharm. Bull.* (*Tokyo*), **16**, 1864 (1968).
41. J. C. M. Schogt, P. H. Begemann, and J. H. Recourt, *J. Lipid Res.*, **2**, 142 (1961).
42. P. V. Rao, S. Ramachandran, and D. G. Cornwell, *J. Lipid Res.*, **8**, 380 (1967).
43. M. M. E. Metwally, C. H. Amundson, and T. Richardson, *J. Am. Oil Chemists' Soc.*, **44** (3), 136A (1967).
44. A. Karmen, *J. Lipid Res.*, **8**, 234 (1967).
45. J. W. Ralls, *Anal. Chem.*, **36**, 946 (1964).
46. L. A. Jones and R. J. Monroe, *Anal. Chem.*, **37**, 935 (1965).
47. S. Nishi, *Tokyo Kogyo Shikensho Hokoku*, **60**, 344 (1965).
48. I. R. Hunter and M. K. Walden, *J. Gas Chromatog.*, **4**, 246 (1966).
49. D. F. Gadbois, P. G. Scheurer, and F. J. King, *Anal. Chem.*, **40**, 1362 (1968).
50. W. L. Stanley, R. M. Ikeda, S. H. Vannier, and L. A. Rolle, *J. Food Sci.*, **26**, 43 (1961).
51. P. Ronkainen and S. Brummer, *J. Chromatog.*, **28**, 253 (1967).
52. P. Ronkainen, S. Brummer, and H. Suomalainen, *J. Chromatog.*, **28**, 270, 443 (1967).
53. M. Beroza and B. A. Bierl, *Anal. Chem.*, **38**, 1976 (1966); *Mikrochim. Acta*, **1969**, 720.
54. R. D. Stevens and C. H. Van Middelem, *J. Agr. Food Chem.*, **14**, 149 (1966).
55. H. M. Fales and T. Luukkainen, *Anal. Chem.*, **37**, 955 (1965).
56. M. G. Horning, E. A. Boucher, A. M. Moss, and E. C. Horning, *Anal. Letters*, **1**, 713 (1968).

57. E. M. Chambaz and E. C. Horning, *Anal. Biochem.*, **30**, 7 (1969).

58. V. Medonos, V. Ruzicka, J. Kalina, and A. Marhoul, *Collection Czech. Chem. Commun.*, **33**, 4393 (1968).

59. T. Balint and L. Szepesy, *J. Chromatog.*, **30**, 433 (1967).

60. M. Senn, W. J. Richter, and A. L. Burlingame, *J. Am. Chem. Soc.*, **87**, 680 (1965).

Electroanalytical

61. E. V. Zobov and Yu. S. Lyalikov, *Chem. Abstr.*, **52**:14431h (1958).

62. A. Berka, J. Dolezal, J. Janata, and J. Zyka, *Anal. Chim. Acta*, **25**, 379 (1961).

63. H. A. Laitinen and L. W. Burdett, *Anal. Chem.*, **22**, 833 (1950).

64. I. M. Kolthoff and J. J. Lingane, *Polarography*, 2nd Ed., Interscience, New York, 1952; L. Meites, *Polarographic Techniques*, 2nd Ed., Interscience, New York, 1965.

65. P. J. Elving and E. Rutner, *Ind. Eng. Chem., Anal. Ed.*, **18**, 176 (1946).

66. R. E. Van Atta and D. R. Jamieson, *Anal. Chem.*, **31**, 1217 (1959).

67. A. F. Krivis and G. R. Supp, *Anal. Chem.*, **35**, 1411 (1963).

Nuclear Magnetic Resonance

68. R. E. Klinck and J. B. Stothers, *Can. J. Chem.*, **40**, 1071 (1962).

69. R. E. Klinck and J. B. Stothers, *Can. J. Chem.*, **44**, 45 (1966).

70. D. W. Boykin, A. B. Turner, and R. E. Lutz, *Tetrahedron Letters*, 817 (1967).

71. G. J. Karabatsos, F. M. Vane, R. A. Taller, and N. Hsi, *J. Am. Chem. Soc.*, **86**, 3351 (1964).

Radiochemical

72. W. M. Stokes, W. A. Fish, and F. C. Hickey, *Anal. Chem.*, **27**, 1895 (1955).

73. K. Nitzl, *Papier*, **21**, 393 (1967).

74. B. Y. Cobb, *Proc. 5th Natl. Peanut Res. Conf.*, 1968, pp. 7–17.

75. Y. Nagase, S. Baba, and T. Ido, *J. Pharm. Soc. Japan*, **84**, 202 (1964); *Chem. Abstr.*, **61**:10279h.

76. J. M. Steim and A. A. Benson, *Anal. Biochem.*, **9**, 21 (1964).

77. P. Rochas and L. Gavet, *Melliand Textilber.*, **49**, 1123 (1968).

78. A. Riondel et al., *Proc. Endocrinol. Soc.*, Chicago, June 21–23, 1962, p. 16.

79. J. F. Tait, et al., *in* S. Rothchild, Ed., *Advances in Tracer Methodology*, Vol. 2, Plenum, New York, 1965, pp. 227–235.

80. J. K. Weltman and D. W. Talmage, *Intern. J. Appl. Radiation Isotopes*, **14**, 11 (1963).

81. W. S. Ruliffson, H. L. Lang, and J. P. Hummel, *J. Biol. Chem.*, **201**, 839 (1953).

82. M. Jaarma, *Acta Chem. Scand.*, **8**, 860 (1954).

83. J. Z. Beer, *Talanta*, **8**, 809 (1961).

84. J. Z. Beer, *J. Chromatog.*, **11**, 247 (1963).

85. J. Z. Beer, *Carbohydrate Res.*, **1**, 297 (1965).

86. H. S. Isbell, *Science*, **113,** 532 (1951).

87. H. S. Isbell, J. V. Karabinos, H. L. Frush, N. B. Holt, A. Schwebel, and T. T. Galkowski, *J. Res. Natl. Bur. Std.*, **48,** 163 (1952).

88. J. D. Moyer and H. S. Isbell, *Anal. Chem.*, **29,** 393 (1957).

89. J. D. Moyer and H. S. Isbell, *Anal. Chem.*, **30,** 1975 (1958).

90. J. D. Moyer and H. S. Isbell, *Anal. Chem.*, **29,** 1862 (1957).

3 Carboxylic acids, salts, esters, amides, chlorides, anhydrides, and nitriles

I. ABSORPTION SPECTROPHOTOMETRIC METHODS*

CARBOXYLIC ACIDS, SALTS, ESTERS, AMIDES, CHLORIDES, ANHYDRIDES, AND NITRILES

A. Direct Infrared Measurement

Initial evaluation of the infrared spectra of this series of compounds could give the impression that this approach is generally feasible as a basis for quantitative determinations. Several strong characteristic infrared absorption bands are available for each of these compounds. For example, the C=O stretching vibrations of these compounds are shown in Table 3-1. In addition other bands are available, such as the C—O and O—H stretching bands in acids at 6.94 to 7.17 μ and 2.8 to 3 μ, respectively. Esters exhibit intense C—O stretching vibrations at 7.81 to 9.09 μ.

However, the fact that many absorption bands are exhibited by these compounds results in crowded spectra, especially for complex mixtures in which several of the compounds are present. Multicomponent mixtures often show high overall absorption throughout the commonly used infrared region. As with alcohols, intermolecular association through hydrogen bonding presents difficulties in arriving at analytical conditions of general usefulness for carboxylic acids and amides.

In the case of carboxylic acids more success is experienced if they are first converted to the esters or salts. This approach eliminates the association problems. For example, Childers and Struthers (1) were able to analyze mixtures of acids in the form of their salts in mineral oil mulls.

---

* Written by J. Gordon Hanna.

Table 3-1.   C=O Stretching Vibrations for Carboxyl Compounds

Compound type	C=O stretching, wavelength, μ
Carboxylic acids	5.65–5.80
Salts	6.20–6.40
Esters	5.75
Acid chlorides	5.55
Acid anhydrides (doublet)	
Open chain	5.35–5.49 and 5.55–5.71
Cyclic	5.40–5.55 and 5.59–5.75
Amides	5.90–6.10

For routine analyses in situations where the compound matrix is well defined and the exact position of the most intense absorption is known, a direct infrared determination should be considered. Each specific situation requires evaluation of the applicability of the technique.

B.   Direct Ultraviolet Measurement

Ultraviolet measurements are of minor importance for the determination of these compounds. The carboxyl group itself has only weak absorptions except when supported by conjugated unsaturation or an aromatic nucleus. In some situations the region 250 to 280 $m\mu$ can be used for aromatic carboxylic compounds.

C.   Determination after Chemical Reaction

Ferric Hydroxamate Reaction.   The strong absorption of the red ferric chelates of hydroxamic acids is the basis for the most important spectrometric methods for the determination of these compounds. Carboxylic esters, acid chlorides, and anhydrides treated with hydroxylamine produce hydroxamic acids by the reations shown in equations 1 to 3:

$$RCOOR' + NH_2OH \xrightarrow{OH^-} RCONHOH + R'OH \qquad (1)$$

$$RCOCl + NH_2OH \longrightarrow RCONHOH + HCl \qquad (2)$$

$$(RCO)_2O + NH_2OH \longrightarrow RCONHOH + RCOOH \qquad (3)$$

The formation of the ferric hydroxamate complex is assumed to proceed according to the reaction shown in equation 4:

$$3RCONHOH + 3H_2O + Fe^{3+} \rightleftharpoons Fe(RCONHO)_3 + 3H_2O + 3H^+ \quad (4)$$

Feigl and his coworkers (2) first introduced the use of the ferric-hydroxamic acid reaction as a spot test for carboxylic esters. Esters, carboxylic acids, acid halides, anhydrides, and amides can be determined directly or indirectly on the basis of this reaction.

METHOD FROM GODDU, LEBLANC, AND WRIGHT (3)

Apparatus.   Beckman Model B spectrophotometer with 1-cm cells.

Reagents.   Hydroxylamine hydrochloride, 12.5%. Reflux 12.5 grams of the solid material with 100 ml of methanol.

Sodium hydroxide, 12.5%. Reflux 12.5 grams of sodium hydroxide with 100 ml of methanol.

Ferric perchlorate stock solution. Dissolve 5.0 grams of ferric perchlorate in 10 ml of 70% perchloric acid and 10 ml of water. Dilute to 100 ml with anhydrous 2B alcohol. Cool under a tap as the alcohol is added. Alternatively, weigh 0.8 gram of iron wire in a 50-ml beaker. Add 10 ml of 70% perchloric acid and heat until the iron dissolves. Be careful with this reaction as the iron dissolves rapidly when the acid is hot. Cool the beaker and transfer the contents to a 100-ml volumetric flask with 10 ml of water and dilute to volume with anhydrous 2B alcohol. Cool under at tap as the alcohol is added.

Reagent solution. Add 40 ml of stock solution to a 1-liter volumetric flask, introduce 12 ml of 70% perchloric acid, and dilute to volume with anhydrous 2B alcohol. Add the alcohol in 50- to 100-ml increments, cooling between each addition, until the perchloric acid has been diluted to about 10% of its original concentration. The ferric ion concentration of this solution is 5.7 mM, and the acid concentration is 0.16 M.

Procedure for the Determination of Esters.   To prepare the alkaline hydroxylamine reagent, mix equal volumes of 12.5% hydroxylamine hydrochloride solution and 12.5% methanolic sodium hydroxide solution and filter the precipitated sodium chloride on Whatman No. 40 filter paper. The clear filtered reagent solution is usable for 4 hours.

Dissolve the sample in anhydrous 2B alcohol, isopropanol, or benzene (water solution may also be used but produces a slight loss of sensitivity), to obtain a concentration between 0.01 and 0.001 M. Pipet 5 ml of the sample solution into a 25-ml flask with a ground-glass joint. (If a more concentrated solution is to be analyzed, or if a calibration curve is to be prepared, a smaller volume of sample should be used and enough solvent added to make the total volume 5 ml in the flask—e.g., 2 ml of sample and 3 ml of solvent.) Add 3 ml of the filtered alkaline reagent solution to each sample flask and to a blank that contains 5 ml of solvent. Add a boiling chip to each flask; then place the flasks on a hot plate on low heat and attach reflux condensers. Reflux the samples for 5 minutes. Remove the flasks from the hot plate, do not wash down the condensers, cool to room temperature, and wash the contents into a 50-ml volumetric flask with the ferric perchlorate reagent. Dilute to volume with the reagent. Shake the flasks to ensure complete solution of the initially precipitated ferric hydroxide. After several minutes read the absorbance of the samples against the reagent blank at the maximum absorbance wavelength (Table 3-2). Prepare a calibration curve based on the same ester as that determined or, less preferably, on an ester of the same acid as the acid portion of the ester to be determined. Use the same solvent in all cases.

Table 3-2.   Absorption Maxima and Molar Absorptivities of Ferric Hydroxamates Formed from Different Esters (3)

Ester	Wavelength, $m\mu$	Molar absorptivity $\times 10^3$
Ethyl formate	520	1.06
Ethyl acetate	530	.1.10
n-Butyl acetate	530	1.06
n-Amyl acetate	530	1.05
Phenyl acetate (impure)	530	0.99
Triacetin	530	3.33
Ethyl propionate	530	1.02
γ-Butyrolactone	530	1.11
Methyl *n*-butyrate	530	1.06
n-Butyl *n*-butyrate	530	1.05
n-Amyl *n*-butyrate	530	0.91
Dimethyl malonate	520	1.72
Dimethyl maleate	520	1.53
Dimethyl adipate	530	2.04
Pentaerythritol tetracaproate	530	3.89
Methyl oleate	530	1.00
Methyl benzoate	550	1.13
n-Butyl benzoate	550	1.08
Benzyl benzoate	550	1.13
Methyl *p*-toluate	550	0.94
Dimethyl *o*-phthalate	540	1.48
Dimethyl isophthalate	540	2.42
Dimethyl terephthalate	540	2.37

To analyze a mixture of anhydride and ester for both constituents, a calibration curve for the anhydride, based on the conditions described for the ester determination, must also be prepared. The concentration of anhydride is determined under neutral hydrolysis conditions. Then the absorbance due to that concentration of anhydride under basic hydrolysis conditions is subtracted from the observed absorbance before the ester content is calculated.

Procedure for Anhydrides and Lactones in the Presence of Esters. To prepare the hydroxylamine reagent, neutralize a portion of the methanolic hydroxylamine hydrochloride to the phenolphthalein end point with the 12.5% methanolic sodium hydroxide. Filter the precipitated sodium chloride on Whatman No. 40 filter paper. The clear filtered reagent is usable for at least 4 hours.

If the sample contains an anhydride, dilute it with benzene to obtain a concentration of 0.01 to 0.001 M. Dry the benzene before use—for example, over anhydrous calcium sulfate for 24 hours. The solvents used for the ester determination can serve also for lactones.

Perform the analysis in the same manner as that described for the ester analysis but use a 10-minute reflux time. Prepare calibration curves based on the analysis of known concentrations of anhydride or lactone. Under these neutral conditions the calibration curves are not always straight lines.

The reaction of anhydrides of carboxylic acids is similar to that of esters under alkaline conditions, forming 1 mole of hydroxamic acid per mole of anhydride. With neutral conditions, hydroxylamine is a strong enough base to react with the anhydride, yet no reaction occurs with esters. This permits the selective determination of anhydrides. About 65% of the color obtained under alkaline conditions develops when a 10-minute reflux time is used with the neutral reagent. Of the esters tested, only phenolic esters, polyesters, lactones, and formates react with the neutral reagent. Acid chlorides react in both procedures and can be determined.

Acids, most amides, and nitriles do not interfere with these procedures. High concentrations of carbonyls will consume much of the reagent and necessitate the use of higher concentrations of hydroxylamine.

According to Cheronis and Ma (4), carboxylic acids can be determined colorimetrically by conversion to the corresponding acid chlorides and measured as the ferric hydroxamate complexes.

Bergman (5) studied the amide-hydroxylamine reaction and used it as a basis for the colorimetric determination of amides of hydroxamic acids:

$$RCONH_2 + NH_2OH \rightarrow RCONHOH + NH_3 \qquad (5)$$

Reagent. Ferric chloride, 0.74 M, in 0.1 N hydrochloric acid.

Apparatus. Klett-Summerson photoelectric colorimeter with a No. 54 filter (spectral range, 500 to 570 mμ).

Procedure. Dissolve the sample in water to obtain an amide concentration of 5 to 10 $\times$ 10^{-3} M. To 1 ml or less of the sample solution add 2 ml of the alkaline hydroxylamine reagent prepared from equal volumes of 2 N hydroxylamine sulfate and 3.5 N sodium hydroxide. Add sufficient water to give a total volume of 3 ml, if necessary. Keep at the temperature and for the time indicated in Table 3-3. Cool rapidly to room temperature and add 1 ml each of 3.5 N hydrochloric acid and ferric chloride reagent. Read the absorbance within 5 minutes and obtain the concentration of amide from a calibration curve relating concentration to absorbance.

The relationship between absorbance and concentration units was found to be linear for all compounds tested. A No. 50 filter (spectral range, 470 to 530 mμ) was found necessary for fluoroacetamide because fluoroacethydroxamic acid has its maximum absorbance near 500 mμ. Amides react much more slowly than the corresponding esters. In many cases this difference may be used to analyze for both types of compounds successively by the same method.

NITRILES

A. Direct Infrared Determination

The nitrile group, C$\equiv$N, has characteristic absorptions in the infrared region at 4.35 to 4.55 μ. Although the intensities of the absorptions are relatively low, the nitrile peaks appear in a region of the spectrum essentially free from interference by other functional groups. The dearth of other general methods for the determination of the nitrile group makes the direct infrared approach particularly useful. For example, Dinsmore and Smith (6) determined the acrylonitrile content of Buna-N copolymers by comparing the nitrile absorption at about 4.47 μ with the absorptions of known nitrile concentrations.

B. Measurement after Chemical Reaction

Because of the various reaction characteristics exhibited by different nitrile compounds, no satisfactory general chemical method for their deter-

Table 3-3. Optimal Conditions for Conversion of Amides to Hydroxamic Acids (5)

Compound	Temperature, °C	Reaction time, minutes	Klett reading, units/μM
Acetamide	60	120	90
	26	480	103
N-Methylacetamide	60	420	57
Acetanilide	60	180	70
N¹-Acetylsulfanilamide	60	240	70
Acetylglycine	60	240	35
Fluoroacetamide	26	60	62
Formamide	26	60	80
	60	10	75
Dimethylformamide	26	240	45
Succinamide	60	120	85
Caprolactam	60	420	41
Asparagine	60	180	38
Glutamine	60	180	35
Glutathione	60	120	48
Glycylglycine	60	120	25
Nicotinamide	26	480	45
N¹-Methylcicotinamide methosulfate(I)	26	360	45
Nicotinic acid methylamide	60	240	30
Coramine (nicotinic acid diethylamide)	60	480	6
Pantothenic acid, calcium salt	26	300	89
Barbitone	100	45	1.7
Pentobarbitone	60	300	1.5
Phenobarbitone	100	120	7.5
Evipan, sodium	100	30	9

mination has been developed, regardless of the type of end measurement applied. Methods for individual nitriles based on spectrometric end measurements are about as scarce.

One rather indirect method of this kind has been published by Haslam and Newlands (7) for the determination of acrylonitrile in air at concentrations of about 0 to 150 mg/m³. A known volume of air is passed through a solution of lauryl mercaptan in isopropanol. Ethanol is added to catalyze the reaction of acrylonitrile and mercaptan:

$$CH_2{=}CHCN + C_{12}H_{25}SH \rightleftharpoons C_{12}H_{25}SCH_2CH_2CN \qquad (6)$$

After the mixture is acidified with acetic acid, a volume of standard iodine solution equivalent to the lauryl mercaptan originally used is added. The excess iodine, which is proportional to the amount of acrylonitrile present in the original sample of air, is then determined spectrophotometrically. The mercaptan-iodine reaction is shown in equation 7:

$$2C_{12}H_{25}SH + I_2 \rightarrow C_{12}H_{25}SSC_{12}H_{25} + 2HI \qquad (7)$$

α,β-Unsaturated nitriles, esters, and aldehydes, if present, are expected to interfere.

METHOD FROM HÁSLAM AND NEWLANDS (7)

Pipet 10 ml of mercaptan solution (1% lauryl mercaptan in isopropanol) and 5 ml of isopropanol into each of two absorbers. Place the absorbers in crushed solid carbon dioxide. Connect in series the two absorbers, a gas meter, and a source of suction. Regulate the flow rate at about 2 liters/minute and draw about 40 liters of the atmosphere to be tested through the absorbers. Disconnect the absorbers and allow them to come to room temperature.

To prepare the blank, pipet 10 ml of the mercaptan solution, 5 m. of isopropanol, and 1 ml of 5% ethanolic potassium hydroxide into a 50-ml volumetric flask, mix, and set aside for 4 minutes. Then pipet 2 ml of acetic acid into this solution, mix, titrate with 0.03 N iodine solution to a faint yellow end point, and dilute to 50 ml with isopropanol.

Pipet 1 ml of 5% ethanolic potassium hydroxide into each absorber, mix, and allow to stand for 4 minutes. Then pipet 2 ml of acetic acid into each absorber, mix, and add exactly twice the volume of 0.02 N iodine solution used for the blank. Transfer the contents of each absorber to the same 100-ml volumetric flask, wash the absorbers with isopropanol, add the washings to the flask, and dilute to the mark with isopropanal.

Measure the absorbance of the test solution against the blank solution. From a previously prepared calibration curve relating absorbance to concentration of acrylonitrile, determine the concentration of acrylonitrile in the sample solution.

A Spekker absorptiometer with 1-cm cells and Ilford No. 601 spectrum violet filters was recommended by Haslam and Newlands for measurement of the absorbance. Undoubtedly, however, other instruments can be used.

The average recovery reported was about 90% of the theoretical value.

Table 3-4. Typical Procedures for Esterification of Carboxylic Acids for Quantification by Gas Chromatography

Alcohol moiety	Typical reagents	Types of compounds and references
Methyl	Diazomethane	Aromatic acids or herbicides,[1-5] fatty and resin acids,[6] phosphorus acids,[1] unsaturated fatty acids[7]
	Tetramethylammonium hydroxide	Organic and fatty acids,[8-10] aliphatic and aromatic mono and dibasic acids,[11] indole acids,[12] digestion products of tissues[13]
	BF$_3$-methanol	Herbicides,[14] lower fatty acids[15]
	Dimethyl sulfate-methanol	Herbicides[16]
	Methanol-sulfuric acid	Aliphatic acids,[17] pyridine carboxylic acids[18]
	(1) SOCl$_2$; (2) methanol	Pyridine carboxylic acids[18]
Ethyl	Diazoethane	Aromatic or phosphorus acids[1]
2-Chloroethyl	2-Chloroethanol-H$_2$SO$_4$	Mono- and short-chain dicarboxylic acids[19]
Butyl	BF$_3$-butanol	Fatty acids[20]
	Tetrabutylammonium hydroxide	C_1–C_7 monocarboxylic acids and lactic acid[21]
Pentafluorobenzyl	α-Bromo-2,3,4,5,6-penta-fluorotoluene and K$_2$CO$_3$	C_2–C_{18} monocarboxylic acids[22]
Trimethylsilyl	Bis(trimethylsilyl)acetamide or bis(trimethylsilyl)trifluoroacetamide	Aromatic acids[23]
Halomethyl-dimethylsilyl	Halomethyldimethylchlorosilane and diethylamine	Pesticidal aromatic acids[24]

[1] C. W. Stanley, *J. Agr. Food Chem.*, **14**, 321 (1966).

[2] J. M. Devine and G. Zweig, *J. Assoc. Offic. Anal. Chemists*, **52**, 187 (1969).

[3] E. A. Woolson and C. I. Harris, *Weeds*, **15**, 168 (1967).

[4] D. G. Crosby and J. B. Bowers, *Bull. Environ. Contam. Toxicol.*, **1**, 104 (1966).

[5] G. Yip, *J. Assoc. Offic. Agr. Chemists*, **45**, 367 (1962); **47**, 1116 (1964).

[6] F. H. M. Nestler and D. F. Zinkel, *Anal. Chem.*, **39**, 1118 (1967).

[7] R. T. Holman and J. J. Rahm, *Progress in Chemistry of Fats and Other Lipids*, Vol. IX; *Polyunsaturated Acids*, Part 1, R. T. Holman, ed., Pergamon, New York, 1966.

[8] E. W. Robb and J. J. Westbrook, III, *Anal. Chem.*, **35**, 1644 (1963).

II.   GAS CHROMATOGRAPHIC METHODS*

A.   Esterification of Carboxylic Acids

Although carboxylic acids may be analyzed directly by GC, quantitative analysis by this method is generally avoided. Instead, derivatives are usually formed to reduce polarity and increase volatility, thereby facilitating quantitative analysis by overcoming adsorption difficulties, improving thermal stability, and allowing analyses to be conducted at lower temperatures.

Esters are the derivatives of acids most often used in GC. Table 3-4 lists typical procedures employed for their preparation.

Diazomethane has been widely used to prepare methyl esters quantitatively. (See the discussion in Chapter 1, Section II.B, for the method of use.) Diazoethane may be similarly employed to prepare ethyl esters.

Methyl esters of carboxylic acids have also been formed quantitatively by the pyrolysis of tetramethylammonium salts of mono- and dicarboxylic acids in the hot (ca. 350°C) injection port of the gas chromatograph. The procedure facilitates the analysis of aqueous solutions of acids and, by converting the acids to their salts, avoids the loss of volatile fatty acids during analysis. Polyunsaturated acids may be similarly analyzed if the highly alkaline solution of the salts is made nearly neutral (pH 7.5 to 8.0) with acetic acid before injection (8). The method fails, however, for oxalic, malonic, and hydroxy acids.

Other procedures for methylation employ methanol plus dimethyl sulfate or a catalyst (BF_3, HCl, H_2SO_4). Liliedahl (9) compared a number of

9 D. T. Downing, *Anal. Chem.*, **39**, 218 (1967).

10 D. T. Downing and R. S. Greene, *Anal. Chem.*, **40**, 827 (1968).

11 J. J. Bailey, *Anal. Chem.*, **39**, 1485 (1967).

12 R. H. Leonard, *J. Gas Chromatog.*, **5**, 323 (1967).

13 J. MacGee, *J. Gas Chromatog.*, **6**, 48 (1968).

14 W. H. Gutenmann and D. J. Lisk, *J. Agr. Food Chem.*, **11**, 304 (1963).

15 J. Churacek, M. Drahokoupilova, P. Matousek, and K. Komarek, *Chromatographia*, **2**, 493 (1969).

16 J. E. Scoggins and C. H. Fitzgerald, *J. Agr. Food Chem.*, **17**, 156 (1969).

17 M. Rogozinski, *J. Gas Chromatog.*, **2**, 163 (1964).

18 H. Liliedahl, *Acta Chem. Scand.*, **20**, 95 (1966).

19 A. Karmen, *J. Lipid Res.*, **8**, 234 (1967).

20 E. A. Crowell and J. F. Guymon, *Am. J. Enol. Viticult.*, **20**, 155 (1969).

21 J. W. Schwarze and M. N. Gilmour, *Anal. Chem.*, **41**, 1686 (1969).

22 F. K. Kawahara, *Anal. Chem.*, **40**, 2073 (1968).

23 K. Blau, *Clin. Chim. Acta*, **27**, 5 (1970).

24 C. A. Bache, L. E. St. John, Jr., and D. J. Lisk, *Anal. Chem.*, **40**, 1241 (1968).

* Written by Morton Beroza and May N. Inscoe.

methylation techniques for use in the analysis of pyridine carboxylic acids; although no single method was adequate for all acids under investigation, quantitative results could be obtained by at least one procedure for every acid tried.

Esterification with larger alcohol moieties follows the general pattern used in the methylation procedures. When higher sensitivities are needed, derivatives that give enhanced response with electron capture detection (2-chloroethyl, pentafluorobenzyl, halodimethylsilyl) are useful.

B.   Subtraction of Free Carboxylic Acids by Gas Chromatography

It is sometimes convenient to determine fatty acids directly, that is, without derivatization, and good quantitative data have been obtained for the volatile fatty acids, in the C_1 to C_6 range, for example, in rumen fluid (10–12). Phosphoric acid is added to the packing, and/or the sample is treated with acid to repress ionization, dimerization, or adsorption of the constituent acids during the chromatography (13, 14). With a packing consisting of 0.25% Carbowax 20M—0.4% isophthalic acid on 200-μ glass microbeads, C_2 to C_{18} free fatty acids have been determined (15).

Subtraction of free acids from the carrier gas stream is accomplished by including in the GC pathway a reaction loop (described in Chapter I, Section II.F). The reactor loop (16) is 6 inches long and contains 10% by weight of zinc oxide physically mixed with the packing used to chromatograph the free acids [25% LAC-2R-446 plus 2% phosphoric acid on 60/80 mesh Celite 545 (17)]. Carboxylic acids, except for those considered hindered (α-alkyl substituted), were quantitatively "subtracted" from the chromatogram. Peaks of hindered acids were delayed, were drastically reduced in height, and tailed badly. (This behavior can be used to recognize alpha substitution in unknown acids.) Chromatograms with and without the reaction loop can serve to show which are the acid peaks and, by difference, how much of each acid is present. The reaction loop has a limited capacity and must be checked periodically and replaced when saturated. Zinc oxide has been used to subtract carboxylic acids quantitatively without affecting the chromatography of aldehydes and aldehyde-esters formed in the pyrolysis of ozonized fatty acid methyl esters (18).

C.   Decarboxylation of Carboxylic Acids for GC Determination

Substituted malonic acids may be decarboxylated in the injection port of the gas chromatograph, and the resulting monocarboxylic acids analyzed by GC (19). The injection temperature was 190 to 220°C, the higher temperature being used with disubstituted malonic acids. On-column pyrolytic decarboxylation at 385°C, followed by GC determination of the decarboxylated product, has also been used in a sensitive analytical method for the herbicide Picloram (4-amino-3,5,6-trichloropicolinic acid) (20).

α-Hydroxy acids treated with periodic acid are cleaved to form CO_2 and an aldehyde or ketone which may be determined by GC. (See Chapter 1, Section II.E.) It is also possible to measure the CO_2 generated to determine the amount of α-hydroxy acid in a molecule by the method described in Section D, which was devised primarily for the determination of aromatic and aromatic-type heterocyclic carboxylic acids.

D.   Microdecarboxylation of Aromatic, Aromatic-Type Heterocyclic, and α-Hydroxy Carboxylic Acids for Quantitative Determination by Gas Chromatography

METHOD OF MA, SHANG, AND MANCHE (21)

After heating the carboxylic acid in the presence of quinoline and cupric carbonate in a closed reaction vessel, the CO_2 formed is led into a gas chromatograph and determined by its peak height, which varies linearly with the amount of carboxylic acid in the sample:

$$R{-}COOH \xrightarrow[\Delta]{\text{catalyst}} RH + CO_2$$

Apparatus.   A gas chromatograph equipped with a 2-foot column packed with 25% (w/w) SF 96 silicone oil on 30/60 mesh Chromosorb (type not specified) at 51°C and a thermal conductivity detector is used with helium as the carrier gas; the flow rate is 90 ml/minute. The reaction vessel, shown in Figure 3-1, is inserted into the system with side arm A connected to the helium source and side arm B to the injection port. The central portion, C, of the vessel, which is 20 mm o.d. and 90 mm long, has a capacity of 8 ml. Side arms A and B are 8 mm o.d. and 50 mm long and are connected to

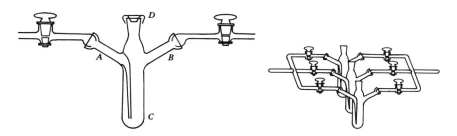

Figure 3–1. Single-reaction-vessel unit.   **Figure 3–2.** Multiple-reaction-vessel units.

stopcocks by means of ball-and-socket joints; side arm *A* has a sealed-in delivery tube. The vessel is closed with the rubber cap stopper, *D*.

A heating block (22) serves to heat the reaction mixture in *C*.

Several reactions may be conducted simultaneously with the multiple-reaction vessels shown in Figure 3-2.

Reagents.   Synthetic quinoline and cupric carbonate (Mallinkrodt Chemical Works) were used without purification.

Procedure.   The sample, containing 0.05 to 0.15 meq of carboxyl group, is weighed and inserted into the bottom of the vessel, *C* (Figure 3-1). Cupric carbonate (10 $\pm$ 0.05 mg) is added, and then 0.20 ml quinoline from a micro-syringe. The vessel is closed with stopper *D*, and side arms *A* and *B* are connected to the stopcocks. Minute amounts of high-vacuum grease are applied to the joints. After helium has been passed through the reaction vessel for 5 minutes to displace the air, the amplifier and recorder are switched on. A steady base line should be obtained; otherwise flushing is continued until the base line becomes steady. The two stopcocks are now turned to stop the passage of the helium through the vessel, and the mixture in the reaction vessel is heated to 225°C for 45 minutes. The reaction vessel is tapped occasionally so that the liquid condensing on the walls drops to the bottom of *C*. The vessel is then cooled to 100°C. When both stopcocks are opened, the helium carrier gas sweeps the CO_2 into the gas chromatograph and the peak height is determined. The peak emerges in 4 to 5 seconds.

Results and Discussion.   Linear relationships were found between the amount of sample (0.05 and 0.15 meq) and the quantity of CO_2 produced for *o*-, *m*-, and *p*-hydroxybenzoic acids, *o*-aminobenzoic acid, *o*-chlorobenzoic acid, 2,5-dihydroxybenzoic acid, 2,4-dichlorobenzoic acid, 2,4- and 3,5-dinitro-benzoic acids, and 2-furoic acid. Calibration curves are constructed for each acid. (They do not pass through zero, probably because the CO_2 formed is the sum of that from the acid and that from the cupric carbonate.) The method is said to be more specific and precise than manometric procedures such as that of Hubacher (23).

Beroza (24) reported that aromatic-type heterocyclic carboxylic acids (e.g., nicotinic, isonicotinic, pyrazine 2,3-dicarboxylic, and quinolinic acids) and α-hydroxy acids are readily decarboxylated by essentially the same reaction procedure but that some of the acids require as much as 2 hours for complete reaction. The procedure outlined should be applicable to these compounds with the longer heating period.

A more suitable chromatographic column—1.5 m, 6-mm-o.d. column containing 35/60 mesh Porapak Q (Waters Associates, Framingham, Mass.) (25)—would improve the accuracy and specificity of this determination.

The apparatus of Norton, Turner, and Salmon (see Chapter 8, Section II) could also be used in place of that described.

E. Hydroxy Acids by Gas Chromatography

The polar OH and COOH groups of hydroxy acids are normally derivatized to secure good quantitative data by GC. Radin (26), in his review of the isolation, structure determination, and quantification of hydroxy fatty acids, considers GC the method of choice for separating mixtures of hydroxy fatty acids into their components for quantification, nonhydroxy fatty acids being first separated by solvent partition, precipitation, or chromatography. Some typical procedures used to analyze hydroxy fatty acids are given in Table 3-5. The procedures, in general, follow those already outlined for the derivatization of the individual functional groups. (See Chapter 1, Section II.A–D for OH and this chapter, Section II.A for COOH.) For various reasons—avoidance of interference, ease or rapidity of analysis, more complete reaction—certain derivatives are preferred to others.

Some reports indicate that the OH group need not be derivatized in order to determine saturated hydroxy fatty acids shorter than 20 carbon atoms (27); the use of all-glass, well-siliconized columns containing lightly loaded packings is believed to help toward this end (26). Quantitative data obtained with the OH group free are probably less reliable than those obtained with the OH derivatized.

Some hydroxylated fatty acids having the OH close to a double bond are decomposed (dehydrated) in GC, and acetylation does not prevent this (28).

The analysis of α-hydroxy acids by periodate oxidation to an aldehyde ($RCHOHCOOH \rightarrow RCHO$), which is readily determined by gas chromatography, has already been discussed. See Chapter 1, Section II.E.

Acidic reaction products from the oxidation of hexoses (polyhydroxy monocarboxylic acids) in alkaline solution were analyzed by GC after conversion to lactones with HCl and then trimethylsilylation, using hexamethyldisilazane and trimethylchlorosilane (29).

F. Determination of Keto Acids by Gas Chromatography

Keto acids may normally be determined by GC after suitable derivatization of the carboxyl group (see Chapter 3, Section II.A). As part of a GC method for acids of the Krebs cycle and related compounds, methoximes of keto acids (as trimethylsilyl esters) proved useful (30). In another procedure, because pyruvic and α-ketoglutaric acids can form two products by acid-catalyzed methanolysis or treatment with diazomethane, these acids were converted to quinoxalones and treated with bis(trimethylsilyl)acetamide (BSA) and pyridine for GC (31):

RCOCOOH + (structure with NH₂ groups) ⟶ (quinoxalinone structure with R) —BSA→ (silylated quinoxalinone structure with Si(CH₃)₃)

Table 3-5. Typical Procedures Used to Determine Hydroxy Acids Quantitatively by Gas Chromatography [a]

Derivatives of			
Carboxyl	Hydroxyl	Reagents	Compounds and references
Ethyl	Acetyl	(1) Ethanolic HCl; (2) Pyridine-acetic anhydride	Panthotenates[1]
Methyl	TMS	(1) Diazomethane; (2) HMDS + TMCS	Phenolic acids in human urine,[2,3] in cerebrosides,[4] in oils[5]
TMS	TMS	BSTFA + TMCS or BSA + TMCS	Homovanillic and vanilmandelic acids,[6] Krebs cycle acids,[7] bile acids[8]
Methyl	Methyl	Diazomethane-methanol	Anacardic (6-alkylsalicylic) acids[9]
Methyl or propyl	Methyl or propyl	Diazomethane or diazopropane	Hydroxydichlorophenoxy-acetic acids[10]

[a] Abbreviations: TMS = trimethylsilyl; HMDS = hexamethylsilazane; TMCS = trimethylchlorosilane; BSTFA = bis(trimethylsilyl)trifluoroacetamide; BSA = bis-(trimethylsilyl)acetamide.

[1] A. R. Prosser and A. J. Sheppard, *J. Pharm. Sci.*, **58**, 718 (1969).
[2] F. Karoum, C. R. J. Ruthven, and M. Sandler, *Clin. Chim. Acta*, **20**, 427 (1968).
[3] J. Gentz, B. Lindblad, S. Lindstedt, and R. Zetterstroem, *J. Lab. Clin. Med.*, **74**, 185 (1969).
[4] P. Capella, C. Galli, and R. Fumagalli, *Lipids*, **3**, 431 (1968).
[5] A. Prevot and C. Barbati, *Rev. Franc. Corps Gras*, **15**, 157 (1968).
[6] T. J. Sprinkle, A. H. Porter, M. Greer, and C. M. Williams, *Clin. Chim. Acta*, **25**, 409 (1969).
[7] M. G. Horning, E. A. Boucher, A. M. Moss, and E. C. Horning, *Anal. Letters*, **1**, 713 (1968).
[8] P. Back, *Z. Klin. Chem. Klin. Biochem.*, **7**, 365 (1969).
[9] J. L. Gellerman and H. Schlenk, *Anal. Chem.*, **40**, 739 (1968).
[10] N. C. Glaze and M. Wilcox, *J. Chromatog.*, **34**, 391 (1968).

G.   Determination of Amino Acids by Gas Chromatography

The great importance of proteins and their constituent amino acids has recently sparked considerable interest in the determination of the amino acids by GC. Although automated analytical techniques using column chromatography are available, potential advantages of the GC approach are more rapid analyses and a reduction in the size of sample needed. Many derivatives have been proposed to allow the quantitative determination of the twenty amino acids commonly found in proteins. Much of the difficulty in choosing the best derivative stems from the fact that the twenty amino acids contain a dozen different functional groups, and a single scheme that will permit quantitative analysis of *all* of the amino acids is the desired objective. Degradative procedures have, in general, been ruled out, since they remove characteristic differences between some of the amino acids, and the major thrust recently has been directed toward a procedure giving complete, or at least reproducible, derivatization of polar or unstable functional groups. Since a recent article by Blau (32) discusses this subject in ample detail, it will not be reviewed here.

Several of the methods advanced for protein amino acids and some related compounds are listed in Table 3-6. The method of Gehrke and Stalling (33a), in which the protein amino acids are determined as trifluoroacetyl *n*-butyl esters, appears to have gained acceptance, and the derivatives are now available commercially (Regis Chemical Co., Chicago, Ill.). Trimethylsilyl ether-esters formed with bis(trimethylsilyl)acetamide in a single-step derivatization of amino acids give peaks except for arginine (34); sixteen of these derivatives are now commercially available (Pierce Chemical Co., Rockford, Ill.). Trifluoroamino acid methyl esters are also being advanced as suitable derivatives. The use of fluorinated derivatives (*N*-pentafluoropropionyl and *N*-heptafluorobutyryl) has been suggested as a means of greatly improving the sensitivity of analysis and decreasing the time involved (35).

H.   Determination of Cyclopropenoid and Cyclopropanoid Acids by Gas Chromatography

The cyclopropene acids are present as glycerides in several edible oils; because they are considered harmful (36), methods are needed to assure the absence of these acids in oils. Schneider et al. (37) treated the oils with sodium methoxide in methanol to form the methyl esters of the fatty acids, which were then converted to an ether and ketone derivative with silver nitrate in methanol for quantitative determination by GC. With oils containing less than 5% cyclopropenoid fatty acids, normal fatty acids were first

Table 3-6.   Some Recent Derivatization Procedures for Determining Amino Acids Quantitatively by Gas Chromatography [a]

Derivatives of		Comments and reference
Carboxyl	Amino	
Butyl	TFA	Methyl ester hydrochlorides interesterified to butyl ester hydrochlorides, and *N*-TFA derivatives formed in sealed tube;[1,2] 20 amino acids analyzed on two separate columns[3]
Methyl	TFA	20 amino acids separated on two separate columns;[4] 20 amino acids separated at three temperatures[5]
TMS	TMS	BSA used to form TMS ether-esters;[6] 12 sulfur-containing amino acids analyzed;[7] iodoamino acids and related compounds reacted with BSA and subnanogram amounts detected with EC detection;[8] BSTFA used in a comprehensive method for 20 amino acids[9]
Methyl	Alkylidene and alkyl	Alkylidene and alkyl amino acid esters are highly stable to moisture and temperature[10]
Propyl	Acetyl	Propanol-HCl used, then pyridine-acetic anhydride[11]

[a] Abbreviations: TFA = trifluoroacetyl; TMS = trimethylsilyl; BSA = bis(trimethylsilyl)acetamide; EC = electron capture; BSTFA = bis(trimethylsilyl)-trifluoroacetamide

[1] C. W. Gehrke and D. L. Stalling, *Separation Sci.*, **2**, 101 (1967).
[2] G. P. James and P. A. Van Dreal, *Clin. Chem.*, **15**, 794 (1969).
[3] C. W. Gehrke, R. W. Zumwalt, and L. L. Wall, *J. Chromatog.*, **37**, 398 (1968).
[4] R. R. Crichton and G. Leaf, *Biochem. J.*, **99** (3), 47P (1966).
[5] S. Makisumi and H. A. Saroff, *J. Gas Chromatog.*, **3**, 21 (1965).
[6] J. F. Klebe, H. Finkbeiner, and D. M. White, *J. Am. Chem. Soc.*, **88**, 3390 (1966).
[7] F. Shahrokhi and C. W. Gehrke, *J. Chromatog.*, **36**, 31 (1968).
[8] K. Funakoshi and J. H. Cahnmann, *Anal. Biochem.*, **27**, 150 (1969).
[9] C. W. Gehrke, H. Nakamoto, and R. Zumwalt, *J. Chromatog.*, **45**, 24 (1969).
[10] J. W. Davis, Jr., and A. Furst, *Anal. Chem.*, **40**, 1910 (1968).
[11] J. R. Coulter and C. S. Hann, *J. Chromatog.*, **36**, 42 (1968).

removed by liquid chromatography on alumina. In another procedure the cyclopropene and cyclopropane fatty acid methyl esters were determined directly by on-column injection, using a silicone liquid phase on Gas Chrom Q or Diatoport S. Under these conditions, isomerization of the cyclopropene group was not observed. The compounds determined were sterculic, 2-hydroxysterculic, dihydrosterculic, and malvalic acids (38).

I.   Release of Carboxylic Acids from Salts for GC Analysis

Several procedures have been used to release the lower carboxylic acids from their sodium or potassium salts before they are passed into a gas chromatograph for analysis. For example, C_2 to C_5 acids were released by reacting the sample with excess phosphoric acid (39) or passing it through packing containing anhydrous $NaHSO_4$ (40) just ahead of the chromatographic column.

J.   GC Analysis of Esters

In the absence of polar groups, esters are readily determined quantitatively by GC. Polyester liquid phases have been widely used in these analyses because they give symmetrical peaks of simple esters and provide, in part, separations on the basis of unsaturation. Nonpolar liquid phases, although giving symmetrical peaks of simple esters and very good quantitative data, do not separate saturated and unsaturated esters. Thus, the nonpolar column can be used to separate esters roughly according to molecular weight (e.g., n-C_{16} from n-C_{18} esters), and the polyester columns can serve to separate each group additionally according to unsaturation (0, 1, 2, or more double bonds). Esters of high molecular weight or those of an involatile complex (e.g., a phospholipid) are usually converted to more volatile derivatives of either the acid or the alcohol moiety or both by transesterification, alcoholysis, or saponification followed by esterification or etherification. Should esters contain polar groups, derivatives of the latter are formed at some point in the determination. Thus, acetylation of mono- and diglycerides assured complete elution of these esters in GC analysis; without acetylation recovery can be incomplete (41, 42). Mono- (C_2 to C_{18}) and diglycerides (C_4 to C_{36}) have also been determined after conversion of their free hydroxy groups to trimethylsilyl ethers with bis(trimethylsilyl)acetamide (43).

Hydrolysis or saponification of esters normally yields carboxylic acid and alcohol or phenol fragments, which may be separated and individually derivatized for GC analysis. (See Chapter 1, Sections II.A.–D. Also this chapter, Section II. for derivatization of these fragments.) The alcohol (and possibly interfering material), if not too polar and free of acidic groups, can be separated or removed by extracting an alkaline saponification mixture with an immiscible solvent (44), or it may be possible to evaporate the alcohols from the mixture *in vacuo* to avoid interference from them. The acid, if not too polar because of other functional groups in the molecule, may be extracted with an immiscible phase after acidifying the saponification mixture.

Occasionally, derivatization of the alcohol or acid fragments is unnecessary or less desirable. In one GC determination of butyric acid in butter fat, the fat is saponified by refluxing with 0.5 N KOH in aqueous isopropyl alcohol and the solution evaporated; the residue is shaken with a methylene chloride-formic acid mixture to make it acid, and an aliquot is injected into the gas chromatograph for determination of the butyric acid in free form. The procedure is said to avoid losses that would normally occur if the acid were methylated (45). A means suggested to overcome losses of lower fatty acids is to form butyl instead of methyl esters (46).

Transesterification, especially methanolysis, is widely used to simultaneously hydrolyze and derivatize esters quantitatively for GC analysis. The procedure is rapid and usually avoids the use of concentrated alkali, which can partially isomerize polyunsaturated acids. Methanolysis is accomplished by reacting the ester with methanol containing an acid or base; the product obtained is the methyl ester of the acid moiety. Methanolic hydrogen chloride has been used to determine fatty acid methyl esters derived from lipids (47) and from wax esters (48). In the latter analysis the alcohols and methyl esters are separated by column chromatography before their analysis by GC, the alcohols being determined as trifluoroacetates. Methanolsulfuric acid has been used to determine C_{14} to C_{20} fatty acid methyl esters from human serum lipids (49); BCl_3-methanol is another reagent employed in the analysis of lipids (50). In a convenient procedure that may be conducted at room temperature with no evaporations, a fat in benzene is allowed to react in a closed flask overnight with 2,2-dimethoxypropane (DMP) and methanolic hydrogen chloride (51); after neutralizing the solution, an aliquot is injected into the gas chromatograph. In addition to the methyl esters, isopropylidene glycerol, formed from DMP and glycerol, appears on the chromatogram and serves as a convenient retention time marker. The DMP also acts as a water scavenger and helps to drive the reaction to completion.

A variety of alkaline agents and conditions have been used for transesterification of esters. Six procedures for obtaining fatty acid methyl esters (C_6 to C_{18}) from oils and fats utilized KOH or sodium methoxide in transesterification or saponification-esterification reactions; results of the methods were similar for the major acids of a sample, so long as precautions to prevent the loss of lower-molecular-weight acids were taken (52). Treatment of milk fat with methanolic KOH or $NaOCH_3$ at room temperature results in almost instantaneous formation of methyl esters from glyceride fatty acids, and the solution is analyzed directly by GC (53). Ester cerebrosides, cholesteryl esters, waxes, and several oils are transesterified to form methyl esters by warming with methanolic sodium methoxide and then treating with methanolic HCl; after evaporating the solution, an aliquot in a suitable solvent is injected into the gas chromatograph (54). Fatty acids in vegetable oils were determined as their methyl esters after the oil was heated first with sodium methoxide in methanol and then with formic acid in a special microreactor made from a soldering gun; the carrier gas sweeps the methyl esters into the gas chromatograph via a needle when the sample is heated to 250°C with the soldering gun (55). The cholesteryl ester fraction of human serum was isolated by thin-layer chromatography and the methyl esters of the fatty acids were prepared by transmethylation with 0.1 N sodium methoxide in absolute methanol for determination by GC (56).

In another simple and convenient method for determining fatty acid content (C_4 to C_{18}) in fats and oils, the sample is heated with potassium methoxide in methanol under nitrogen in a vial for 2 minutes at 65°C, the first half-minute with shaking. A silica gel and calcium chloride mixture is added and mixed, and then CS_2 is added and shaken; after clarifying the solution by centrifugation, an aliquot of the CS_2 is injected into the gas chromatograph. The silica gel causes the reaction mixture to become one phase and facilitates extraction of the methyl esters by CS_2. It also adsorbs small amounts of free fatty acids in the oil and prevents their interference in the analyses. The calcium chloride complexes with enough methanol to prevent it from obscuring the methyl butyrate peak. Finally, CS_2 does not obscure the peaks of the low-molecular-weight fatty acid methyl esters as methanol does in many instances. This rapid method gives results fully comparable to those obtained with more time-consuming procedures (57). Treatment of free fatty acids in an oil sample with potassium methoxide as described does not form methyl esters. The addition of BF_3 and another 2-minute heating at 65°C will methylate them.

K.   Reduction of Ester Groups for Gas Chromatography

It is sometimes convenient or desirable to reduce the acid moieties of esters to alcohols. Glycerol and fatty acids in glycerides can be determined simultaneously by reducing the glycerides to fatty alcohols (C_{14} to C_{18}) and glycerol with lithium aluminum hydride and then acetylating the products with acetic anhydride (58). Fatty acids at the level of 10 to 30 μg have been determined in lipids by this same general procedure (59).

A specific and sensitive method for long-chain S-acyl esters is based on the fact that O-acyl esters are not reduced by $NaBH_4$ whereas S-acyl esters are, yielding the long-chain alcohols (60). The method is limited by the fact that Δ^2-S-acyl esters are reduced to the saturated alcohols, and keto groups to alcohols, by the $NaBH_4$. Dihydroxy compounds should be acetylated for quantitative recoveries.

L.   Determination of Amides and Imides by Gas Chromatography

Although direct determination of amides with no polar groups in the molecule is easily accomplished by GC (61), difficulties are frequently encountered in determining the more polar carboxamides. Carboxamides have been converted quantitatively to their respective nitriles by heating with P_2O_5 in dioxane, and the nitriles determined by GC (62). Similarly, phosgene has been used to convert pyridine carboxamides to their nitriles for GC (63).

Mucopolysaccharides, which contain N-acetyl groups, can be hydrolyzed by heating with 2 N HCl for 3 hours at 100°C; the acetic acid liberated is determined directly by GC on a Chromosorb 102 column (64). Treatment with methanolic HCl has been used to transform nonvolatile substituted benzamides to the methyl esters of the corresponding benzoic acids for GC (65).

Active hydrogen analyses using lithium aluminum hydride have been successfully performed on amides (66). See Chapter 8, Section II.

Other procedures used for the GC determination of amides have included formation of the TMS derivative (67) and reduction of cyclic amides to the corresponding cyclic amines with $LiAlH_4$ (68).

M.   Determination of Nitriles by Gas Chromatography

In the absence of polar groups, nitriles chromatograph well and are readily quantified by GC (69–71). In a method for the determination of pentachloromandelonitrile, the compound is quantitatively converted to pentachlorobenzaldehyde, by loss of HCN, when injected into the gas chromatograph (72.)

III.   ELECTROANALYTICAL METHODS*

A.   Coulometric Titration of Carboxylic Acids

The volumetric determination of carboxylic acids with standard base, in aqueous and nonaqueous media, has been well covered (73). Titrations with standard solutions, however, are not as convenient (and possibly as precise) as coulometric titrations. In the latter case, the titrant is generated at the time of use, *in situ*, by an electrolytic reaction. The quantity of electricity consumed in the electrolysis is measured and used to calculate the unknown con-

* Written by Alan F. Krivis.

centration through Faraday's laws. The coulometric approach discards standard solutions completely, and in many cases does not require standard samples. Furthermore, because electricity (or, actually, the length of time a constant-current power supply is turned on) is the real "titrant," automation of the analysis becomes very easy to carry out; electricity can be transferred and measured much more cheaply and readily than volumes of standard solutions. Detection systems remain the same as in a buret titration, so that potentiometric measurement of the end point of the coulometric titration works very well.

The electrolysis of water serves as the basis for coulometric titration of an acid. If platinum electrodes are used for anode and cathode, water is reduced to OH^- at the cathode and oxidized to H^+ at the anode. This then means that the anode and cathode must be isolated from each other by means of a salt bridge and the sample titrated in the cathode chamber. However, if an attackable anode is employed, both electrodes may be present in the same beaker.

For example, if a silver anode is used and bromide is added to the reaction mixture, the anode reaction will produce silver bromide, which causes no interference in the acid titration (74–76). The procedure of Carson and Ko (76) for aqueous titrations is described below; other solvents have also been used such as 70% isopropanol (76), benzene-methanol (77), acetone (78), and isopropanol (79). The nonaqueous media may require more elaborate procedures, and the reader is referred to the original articles for details.

EXPERIMENTAL METHOD [BASED ON THE STUDY BY CARSON AND KO (76)]*

Apparatus. A constant-current power supply and connected timer, as described in Chapter 1, Section III, is the most convenient generating source (coulometer). Several commercial units are available.

The cell for use with an attackable silver anode consists of a suitable-sized beaker into which is dipped the generating electrodes. A cylindrical gauze platinum cathode surrounding a tight helix of silver wire, which may be made from about 10 inches of No. 22 wire, constitutes a convenient and useful arrangement.

This configuration accomplishes two purposes. First, the electric field developed between anode and cathode is "shielded" from the rest of the solution, and the normal detecting electrodes (glass-calomel) can function properly. Second, a minimum amount of space is taken up by this physical arrangement. This second point can be important because glass and calomel

---

* Reprinted in part with the permission of the copyright owner, the American Chemical Society.

electrodes have to be dipped into the solution and a magnetic stirring bar inserted also.

The cell for isolated-anode titrations is identical to the one shown in Figure 1-4 on p. —, for use in coulometric brominations. The only change is the substitution of a glass-calomel detecting electrode pair for the platinum-platinum pair shown in the figure.

A pH meter serves to monitor the titration and detect the end point. Automatic shut-off at the end point can be arranged by use of a pH meter incorporating a meter-relay circuit or by other types of controllers.

Generating Solvents.   Nonaqueous titrations are carried out in acid-free 70% isopropanol with 0.001 M LiCl as the electrolyte.

Aqueous titrations utilize 0.05 to 0.1 M KBr in carbon dioxide-free water as the medium.

Isolated Anode Procedure.   Dissolve a sample containing *ca.* 0.03 to 0.05 meq of acid in a suitable volume of 70% isopropanol containing 0.001 M LiCl and quantitatively transfer to the titration vessel. Mount the vessel in place, protecting the contents from atmospheric acids and bases. Dip the filled anode chamber, the cathode, and the detecting electrodes in place. Titrate the sample at a current of *ca.* 20 mA. For optimum results, generation of the base should be incremental in the vicinity of the end point, with adequate time allowed for equilibration of the detecting system. Titration may be made to a fixed end-point pH, or a plot of each titration can be prepared.

Attackable Anode Procedure.   Dissolve a sample containing *ca.* 0.03 to 0.05 meq of acid in a suitable volume of aqueous 0.1 M KBr and transfer to the titration vessel. Mount the vessel in place, protecting the contents from atmospheric acids and bases. Dip the generating and detecting electrodes into the solution, add a magnetic stirring bar, and start the stirrer. Titrate the sample at *ca.* 20 mA current and record the time at the end point. For optimum results, generation of the base should be incremental in the vicinity of the end point, with adequate time allowed for equilibration of the detecting electrodes after each increment of titrant is generated. The titration may be carried out to a fixed pH (equivalence point pH), or a plot of each titration may be prepared. The generating anode should be cleaned in concentrated ammonia solution after each titration.

Calculations.   Equation 1 shows the calculation of milliequivalents of acid:

$$\text{milliequivalents of acid} = \frac{\text{current} \times \text{time}}{96500} \tag{1}$$

B. Conductometric Titration of Acids

The determination of acids by potentiometric titration with base is well established—even entrenched. However, a conductometric titration of acids is quite valuable in at least two sorts of problems, although it is not normally contemplated. The areas under discussion include low levels of materials and also weak acids, or even salts.

It is a fact that the accuracy with which high and low concentrations of materials may be conductometrically titrated is the same. In practice this means that dilute solutions may be titrated more accurately by a conductance measurement than by a potential measurement. In many instances, the conductometric titration of weak electrolytes is also far easier than the corresponding potentiometric titration; instrumentation may be simpler and the electrode systems are less troublesome.

In the past, one barrier to the use of conductometric titrations was the lack of automatic or semiautomatic equipment. Manual, point-by-point plotting of titrations is extremely time consuming. With the availability of conductance monitors,* however, some change in attitude may be predicted.

The titration of acids may be carried out in almost any solvent which will dissolve sample and titrant (*cf.* Chapter 1, Section III, for the use of DMSO). The procedure described below (80) is oriented toward the titration of very weak acids. In essence, the sample is added to excess ammonia and the ammonium ion is neutralized with standard lithium hydroxide solution.

The use of ammonia accomplishes two things. For one, the acidity of the sample is enhanced. Additionally, the solubility of acids in the solvent is significantly increased. The end result is that a variety of very weak mono- and dibasic acids can be titrated. Table 3-7 lists the acids which were titrated in this way.

EXPERIMENTAL METHOD [ADAPTED FROM THE STUDY OF GASLINI AND NAHUM (80)]†

Apparatus.   The conductivity apparatus described in Chapter 1, Section III, can be used for this titration. Alternatively, a conductivity monitor, buret, and recorder may be assembled into an automatic titrator. Information

---

* Several firms, two of which are Leeds & Northrup and Beckman Instruments, offer excellent continuous conductivity monitors. An inexpensive but reliable unit is available from ESSE Laboratories, P.O. Box 863, Akron, Ohio 44309.

† Reprinted in part with the permission of the copyright owner, the American Chemical Society.

Table 3-7. Acids Successfully Titrated in Ammonia Solution

Aminobenzoic acid	Pyrogallol
Benzoic acid	Phloroglucinol
Guaiacol	Salicylic acid
Phenol	σ-Coumaric acid
Phenylacetic acid	Alizarin
2-Naphthol	Acetoacetanilide
Vanillin	5,5-Dimethyl-1,3-cyclohexanedione
Resorcinol	Ethylacetodicarboxylate
Catechol	Ethyl malonate
Hydroquinone	Succinimide

about this application should be obtained from the supplier of the components.

Reagents and Solutions. For many acids the optimum concentration of ammonia in the solvent is 0.2 M, but a maximum of 2 M is possible where needed.

The ammonia is obtained, free of carbonate, by heating a reagent-grade concentrated solution in the presence of barium hydroxide and collecting the gas in carbon dioxide-free distilled water.

A solution of 2.5 M lithium hydroxide in carbonate-free water serves as the titrant and can be standardized against potassium acid phthalate.

Procedure. Dissolve a weighed sample of acid (ca. 5 mM) in several milliliters of the 0.2 M ammonia solvent and transfer quantitatively to the cell. Add sufficient ammonia solution to bring the volume to 125 ml and start the stirring apparatus. In most instances, it will not be necessary to thermostat the cell, but this can be done if required. Titrate with the lithium hydroxide solution, using a microburet. Plot the conductance versus the titrant volume and locate the end point of the titration. In the case of a monobasic acid the end point is at the intersection of the two straight lines. Dibasic acids will show two such intersections when plotted. Calculate the amount of acid in the sample.

Calculations.

$$\text{Per cent of acid} = \frac{(\text{ml titrant})\ (\text{molarity titrant})\ (\text{mol. wt. unknown})\ (100)}{(\text{sample weight in grams})\ (1000)}$$

IV. NUCLEAR MAGNETIC RESONANCE METHODS*

A. Carboxylic Acids

A series of unsaturated fatty acids was studied by Purcell, Morris, and Susi (81), using 60 MHz ¹H nmr. These acids were cataloged as those containing isolated double bonds, two conjugated multiple bonds, double bonds separated by one methylene group, and double bonds adjacent to the carboxyl groups. The examples cited were 9-*cis*-octadecenoic acid and 9-*trans*-octadecenoic acid, 9,11-*trans,trans*-octadecadienoic acid, 9,12-*trans,trans*-octadecadienoic acid, and 2-*cis*-octadecenoic acid. The chemical shift data for the various protons were reported.

Polyalcohols, hydroxy acids, and derivatives, of D-gluconic acids were studied by Sawyer and Brannan (82). These compounds were studied in D_2O, using tetramethylammonium chloride as an internal reference. The methyl resonance line is 2.17 ppm from 3-(trimethylsilyl)-1-propanesulfonic acid sodium salt. The polyalcohols reported were ethylene glycol, glycerol,

sorbitol, and erythritol, and the chemical shifts of the —CH_2O and —CHO groups were determined. The chemical shifts of the aliphatic and hydroxyl acids and their salts were obtained. Acetic acid, malonic acid, succinic acid, glycolic acid, tartronic, and tartaric acid all showed a diamagnetic displacement for the salt.

The proton chemical shifts of D-glyceric acid, saccharic acid, and D-gluconic acid were studied as a function of pH. The lactones of these compounds were also reported.

The quantitative determination of the elements hydrogen and fluorine in organic compounds was reported by Paulsen and Cooke (83). The authors present a discussion in which the effects of parameters such as sample spinning and sweep rate are shown. The quantitative determination of acidic protons was carried out, and the effect of water on those exchangeable protons was demonstrated. Compounds such as acetic acid, propionic acid, butyric acid, valeric acid, cyclohexane-carboxylic acid, formic acid, and benzoic acid were reported. The fluorinated compounds studied were *m*-bromo-α,α,α, trifluorotoluene, trifluoroacetic acid, 2,2,3,3,-tetrachloro-hexafluorobutane, 2,3-dibromooctafluorobutane, and a mixture of *cis*- and *trans*-1,2-dichlorotetrafluoropropene. Another study related to the choice of operating parameters such as power levels is that of Jungnickel and Forbes (84).

* Written by Harry Agahigian.

Mixtures of carboxylic acids and anhydrides have been analyzed by means of nmr spectroscopy. Parker (85) utilized the difference of the chemical shifts of the alpha protons between the anhydride and the acid. Generally, the alpha protons of the anhydrides are at lower field than the acids. Aromatic as well as aliphatic acids and anhydrides, such as acetic, propionic, succinic, phthalic, chloroacetic, fumaric, and maleic, were reported.

B.   Aliphatic Esters*

The identification of esters and their determination can be accomplished by nmr, but the effect of the ester functionon the adjacent protons serves as the means of identification. Rosado-Lojo, Hancock, and Danti (86) obtained chemical shift data for 18 acetate esters, CH_3COOR, and 16 methyl esters, $RCOOCH_3$, in carbon tetrachloride solution and neat. The authors correlated the structure of the ester with the rate of hydrolysis, which is determined by steric, polar, and hyperconjugation effects. The chemical shifts of the acetate esters were measured in relation to methyl acetate and exhibited plus or minus shifts ranging from ± 1.0 for the $(CH_3CH_2)_2CH-$ group to 5.6 Hz for the *tert*-butyl group. The chemical shift of the methyl on the carbonyl carbon is 118.5 Hz from TMS. The acetate esters contain resonances in this region.

The methyl esters in carbon tetrachloride had chemical shifts ranging from -2.5 Hz for $(t\text{-}Bu)_2CH-$ to $+0.9$ for $(CH_3CH_2)_2CH-$. In relation to tetramethylsilane the chemical shift of the $-OCH_3$ group is 217.5 Hz. The $-OCH_3$ shifts are approximately ± 3 Hz. Closely resembling the work of Rosado-Lojo et al. is that of Kan (87). He, however, also presented data on the succinates as well as similar compounds. His article emphasized correlation to hydrolysis rates and a mechanistic approach.

C.   Polyester Resins

Percival and Stevens (88) have reported a method for the semiquantitative determination of polyester resins. The samples are measured in acetone or benzene, but these cause problems due to overlapping of the solvent peaks and the sample. The acids used in the polyester preparation were isophthalic, maleic, fumaric, adipic, and empol 1014. The glycols were ethylene, propylene, diethylene, dipropylene, and other commercially available products. Eighteen resins were studied, and the chemical shifts of the components reported. The peak areas were used to obtain the quantitative data.

* Written by Harry Agahigian.

The use of esterification with trifluoroacetic acid was reported by Konishi and Kanoh (89). The compounds studied were 1,2-propanediol monoesters of long-chain fatty acids.

D. Polyurethanes

The identification of eleven polyurethanes and their compositions were reported by Brame, Ferguson, and Thomas (90); the experiments were carried out in solutions of arsenic trichloride at 100°C and 60 MHz. The general reaction is as follows:

$$R(OH)_2 + R(NCO)_2 \rightarrow \text{polyurethane}$$

The polyurethanes were studied as 15% w/v solutions, and tetramethylsilane and hexamethyldisiloxane were used as internal standards. Trifluoracetic acid might also serve as a solvent. The polyethers used for urethane preparation were poly(tetramethylene ether) glycol and poly(propylene ether) glycol. These were reacted both with methylene-bis(4-phenylisocyanate) and toluene diisocyanate. The polyesters used were polyester (adipic acid-diethylene glycol)diol, polyester (adipic acid-1,4-butanediol)diol, polyester (adipic acid-1,2-ethanediol)diol, polyester (adipic acid-1,2-ethanediol and 1,2- propane diol)diol; these were formed into urethanes by reaction both with methylene-bis(4-phenylisocyanate) and toluene diisocyanate.

Although ambiguity in regard to the assignment of the NH resonance may cause some problems with the line integration, generally the quantitative data are good. The molecular weight of the polyether or the polyesterdiol can be estimated. An estimate of the mole fractions of a polyester-based polyurethane shows that the method is in good agreement with the hydrolysis products.

Component	Mole Fraction	
	By nmr	By Hydrolysis
Adipic acid	0.37	0.35
Ethanediol	0.44	0.46
Methylene-bis(4-phenylisocyanate)	0.14	0.04

The chemical shift data for this series are also reported.

E. Amides and Ureas

The method of Leader (see Chapter 1, Section IV.A) has been applied also to amides and ureas. The ^{19}F chemical shifts for amides, monosubstituted ureas, and sulfonamides fall within a range of 0 to 1.8 ppm from the hexa-fluoroacetone H_2O adduct. The related compounds reported were acetamide, acrylamide, chloroacetamide, benzamide, sulfanilamide, p-toluenesulfonamide, hydrazine, 2,4-dinitrophenylhydrazine, methylthiourea, phenylurea, and methylurea. It is possible to establish interferences from other active hydrogen components which might cause problems by noting the presence of resonances at higher field.

Ma and Warnhoff (cf. Chapter 11, Section IV.A) studied and reported nmr data on the following amides:

	Chemical Shift N—CH$_3$
N,N-Dimethylformamide	2.85 2.96
N,N-Dimethylacetamide	2.90 3.02
N-Methylurea	2.77
N-Methylpyrrolidone	2.82
N-Methylphthalimide	3.16
N-Methylthiobenzanilide	3.89
N-Methylbenzanilide	3.48

V. RADIOCHEMICAL METHODS*

The importance of the higher fatty acids in biochemistry and in the chemistry of natural products has provided the impetus for the development of a number of sensitive radioreagent techniques applicable to micro and submicro amounts of these compounds. The most widely used derivatives are the methyl-^{14}C and methyl-^{3}H esters and the soaps of radioactive metals. Many monocarboxylic acids in macro quantities may also be determined by isotope dilution as the anilides or substituted anilides. Carboxyl groups in certain cellulosic materials may be determined as salts of radioactive metals.

A. Conversion to Methyl Esters

Determination of individual carboxylic acids in mixtures by chromatographic separation methods is greatly facilitated if the acids are first converted to their methyl esters. The same approach is also useful for determining the sum of free acids. Quantitative derivatization is effected in radiochemical

* Written by D. Campbell

methods by treating the acids with (a) diazomethane-^{14}C in ether-methanol, (b) tritiated water in ether, followed by nonradioactive diazomethane, (c) a solution of sodium hydroxide and sodium methoxide-^{3}H in methanol-^{3}H, and (d) methanol-^{14}C in the presence of boron trifluoride. Excess radioreagent is easily removed in each case, and the methods are thus highly sensitive.

Diazomethane reacts rapidly and quantitatively with carboxylic acids in a 9:1 ethyl ether-methanol solution at room temperature without significant side reactions (91):

$$R-COOH + {}^{14}CH_2N_2 \rightarrow R-COO^{14}CH_3 + N_2$$

In practice, the radioreagent is generated as needed in a nitrogen atmosphere in a closed system that contains the acid specimen; persistence of a yellow color indicates excess reagent, which is removed in a current of nitrogen. Diazomethane-^{14}C is produced by heating a labeled nitroso derivative, prepared from methylamine-^{14}C, in the presence of a strong base. Although a variety of such precursors are in common use for synthesizing the unlabeled compound, only N-methyl-N-nitrosourea (92), N-nitroso-N-methylamino-isobutyl ketone (93), and N-methyl-N-nitroso-p-toluenesulfonamide (91, 94, 95) have been used as sources of the labeled compound. The nitrososulfonamide (NMSA) is preferable, since it is reasonably stable at room temperature, produces $^{14}CH_2N_2$ in yields of up to 85%, and is commercially available.

A study of esterification conditions revealed that conversion of fatty acids, for example, palmitic acid, was complete in 10 minutes in a 9:1 (v/v) mixture of ethyl ether and methanol (91). The latter evidently functions solely as a catalyst, since the molar specific activity of the ester agreed well with that of the labeled NMSA. This agreement is a requisite condition for the use of $^{14}CH_2N_2$ as a quantitative radioreagent in the presence of methanol. Reactions were not complete in ether or methanol alone in 30 minutes. Prolonged reactions, in general, are undesirable, since they lead to impurities such as the polymer ($^{14}CH_2)_x$, which is identifiable as turbidity or flakes in the solution.

Generation of diazomethane-^{14}C and subsequent esterification can be accomplished in an experimental arrangement as simple as a series of three side-arm test tubes (91). In the first, nitrogen used as carrier gas is saturated with ether. In the second, $^{14}CH_2N_2$ is formed by treating the NMSA-^{14}C (ca. 2 mM per millimole of monofunctional acid), dissolved in a mixture of ether and diethylene glycol monoethyl ether, with aqueous KOH. The third tube contains 5 to 30 mg of acids in 9:1 ether-methanol. Derivatization and removal of unconsumed reagent are performed in 10 to 12 minutes. Another apparatus (92, 93) employs 5-ml flasks, stopcocks, and a series of absorbers for collecting excess reagent; sample weights of 3 to 6 mg are used with reaction periods of 1 to 4 hours at room temperature. Completion of esterifications with CH_2N_2 is indicated by a yellow tinge in the derivative solution.

Radioassays of esters of fatty and bile acids prepared with $^{14}CH_2N_2$ and separated by paper chromatography have been made by direct scanning of the paper strip (91, 94, 95) or by preparing from the latter an autoradiogram, which is then scanned with a recording microphotometer (92, 93). The liquid scintillation counting method, which is more advantageous, has been employed after column chromatography (96). When liquid scintillation counting was combined with thin-layer chromatography, the sensitivity of a method in which $^{14}CH_2N_2$ was used for determining dinitrophenyl derivatives of amino acids (97) was increased one hundredfold (98) to 1 pM with a reproducibility of $\pm 6\%$.

Individual compounds can be identified and estimated by similarly analyzing acid mixtures of known composition. Absence of steric hindrance effects in the derivatizing reaction is a definite advantage of diazomethane.

Diazomethane-^{14}C has also been used to determine gibberellic acid, $C_{19}H_{22}O_6$, and its dihydro derivative. These were present in the range of 10 to 50 $\mu g/ml$ in commercial fermentation liquors (99–102).

Analysis for low concentrations of carboxyl groups on the surface of carbon black particles is also possible by a procedure which combines $^{14}CH_2N_2$, oxygen flask combustion, and liquid scintillation counting (103). In this work, 2 grams of carbon black was stirred in 100 ml of a toluene solution of the radioreagent for 20 hours at room temperature. After filtration, extraction, and drying, Zeisel's method was used to remove the activity present in the methyl ester and ether groups, Subsequent treatment of a separate specimen with ca. 6 N HCl hydrolyzed the esters, which were determined by difference.

If a suitable precursor of diazomethane-^{14}C is unavailable, methyl-^3H esters of carboxylic acids may be prepared by treating a diethyl ether solution of tritiated water and the acids of interest with nonradioactive diazomethane. The carboxyl-^3H acids formed by exchange with the ^3H_2O are converted to the corresponding tritiated methyl esters:

$$R—COO^3H + CH_2N_2 \rightarrow R—COOC^3H_3 + N_2$$

This reaction, which is often termed "tritiating methylation," was used initially for the synthesis of methyl-^3H esters (104, 105). The attractiveness of the analytical method lies in the fact that tritiated water is inexpensive and easily obtained. Applicability of the method to mixtures is based on the nondiscriminating incorporation of tritium into the acids and esters.

In the quantitative procedure developed (106), 0.2 ml of a solution of ^3H_2O in dry diethyl ether (1 λ of water per milliliter of ether, 6.67 μCi of total activity) is added to $\not> 0.033$ meq of the acid mixture in dry diethyl ether, acetone, or dioxane. After 15 minutes, the acids are quantitatively esterified by adding an ether solution of diazomethane. The products are separated by thin-layer chromatography (TLC) on silica gel, and the activity in the isolated esters, which are made visible by suitable spray reagents, is determined by liquid scintillation counting. The radioassays may be made either on esters extracted from the silica gel by diethyl ether-hexane, 7:3 (v/v), followed by evaporation of the solvents and addition of scintillator solution, or by suspension counting of the zones scraped from the TLC plates. Suspension counting is often preferable with highly polar acids. If the former method is used, the extraction from the adsorbent must be quantitative. When the sample contains less than 1 to 5 mg of fatty acids, inactive methyl esters of the compounds to be determined should be added before chromatography. In this way, the method may be extended to microgram amounts of sample. The technique is standardized by adding definite amounts of the acids of interest to the sample and repeating the analysis. Successful analyses of a known mixture of stearic, benzoic, and phthalic acids, as well as of stearic and dl-tartaric acids, indicated that tritium activity was incorporated into the esters in a nondiscriminating manner. This fact was further verified with ^{14}C-labeled stearic acid and benzoic acid. Reproducibility of the method was within $\pm 5\%$.

Tritiated water has also been employed in conjunction with nonradioactive diazomethane for determining the ratio of gibberellic acid to its dihydro derivative in commercial fermentation liquors (102). Although the use of tritiated diazoethane and diazopropane has been suggested (102) for the preparation of ethyl- and propyl-^3H esters in the quantitative analysis of carboxylic acids, the actual application of these reagents has not been reported.

Methanol, which may be labeled with either ^{14}C or ^3H, is an attractive reagent for preparing esters, since it is relatively inexpensive, is easily asasyed by liquid scintillation counting, and has an extensive history of use in unlabeled form. In a procedure suitable for direct determination of the sum of both free and combined fatty acids, esterification, as well as transesterification, is performed by treating the specimen in chloroform with an equal volume of a solution of sodium hydroxide and sodium methoxide-^3H in methanol-^3H (107). The reagent is prepared by dissolving clean sodium metal (4 mg/ml of solution) in a 0.5 N solution of sodium hydroxide in methanol-^3H. The preparation is stored in a polyethylene bottle in a refrigerator to reduce the rate of deterioration; careful exclusion of air is also necessary, since exposure of the reagent to the atmosphere for 4 hours resulted in substantial deactivation. The dry specimen is dissolved in 0.1 to 1 ml of chloroform, and the same volume of reagent is added. The container is stoppered tightly with a cork and shaken for 1 hour at room temperature. Immediately after shaking, 1 drop of 0.1% phenolphthalein solution in absolute methanol is added and the sample is neutralized with glacial acetic acid. The solution is then transferred quantitatively with chloroform to a separatory funnel, and the resulting solution is washed twice with ten times its volume of distilled water; additional washing is advisable in determining < 1 μM of acid. The chloroform layer is transferred quantitatively to another container, and the solvent is carefully removed in vacuum. Methanol is added, the container is swirled, and the methanol is evaporated; the step is repeated several times for less than 1 μM of acid. The residue is then dissolved in a toluene scintillator solution for radioassay by liquid scintillation counting. The absence of nonvolatile labeled impurities in the methanol should be verified by a blank analysis. The acid contents as calculated from direct assay of the methanol were in good agreement with those obtained by a secondary standardization in which known amounts of acid were used. If the sample contains glycerides or other esters, these must be removed before free acids can be determined. Speed, simplicity, and good accuracy in the nanomole range make this procedure very attractive for determining total fatty acid content. The mild esterification conditions and the quantitative transesterification catalyzed by sodium hydroxide are also advantageous.

Esterification by methanol-^{14}C with boron trifluoride as catalyst has proved useful for determining carboxyl end groups in nylon polymers (108). The reagent was prepared by adding 7.5 grams of boron trifluoride to ca. 39.5 ml of methanol. Approximately 30 mg of polymer was dissolved in 0.3 ml of the reagent at 60°C, and the solution was held at this temperature for 10 minutes. Radioassays of the polymer, after its isolation by precipitation in water, were made by oxygen flask combustion and liquid scintillation counting. The MeOH·BF$_3$ reagent quantitatively esterifies fatty acids in 2 minutes at the boiling point (109).

Gas chromatographic separation of methyl-^{14}C esters of C$_{12}$ to C$_{20}$ fatty acids, followed by determination of radioactivity in the gas phase, has been used in the analysis of mixtures of such acids (110).

In isolating methyl esters, precautions must be observed to prevent losses of low-molecular-weight derivatives by volatilization.

B.   Conversion to *p*-Bromophenacyl Esters

The reaction of *p*-bromophenacyl bromide with alkali metal salts of carboxylic acids in water-ethanol solutions, although not usually quantitative, yields derivatives widely employed for identification purposes:

A radioreagent-reverse isotope dilution method that employs p-bromo-phenacyl-^{14}C bromide has been developed for determining myristic acid (111) and may be applicable to other fatty acids as well. Esterification of this acid in 95% ethanol above pH 8.5 at reflux temperature reaches a maximum, determined initially by bromine analysis, of about 60% in 50 minutes and remains constant for at least 30 minutes. The proposed method is based on treatment of the unknown mixture with labeled reagent under conditions identical to those indicated above, followed by addition of a much larger, known quantity of nonradioactive ester. After purification of the myristate to constant specific activity, the amount of ester in the specimen is calculated by equation 2 of Section V, Chapter 11, and from this result the weight of myristic acid present is computed by the factor of ca. 0.60. The average value of 0.60 was verified by applying the radioreagent to known amounts of myristic acid. Application of the method to other fatty acids requires preliminary evaluation or the reproducibility of the esterification reaction.

Derivative activation chromatography (cf. Chapter 2, Section V) has been proposed for determining micro and submicro amounts of the p-bromo-phenacyl esters of carboxylic acids after separation of the derivatives on paper chromatograms (112). Use of the method for determining acids requires a quantitative or reproducible procedure for the preparation of esters. The technique is standardized by concurrent activation and radioassay of a known amount of a bromine compound. For greatest sensitivity, a high-flux source of thermal neutrons, such as a nuclear reactor, is necessary.

C. Conversion to Anilides and Substituted Anilides

p-Chloroaniline-^{36}Cl is useful for determining macro amounts of carboxylic acids, anhydrides, and chlorides by direct isotope dilution. Certain chloro-phenoxyacetic-^{36}Cl acids have also been employed in estimating the un-labeled compounds as the anilides. The derivatives, in general, melt sharply and can be purified by crystallization. p-Iodoaniline-^{131}I is a promising radioreagent for analyzing smaller quantities of material by solid scintillation scanning of a chromatography column on which the labeled p-iodoanilides have been separated. An advantage of primary aromatic amines is that carboxylic acid anhydrides and chlorides usually react quantitatively with them under mild conditions.

The method in which p-chloroaniline-^{36}Cl is employed is direct isotope dilution after complete conversion of the carboxylic acid, anhydride, or chloride to the nonradioactive p-chloroanilide (113). The labeled p-chloroaniline is prepared by chlorinating acetanilide with elemental ^{36}Cl and hydrolyzing the p-chloroacetanilide-^{36}Cl with concentrated hydrochloric acid. The pure labeled derivative of the acid of interest is then prepared. Acids in the sample are quantitatively converted to nonradioactive anilides by treating them with p-chlorophenylphosphazo-p-chloroanilide:

$$Cl-\!\!\bigcirc\!\!-N=P-NH-\!\!\bigcirc\!\!-Cl$$

in the presence of nonradioactive p-chloroaniline. A solution of 1 to 2 mM of the acid, 0.8 gram of p-chloroaniline per millimole of acid, and 0.4 gram of p-chlorophenylphosphazo-p-chloroanilide in 10 ml of toluene is refluxed for 90 minutes. (A 1:1 toluene-benzene mixture must be used with acetic acid.) The solution is cooled, and 1.5 ml of a dioxane solution of the p-chloroanilide-^{36}Cl of interest, which should have a specific activity of ca. 1 μCi/mM, is added. The amount of activity used is approximately 0.01 μCi. After the addition of 10 ml of ethanol, the solution is boiled until nearly clear. The organic solvents are removed with water vapor, 10 ml of 4 N HCl is added, and the precipitate is removed by filtration. The crude p-chloroanilide-^{36}Cl is recrystallized from a suitable solvent until the melting point differs by less than 1°C from that of the pure derivative. Chemical purity is determined from the observed melting point and the known molar melting point depression constant. Purification of the derivative until the specific activity is constant (or, more rigorously, until the specific activity of the derivative isolated from the filtrate agrees with that of the recovered product) may be employed if sufficient compound is available. The use of more than one solvent for the purification is also desirable.

Preparation of a standard is recommended by mixing a known amount of the labeled anilide with a definite quantity of pure unlabeled anilide approximately equal to that obtained from the sample (113); the specific activity of the mixture is then determined by the same radioassay method employed with the sample. In this variation, which obviates measurement of the specific activity of the labeled p-chloroanilide added, the weight of acid present in the specimen is calculated by equation 5 of Section V, Chapter 1. Alternatively, the specific activity of the added p-chloroanilide may be established; the weight of the acid of interest is then calculated by equation 6 of Section V, Chapter 1. The procedure was studied extensively with the former method of calculation, end-window G-M counting, and purity determinations by melting point measurements. The recoveries obtained for acetic and benzoic acids, both alone and in the presence of significant amounts of water, for acetic anhydride, and for stearic acid are indicated in Table 3-8.

The purity of the analytical-grade samples of acetic acid, benzoic acid, and acetic anhydride used in the procedure, as determined by independent chemical methods, was $100.0 \pm 0.2\%$, $100.1 \pm 0.2\%$, and $100.0 \pm 0.4\%$, respectively. The purity of the stearic acid was estimated at 99%. In Procedure I for the analysis of stearic acid, the latter was refluxed with p-chlorophenylphosphazo-p-chloroanilide and p-chloroaniline in toluene. In Procedure II, which is applicable to acids that form nonvolatile chlorides, thionyl chloride was used to first convert the acid quantitatively to the acid chloride. After removal of excess thionyl chloride, the stearoyl chloride was heated gently with an acetone solution of p-chloroaniline. The derivative obtained with Procedure II, however, was more difficult to purify than that resulting from Procedure I.

An important advantage of the method is indicated by its successful application to the determination of stearic acid in an approximately $5:1$ stearic acid-palmitic acid mixture. The procedure is useful when the quantity of acid present yields adequate derivative for purification by crystallization. It is not applicable to dicarboxylic acids because of the possible formation of monoanilides or imides. The conversion of succinic acid, as determined by Procedure I, was ca. 60%. α-Amino acids cannot be determined by the method.

Use of the phosphazo-^{14}C compound has been proposed for the analysis (114) on the basis of its quantitative reaction with monocarboxylic acids that do not contain amino groups.

Isotope dilution that employs the carboxylic acid to be determined as a radioreagent and a nonradioactive substance for derivatization is also a valuable approach, since quantitative derivative formation is not required. This technique is especially useful when simple isotope dilution with the labeled acid alone cannot be conveniently applied because of difficulties in separating homologous or otherwise similar acids. Its application is illustrated by the analysis of mixtures of chlorinated phenoxyacetic acids (115). The acids of interest (e.g., 2,4-dichlorophenoxyacetic and 2,4,5-trichlorophenoxyacetic acid) were prepared labeled with ^{36}Cl, and the pure nonradioactive acids were similarly synthesized. A known weight of the labeled acid of interest, in solution, was then added to the specimen to be analyzed,

Table 3-8. **Determination of Carboxylic Acids and Anhydrides as the p-Chloroanilides-^{36}Cl**

Compound determined	Weight taken, mg	Weight recovered, mg	Per cent recovered
Acetic acid	126.6	126.8	100.2
	127.8	127.3	99.6
	125.0[a]	126.8	101.4
	125.5[b]	127.2	101.3
	122.2[b]	121.4	99.3
	123.5[c]	122.7	99.3
	128.2[d]	127.0	99.0
	128.6[d]	130.4	101.4
	128.0[d]	127.2	99.4
Acetic anhydride	114.2	114.3	100.1
	117.3	118.2	100.8
Benzoic acid	131.2	131.1	99.9
	134.0	134.3	100.2
	134.7	133.5	99.1
	132.0	132.3	100.2
	134.9	135.3	100.3
	126.0[a]	127.1	100.9
	132.4[b]	131.9	99.6
	130.9[b]	128.3	98.0
Stearic acid, Procedure I	251.9	247	98.1
	255.0	250	98.0
	259.3	254	98.0
	248.1	247	99.6
	246.3[e]	245	99.5
	251.4[e]	244	97.1
	250.6[f]	248	99.0
	254.7[f]	253	99.3
Stearic acid, Procedure II	250.1	248	99.2
	255.4	258	101.0
	244.5	244	99.8
	264.5	264	99.8

Reprinted from *Anal. Chem.*, **28,** 1320 (August 1956). Copyright 1956, by the American Chemical Society. Reprinted by permission of the copyright owner.

[a] 10 mg of water added to sample.
[b] 25 mg of water added to sample.
[c] 50 mg of water added to sample.
[d] 10 mg of formic acid and 10 mg of propionic acid added to sample.
[e] 25 mg of palmitic acid added to sample.
[f] 50 mg of palmitic acid added to sample.

as well as to a definite quantity of the same pure nonradioactive acid. The solution that contained the radioactive acid also contained significant amounts of the most probable labeled contaminants; these intentionally added impurities, however, were in nonradioactive form. The respective samples were then dissolved in aqueous sodium hydroxide solution, precipitated with hydrochloric acid, and dried. The derivatives formed by refluxing the isolated acids with aniline were purified by crystallization under specified conditions. Radioassays were made by the technique that does not require knowledge of the specific activity of the acid. Equation 6 of Section V, Chapter 1, can also be used, provided that weights are expressed in grammoles and specific activities are in radioactivity units per mole. Because extremely pure derivatives of the acids of interest in the mixtures encountered were usually difficult to obtain, the chemical purity of sample anilides was calculated from melting point determinations and known melting point depression constants.

The labeled acids used in this method need not possess high chemical or radiochemical purity, since substantial amounts of known impurities in nonradioactive form are intentionally added with the radioactive acid. Purification then removes radioactive impurities with their unlabeled counterparts.

Automatic scanning of a liquid chromatography column on which p-iodoanilides-^{131}I have been separated has been shown to be feasible for resolution and quantitative analysis of mixtures of these derivatives (116). The apparatus makes use of a scintillation detector connected to a recording count-rate meter for assay of the gamma radiation on the column from top to bottom (cf. Section V, Chapter 1). Although the technique was used only for a mixture of the p-iodoanilides of acetic and propionic acids, it is applicable in principle to any p-iodoanilide, or mixture of these derivatives, that can be isolated or separated on a column. Full realization of the potential of the method, however, requires a quantitative procedure for forming p-iodoanilidies from carboxylic acids.

D. Conversion to Radioactive Metal Derivatives

Formation of the ^{60}Co, [110m]Ag, and [203]Hg soaps of fatty acids, usually after separation of the latter by reversed-phase chromatography on paper, is a technique applicable to micro amounts of sample. Although determinations of total higher fatty acids may be made fairly rapidly with either [110m]AgNO$_3$ or [203]Hg(OAc)$_2$ alone, procedures employed with [60]Co(OAc)$_2$ and with chromatograms involve the use of at least two reagents. The methods are standardized with known amounts of the pure acids of interest.

Cobalt-60 acetate, first proposed for determining oleic acid (117, 118) after direct application of the acid to paper, was subsequently used with the saturated fatty acids separated on paper impregnated with a hydrophobic medium (119). Since the free acids react only slowly with the aqueous reagent, the spots are exposed initially to gaseous ammonia in a closed vessel for periods of up to 8 hours to form ammonium salts. In determining oleic acid (117) the spot of the ammonium soap was then treated with a few drops of 2.5% ^{60}Co(OAc)$_2$ solution that had an activity of 10 μCi/ml; excess reagent was washed out with water, and the excised spot was assayed by end-window counting. The procedure appears applicable as well to estimation of the total content of free higher fatty acids.

Lauric, myristic, palmitic, and stearic acids in amounts of 40 μg each were separated by reversed-phase chromatography on paper saturated with a mixture of low-molecular-weight hydrocarbons and then determined with an accuracy that ranged from -1.3 to $+1.4\%$ (119). In this method, orientation, calibration, and sample chromatograms were developed simultaneously and dried. After exposure to ammonia for 6 to 8 hours, the orientation chromatogram was immersed in 0.03% aqueous nonradioactive Co(OAc)$_2$ and the well-separated spots were located with a 1% solution of ammonium sulfide. After treatment with ammonia, the other chromatograms were cut into 3-cm segments. Segments that contained portions of identified single spots were immersed in labeled reagent and washed, and ^{60}CoS was formed. The integrated activity of the ^{60}CoS derived from each acid was then determined. In calculating the amount of each acid in the sample, use was made of the total net activity in the respective sample spot and in the corresponding calibration spot, and of the fact that this integrated activity is proportional to the quantity of acid present. Precise pH control is necessary to avoid errors: at pH < 5.8 soap formation is incomplete, and at pH > 6.0 difficultly removable cobalt ions are adsorbed on the paper.

If the methods that use ^{60}Co(OAc)$_2$ are to be applied to samples that contain saturated fatty acids with fewer than 12 carbon atoms, possible loss of the soaps during washing must be assessed initially by means of the pure acids.

A recent advance in the use of ^{60}Co for determining the total of higher, free, saturated fatty acids is measurement of the activity of the soaps after their formation in a mixture of chloroform and heptane (120). In practice, 50 λ of heptane that contains 1 to 40 nM of the acid is combined with 100 λ of a 4:1 (v/v) chloroform-heptane mixture in a small test tube. An aliquot of 10 λ of a fresh solution composed of ^{60}Co(NO$_3$)$_2$ stock solution-saturated aqueous K$_2$SO$_4$-triethanolamine (10:9:1, v/v), is then added; the aliquot contains 100 nM of cobalt nitrate. The stock solution is prepared by adding 20 mCi of ^{60}Co(NO$_3$)$_2$, enough nonradioactive Co (NO$_3$)$_2$ to give a total of 2 mM of the salt, and 0.8 ml of glacial acetic acid to a 100-ml volumetric flask and diluting to volume with saturated aqueous sodium sulfate solution. The stoppered tube is placed for 30 seconds on a commercial instrument that will vigorously mix the contents. After centrifugation, an aliquot of 100 λ of the organic (upper) layer is taken for radioassay. For highest accuracy, a calibration curve should be prepared from known amounts of the acid to be determined.

When reagent that had a specific activity of 10 mCi/mM was used, the sensitivity of the above procedure was 0.08 nM; the counting method employed, however, was not given. Use of palmitic-^{14}C acid and ^{60}Co(NO$_3$)$_2$ in separate experiments showed that the molar ratio of acid to cobalt in the organic phase was 2.03 $\pm$ 0.16, although only 94% of the palmitic acid activity appeared in this phase. The recoveries observed were significantly dependent on molecular weight for acids that contained fewer than 12 carbon atoms; the recoveries for octanoic acid and decanoic acid were 70% and 90%, respectively. Unsaturated fatty acids were not included in the study.

Nonradioactive silver soaps of the higher saturated fatty acids have been used as intermediates in a method that employs ^{131}I (121, 122). After separation of the acids by reversed-phase chromatography, the chromatograms were immersed in a saturated (ca. 0.05 M) solution of silver acetate for 15 minutes, and then washed thoroughly with water. Treatment for 15 minutes with a 0.01 M K^{131}I solution that had an activity of 3 to 4 μCi/ml was used to form Ag^{131}I. Excess reagent was removed by further water washing, and the strips were dried in the dark. The activity was assayed by means of a strip scanner with a 2-mm slit. Lauric, myristic, palmitic, and stearic acids in amounts of 50 to 400 μg were determined with an accuracy in the range of 4.6 to 7.1%. The direct proportionality observed between integrated spot activity and spot content is useful if the components are well separated; where the peak area is difficult to determine, use can be made of the relationship between maximum spot activity and spot content. Simultaneous development of sample and reference chromatograms is necessary for highest accuracy. The technique cannot be applied to fatty acids that have fewer than 12 carbon atoms, because of the significant solubility of the silver soaps of such acids in water.

Direct determination of total higher fatty acids as the ^{110m}Ag or ^{203}Hg soaps on filter paper has been proposed as a quantitative method (123). The conditions that govern the use of ^{110m}AgNO$_3$, in particular, are well defined. After deposition of $\not> 1000$ μg of the acid on filter paper as a uniform spot approximately 1 cm in diameter, the spot is immersed in 1% ^{110m}AgNO$_3$ at a pH well below 4 for 1 hour. Excess reagent is removed by washing the paper four times with water. If the sample contains octanoic or decanoic acid, the wash water must be saturated with the silver salt of the acid to preclude leaching. Care is also necessary in handling the treated spots to prevent reduction of silver ion by the paper. The use of ^{203}Hg(OAc)$_2$ is similar, although unsaturated acids give addition products as well as soaps with this reagent.

A method based on the differential solubilities of ^{110m}Ag soaps has been suggested for the indirect determination of unsaturated fatty acids in the presence of saturated acids (124). Two separate chromatograms that contain the same amount of the sample are prepared and developed in an identical manner. After the chromatograms have been dried, one is treated with an aqueous solution of ^{110m}AgNO$_3$ at pH 7.6 to 8.0 and the other with a 1:1 ethanol-water solution of ^{110m}AgNO$_3$ at pH 4.3. The soaps of both groups of acids are retained on the paper by the aqueous reagent of higher pH, whereas the soaps of the unsaturated acids are washed out by the mixed solvent of lower pH; the unsaturated compounds are then estimated by difference.

Other radioactive cations that have been used in forming insoluble soaps of fatty acids are those of ^{144}Ce, ^{90}Sr, ^{204}Tl, and ^{95}Zr (121).

The need for determining low concentrations of carboxyl groups in the various forms of chemically modified cellulose has produced a number of methods that employ radioactive metals. Radioassays are made either on the separated cellulose derivative or on the supernatant liquid; precise pH control during both reaction and washing is usually a requisite for accuracy. The radioreagents that have been used for the analysis of various substrates are indicated in Table 3-9.

Table 3-9.   Radiochemical Determination of Carboxyl Groups in Cellulosic Materials

Cellulose form	Radioreagent	Phase monitored	Reference
Carboxymethyl cellulose	Th or U salts	Solid	125
Cellulose	^{144}Ce(OAc)$_3$	Solid	126
Oxidized cellulose	^{60}Co salt	Solid	127
Oxidized cellulose	^{60}Co(NO$_3$)$_2$, ^{110m}AgNO$_3$	Solid	128
Cellulose	^{144}Ce(OAc)$_3$	Solid	129
Cellulose	^{45}Ca(OAc)$_2$	Liquid	130
Oxidized cellulose	^{110m}AgNO$_3$	Solid	131

The feasibility of radiometric titration (132, 133) of organic acids and their soluble salts with ^{110m}AgNO$_3$ has been demonstrated by the analysis of oxalic acid, ammonium oxalate, citric acid, sodium citrate, and sodium *p*-aminosalicylate (134–136). The procedure used with acids and their salts involves detection of excess silver ion in the supernatant liquid after quantitative precipitation of the silver salts. A recent advance in the radiometric determination of oxalic acid is titration with 0.1 N or 0.01 N ^{45}Ca chloride solution (137) in an apparatus designed specifically for detecting weak beta emitters by a scintillation technique (138). Ascorbic acid has been determined by a radiometric redox titration in which ferric chloride is used as titrant and ^{65}Zn amalgam as indicator (139).

E. Other Methods

Radiochemical methods are useful for qualitative and quantitative characterization of carboxyl-containing materials in a variety of other ways and have found extensive application in biochemical analysis. Ammonium thiocyanate-^{35}S (140–142) and phenyl-^{3}H isothiocyanate (143) are valuable radioreagents for quantitative determination of terminal carboxyl (C-terminal) groups in polypeptides as the thiohydantoin and phenylthiohydantoin derivatives, respectively, of the corresponding amino acids. Identification of C-terminal amino acids in proteins is often possible by a selective tritiation reaction that involves treatment of the sample with ^{3}H$_2$O in the presence of acetic anhydride (144–146); dissolution of the protein in ^{3}H$_2$O, followed by addition of pyridine and the anhydride, was recommended on the basis of a critical study (146) of the technique. Lactic acid can be determined by combining isotope dilution with ceric sulfate oxidation (147). Oxalic-^{14}C acid has been used for determining urinary oxalate as the calcium salt (148) and has been employed also in deriving a correction factor for an oxidimetric method for the acid (149).

REFERENCES

Absorption Spectrophotometric

1. E. Childers and G. W. Struthers, *Anal. Chem.*, **27**, 737 (1955).
2. F. Feigl and V. Anger, *Mikrochemie*, **15**, 23 (1934); V. Anger and O. Frehden, *Mikrochemie*, **15**, 9 (1934).
3. R. F. Goddu, N. F. LeBlanc, and C. M. Wright, *Anal. Chem.*, **27**, 1251 (1955).
4. N. D. Cheronis and T. S. Ma, *Organic Functional Group Analysis by Micro and Semimicro Methods*, Interscience, New York, 1964, p. 169.
5. F. Bergman, *Anal. Chem.*, **24**, 1367 (1952).
6. H. L. Dinsmore and D. C. Smith, *Anal. Chem.*, **20**, 14 (1948).
7. J. Haslam and C. Newlands, *Analyst*, **80**, 50 (1955).

Gas Chromatographic

8. D. T. Downing and R. S. Greene, *Anal. Chem.*, **40**, 827 (1968).

9. H. Liliedahl, *Acta Chem. Scand.*, **20**, 95 (1966).

10. T. Hamada, S. Omori, K. Kameoka, S. Horii, and H. Morimoto, *J. Dairy Sci.*, **51**, 228 (1968).

11. B. G. Cottyn and C. V. Boucque, *J. Agr. Food Chem.*, **16**, 105 (1968).

12. D. W. Kellogg, *J. Dairy Sci.*, **52**, 1690 (1969).

13. R. N. Shelley, H. Salwin, and W. Horwitz, *J. Assoc. Offic. Agr. Chemists*, **46**, 486 (1963).

14. R. A. Ledford, *J. Dairy Sci.*, **52**, 949 (1969).

15. J. G. Nikelly, *Anal. Chem.*, **36**, 2244 (1964).

16. B. A. Bierl, M. Beroza, and W. T. Ashton, *Mikrochim. Acta*, **1969**, 637.

17. L. D. Metcalfe, *Nature*, **188**, 142 (1960).

18. V. L. Davison and H. J. Dutton, *Anal. Chem.*, **38**, 1302 (1966).

19. N. E. Hoffman and I. R. White, *Anal. Chem.*, **37**, 1541 (1965).

20. R. C. Hall, G. S. Giam, and M. G. Merkle, *Anal. Chem.*, **42**, 423 (1970).

21. T. S. Ma, C. T. Shang, and E. Manche, *Mikrochim. Acta*, **1964**, 571.

22. T. S. Ma and R. T. E. Schenck, *Mikrochemie*, **40**, 245 (1953).

23. M. H. Hubacher, *Anal. Chem.*, **21**, 945 (1949).

24. M. Beroza, *Anal. Chem.*, **25**, 177 (1953).

25. A. Di Lorenzo, *J. Chromatog. Sci.*, **8**, 224 (1970).

26. N. S. Radin, *J. Am. Oil Chemists' Soc.*, **42**, 569 (1965).

27. J. S. O'Brien and G. Rouser, *Anal. Biochem.*, **7**, 288 (1964).

28. L. J. Morris, R. T. Holman, and K. Fontell, *J. Am. Oil Chemists' Soc.*, **37**, 323 (1960); *J. Lipid Res.*, **1**, 412 (1960).

29. L. A. T. Verhaar and H. G. J. de Wilt, *J. Chromatog.*, **41**, 168 (1969).

30. M. G. Horning, E. A. Boucher, A. M. Moss, and E. C. Horning, *Anal. Letters*, **1**, 713 (1968).

31. N. E. Hoffman and T. A. Killinger, *Anal. Chem.*, **41**, 162 (1969).

32. K. Blau, *in* H. A. Szymanski, ed., *Biomedical Applications of Gas Chromatography*, Vol. 2, Plenum, New York, 1968, pp. 1–52.

33. C. W. Gehrke and D. L. Stalling, *Separation Sci.*, **2**, 101 (1967).

33a. R. W. Zumwalt, D. Roach, and C. W. Gehrke, *J. Chromatog.*, **53**, 171 (1970).

34. J. F. Klebe, H. Finkbeiner, and D. M. White, *J. Am. Chem. Soc.*, **88**, 3390 (1966).

35. G. E. Pollock, *Anal. Chem.*, **39**, 1194 (1967).

36. F. S. Shenstone, J. R. Vickery, and A. R. Johnson, *J. Agr. Food Chem.*, **13**, 410 (1965).

37. E. L. Schneider, S. P. Loke, and D. T. Hopkins, *J. Am. Oil Chemists' Soc.*, **45**, 585 (1968).

38. J. H. Recourt, G. Jurriens, and M. Schmitz, *J. Chromatog.*, **30**, 35 (1967).

39. I. J. Krchma, *J. Gas Chromatog.*, **6**, 457 (1968).

40. P. R. Monk and W. W. Forrest, *J. Chromatog.*, **30**, 203 (1967).

41. V. R. Huebner, *J. Am. Oil Chemists' Soc.*, **36**, 262 (1959).

42. A. Kuksis, W. C. Breckenridge, L. Marai, and O. Stachnyk, *J. Am. Oil Chemists' Soc.*, **45**, 537 (1968).
43. R. Watts and R. Dils, *J. Lipid Res.*, **10**, 33 (1969).
44. J. Hradec, *J. Chromatog.*, **32**, 511 (1968).
45. H. Hadorn and K. Zuercher, *Mitt. Gebiete Lebensm. Hyg.*, **59**, 369 (1968).
46. J. Bezard and M. Bugaut, *J. Chromatog. Sci.*, **7**, 639 (1969).
47. M. Kates, *J. Lipid Res.*, **5**, 132 (1964).
48. R. Kleiman, F. R. Earle, and I. A. Wolff, *J. Am. Oil Chemists' Soc.*, **46**, 505 (1969).
49. F. M. Antonini, P. Tinti, E. Petruzzi, A. Bucalossi, G. Pazzagli, and A. D'Alessandro, *Riv. Ital. Sostanze Grasse*, **46**, 144 (1969).
50. J. I. Peterson, H. de Schmertzing, and K. Abel, *J. Gas Chromatog.*, **3**, 126 (1965).
51. M. E. Mason and G. R. Waller, *Anal. Chem.*, **36**, 583 (1964).
52. G. R. Jamieson and E. H. Reid, *J. Chromatog.*, **17**, 230 (1965).
53. S. W. Christopherson and R. L. Glass, *J. Dairy Sci.*, **52**, 1289 (1969).
54. K. Oette and M. Doss, *J. Chromatog.*, **32**, 439 (1968).
55. V. L. Davison and H. J. Dutton, *J. Lipid Res.*, **8**, 147 (1967).
56. C. Alling, L. Svennerholm, and J. Tichy, *J. Chromatog.*, **34**, 413 (1968).
57. F. E. Luddy, R. A. Barford, S. F. Herb, and P. Magidman, *J. Am. Oil Chemists' Soc.*, **45**, 549 (1968).
58. L. A. Horrocks and D. G. Cornwell, *J. Lipid Res.*, **3**, 165 (1962).
59. F. M. Archibald and V. P. Skipski, *J. Lipid Res.*, **7**, 442 (1966).
60. E. J. Barron and L. A. Mooney, *Anal. Chem.*, **40**, 1742 (1968).
61. F. Acree, Jr., and M. Beroza, *J. Econ. Entomol.*, **55**, 619 (1962).
62. M. Malaiyandi, J. P. Barrette, A. S. Y. Chau, and S. A. MacDonald, *Abstracts of Papers*, B-65, 154th Meeting, American Chemical Society, Chicago, Ill., 1967.
63. V. I. Trubnikov, L. M. Malakhova, and E. S. Zhdanovich, *Zh. Anal. Khim.*, **23**, 1546 (1968).
64. B. Radhakrishnamurthy, E. R. Dalferes, Jr., and G. S. Berenson, *Anal. Biochem.*, **26**, 61 (1968).
65. R. A. Hoodless and R. E. Weston, *Analyst*, **94**, 670 (1969).
66. M. N. Chumachenko, L. B. Tverdyukova, and F. G. Leenson, *Zh. Anal. Khim.*, **21**, 617 (1966).
67. L. T. Sennello and C. J. Argoudelis, *Anal. Chem.*, **41**, 171 (1969).
68. S. Mori, M. Furusawa, and T. Takeuchi, *Anal. Chem.*, **42**, 661 (1970).
69. G. G. Briggs and J. E. Dawson, *J. Agr. Food Chem.*, **18**, 97 (1970).
70. W. Kiessling and K. K. Moll, *J. Prakt. Chem.*, **311**, 876 (1969).
71. M. Taramasso and A. Guerra, *J. Gas Chromatog.*, **3**, 138 (1965).
72. M. Oda, K. Suzuki, T. Kashiwa, K. Ikeda, and T. Hattori, *Bunseki Kagaku*, **18**, 1267 (1969).

Electroanalytical

73. S. Siggia, *Quantitative Organic Analysis via Functional Groups*, 3rd Ed., John Wiley, New York, 1963, pp. 130–202.

150   Carboxylic groups

74. S. L. Szebelledy and S. F. Somogyi, *Z. Anal. Chem.*, **112**, 400 (1938).
75. J. J. Lingane and L. A. Small, *Anal. Chem.*, **21**, 1119 (1949).
76. W. N. Carson, Jr., and R. Ko, *Anal. Chem.*, **23**, 1019 (1951).
77. R. O. Crisler and R. D. Coulson, *J. Am. Oil Chem.' Soc.*, **39**, 470 (1962).
78. C. A. Streuli, J. J. Cincotta, D. L. Maricle, and K. K. Mead, *Anal. Chem.*, **36**, 1371 (1964).
79. G. Johansson, *Talanta*, **11**, 789 (1964).
80. F. Gaslini and L. Z. Nahum, *Anal. Chem.*, **31**, 989 (1959).

Nuclear Magnetic Resonance

81. J. M. Purcell, S. G. Morris, and H. Susi, *Anal. Chem.*, **38**, 588 (1966).
82. D. T. Sawyer and J. R. Brannan, *Anal. Chem.*, **38**, 192 (1966).
83. P. J. Paulsen and W. D. Cooke, *Anal. Chem.*, **36**, 1713 (1964).
84. J. L. Jungnickel and J. W. Forbes, *Anal. Chem.*, **35**, 938 (1963).
85. J. R. Parker, *Anal. Chem.*, **41**, 1103 (1969).
86. O. Rosado-Lojo, C. K. Hancock, and A. Danti, *J. Org. Chem.*, **31**, 1899 (1966).
87. R. O. Kan, *J. Am. Chem. Soc.*, **86**, 5180 (1964).
88. D. F. Percival and M. P. Stevens, *Anal. Chem.*, **36**, 1574 (1964).
89. K. Konishi and Y. Kanoh, *Anal. Chem.*, **40**, 1881 (1968).
90. E. G. Brame, Jr., R. Ferguson, and G. J. Thomas, Jr., *Anal. Chem.*, **39**, 517 (1967).

Radiochemical

91. H. Schlenk and J. L. Gellerman, *Anal. Chem.*, **32**, 1412 (1960).
92. B. P. Smirnov, R. A. Popova, and R. A. Niskanen, *Biokhimya*, **25**, 368 (1960).
93. B. P. Smirnov, *Tr. Komis. po Analit. Khim., Akad. Nauk SSSR*, **13**, 435 (1963).
94. H. K. Mangold, J. L. Gellerman, and H. Schlenk, *Federation Proc.*, **17**, 268 (1958).
95. H. K. Mangold, *Fette, Seifen, Anstrichmittel*, **61**, 877 (1959).
96. P. D. S. Wood and H. S. Sohdi, *Proc. Soc. Exptl. Biol. Med.*, **118**, 590 (1965).
97. K. Heyns, H. Heinecke, and G. Grimmer, *Hoppe-Seylers Z. Physiol. Chem.*, **343**, 116 (1965).
98. K. Heyns and R. Hauber, *Hoppe-Seylers Z. Physiol. Chem.*, **348**, 357 (1967).
99. W. E. Baumgartner, *Proc. Sym. Advan. Tracer Appl. Tritium*, Boston, 1958, p. 33.
100. W. E. Baumgartner, L. S. Lazer, A. M. Dalziel, E. V. Cardinal, and E. L. Varner, *J. Agr. Food Chem.*, **7**, 422 (1959).
101. L. S. Lazer, W. E. Baumgartner, and R. V. Dahlstrom, *J. Agr. Food Chem.*, **9**, 24, (1961).
102. W. E. Baumgartner, in S. Rothchild, ed., *Advances in Tracer Methodology*, Vol. 1, Plenum, New York, 1962, pp. 257–262.
103. E. Papirer and J.-B. Donnet, *Bull. Soc. Chim. France*, **1966**, 2033.
104. L. Melander, *Arkiv Kemi*, **3**, 525 (1951).
105. W. G. Verly, J. R. Rachele, V. du Vigneaud, M. L. Eidinof, and J. E. Knoll, *J. Am. Chem. Soc.*, **74**, 594 (1952).

106. G. K. Koch and G. Jurriens, *Nature*, **208**, 1312 (1965).

107. G. A. Fischer and J. J. Kabara, *Anal. Biochem.*, **25**, 432 (1968).

108. R. G. Garmon and M. E. Gibson, *Anal. Chem.*, **37**, 1309 (1965).

109. L. D. Metcalfe and A. A. Schmitz, *Anal. Chem.*, **33**, 363 (1961).

110. D. Eberhagen, B. Wittman, and W. Seitz, *Z. Anal. Chem.*, **237**, 17 (1968).

111. A. C. Kibrick and S. J. Skupp, *Anal. Chem.*, **31**, 2057 (1959).

112. J. M. Steim and A. A. Benson, *Anal. Biochem.*, **9**, 21 (1964).

113. P. Sorenson, *Anal. Chem.*, **28**, 1318 (1956).

114. V. F. Raaen, G. A. Ropp, and H. P. Raaen, *Carbon-14*, McGraw-Hill, New York, 1968, p. 17.

115. P. Sorenson, *Anal. Chem.*, **26**, 1581 (1954).

116. W. M. Stokes, W. A. Fish, and F. C. Hickey, *Anal. Chem.*, **27**, 1895 (1955).

117. H. P. Kaufmann and J. Budwig, *Fette und Seifen*, **53**, 69 (1951).

118. H. P. Kaufmann, *Fette, Seifen, Anstrichmittel*, **58**, 492 (1956).

119. R. Otto, *Atompraxis*, **7**, 209 (1961); *Isotopentechnik*, **1**, 184 (1961).

120. R. J. Ho and H. C. Meng, *Anal. Biochem.*, **31**, 426 (1969).

121. A. Z. Budzynski, Z. J. Zubrzycki, and I. G. Campbell, *Proc. 2nd Intern. Conf. Peaceful Uses At. Energy*, Geneva, 1958, Vol. 24, p. 274.

122. Z. J. Zubrzycki, A. Z. Budzynski, and I. G. Campbell, *Talanta*, **2**, 165 (1959).

123. M. Jaky and K. Kaffka, *Fette, Seifen, Anstrichmittel*, **62**, 682 (1960).

124. Sh. Akramov and A. L. Markman, *Dokl. Akad. Nauk Uz. SSR*, **24**, 25 (1967); *Chem. Abstr.*, **68**:101682h.

125. C. Simionescu and N. Asandei, *Chem. Anal.*, **40**, 204 (1958); *Chem. Abstr.*, **52**:19121e.

126. P. Daudel, *Chem. Anal.*, **40**, 325 (1958); *Chem. Abstr.*, **52**:130h.

127. P. Rochas, R. Bussiere, and L. Gavet, *Compt. Rend.*, **248**, 3436 (1959).

128. P. Rochas and L. Gavet, *Bull. Inst. Textile France*, No. 87, 19 (1960).

129. P. Valls, A. M. Venet, and J. Pousadier, *Bull. Soc. Chim. France*, C106-9 (1953).

130. J. Hostomsky, J. Tölgyessy, and V. Krivan, *Chem. Zvesti*, **14**, 290 (1960); *Chem. Abstr.*, **54**:25787g.

131. P. Rochas and L. Gavet, *Melliand Textilber.*, **49**, 1123 (1968).

132. T. Braun and J. Tölgyessy, *Talanta*, **11**, 1277 (1964).

133. T. Braun and J. Tölgyessy, *Radiometric Titrations*, Pergamon, New York, 1967.

134. P. Bebesel and I. Sirbu, *Rev. Chim. (Bucharest)*, **11**, 288 (1960); *Chem. Abstr.*, **58**:5034a.

135. G. Straub and Z. Czapo, *Acta Chim. Acad. Sci. Hung.*, **26**, 267 (1961); *Chem Abstr.*, **55**:21963c.

136. J. Tölgyessy and M. Sarsunova, *Z. Anal. Chem.*, **196**, 192 (1963).

137. C. Ömböly, T. Szarvas, and L. Horvath, *Radiochem. Radioanal. Letters*, **1**, 149 (1969).

138. C. Ömböly, T. Szarvas, and G. Vegh, *J. Radioanal. Chem.*, **4**, 215 (1970).

139. T. Braun and E. Körös, *Proc. Intern. At. Energy Agency Symp. Radiochem. Methods Anal.*, Salzburg, 1964, Vol. 2, p. 213.

140. E. Scoffone, A. Turco, D. Chellemi, and M. Scatena, *Proc. 15th Intern. Congr. Pure Appl. Chem.*, Lisbon, 1956, Vol. 2, pp. 750–752.

141. E. Scoffone and A. Turco, *Ric. Sci.*, **26**, 865 (1956).

142. E. Scoffone, A. Turco, and M. Scatena, *Ric. Sic.*, **27**, 1193 (1957).

143. R. A. Laursen, *Biochem. Biophys. Res. Commun.*, **37**, 663 (1969).

144. H. Matsuo, Y. Fujimoto, and T. Tatsuno, *Tetrahedron Letters*, **39**, 3465 (1965).

145. H. Matsuo, Y. Fujimoto, and T. Tatsuno, *Biochem. Biophys. Res. Commun*, **22**, 69 (1966).

146. G. N. Holcomb, S. A. James, and D. N. Ward, *Biochemistry*, **7**, 1291 (1968).

147. R. L. A. Leunissen and D. A. Piatnek-Leunissen, *Anal. Biochem.*, **15**, 409 (1966).

148. B. M. Dean and W. J. Griffen, *Nature*, **205**, 598 (1965).

149. G. H. Koch and F. M. Strong, *Anal. Chem.*, **37**, 1092 (1965).

4 Alkoxyl and oxyalkylene groups

I. ABSORPTION SPECTROPHOTOMETRIC METHODS*

Measurement after Chemical Reaction

1. Hydrolysis of Methoxy Groups to Methanol and Oxidation to Formaldehyde.
Pavolini and Malatesta (1) determined methylenedioxyl and methoxyl in nine alkaloids and six phenols, using warm 80% phosphoric acid to convert the methylenedioxyl to formaldehyde and warm concentrated sulfuric acid for the hydrolysis of the methoxyl groups to methanol. They oxidized the methanol to formaldehyde with potassium dichromate and determined the formaldehyde in both cases, using either Nessler's or Tollen's reagent.

Mathers and Pro (2) cleaved methoxyl hydrolytically to methanol, oxidized the methanol to formaldehyde, and determined the formaldehyde colorimetrically after condensation with chromotropic acid.

METHOD FROM MATHERS AND PRO (2)

Apparatus. Spectrophotometer. Beckman Model DU with 1-cm square cuvettes.

Distillation unit for microdetermination. Total condensation, variable-takeoff type of distillation column with a packed section 1 × 20 cm, lagged with a silver vacuum jacket and single-turn glass helices as packing material.

Reagents. Standard methyl alcohol solution, 20 mg/100 ml in 5.5% (v/v) ethyl alcohol.

Ethyl alcohol solution, 5 to 6% (v/v).

Chromotropic acid solution, 1 gram/25 ml of water. Prepare fresh daily.

---

* Written by J. Gordon Hanna.

153

Potassium permanganate solution. Dilute 3 grams plus 15 ml of 85% phosphoric acid to 100 ml with distilled water.

Procedure. Introduce a weighed sample (approximately 0.1 gram) into a 250-ml flask and attach an efficient reflux condenser. Through the condenser add about 10 ml of concentrated sulfuric acid (double the quantity of sulfuric acid for larger samples) and heat to fumes of sulfur trioxide for 5 minutes. Cool the reaction mixture and dilute with 75 ml of water, added through the condenser. Again cool the solution, remove the condenser, attach a simple distilling head, distill about 45 ml of the liquid into a 50-ml volumetric flask containing 3 ml of 95% ethanol, and make to the mark with water. Pipet 1 ml of this solution into a 50-ml volumetric flask, set in an ice bath, and add 2 ml of chilled permanganate solution. Allow oxidation to take place for 30 minutes at ice-bath temperature; then destroy the excess oxidant with approximately 0.2 to 0.3 gram of sodium bisulfite. To the clear solution add 1 ml of chromotropic acid solution, followed by the slow addition of 15 ml of concentrated sulfuric acid with swirling. Set the open flask in a 55 to 65°C water bath for 30 minutes. Dilute the solution with water, cool to room temperature, and dilute to volume with water. Prepare the reference blank and standard methanol color by treating, respectively, 1 ml of 5.5 to 6% ethanol and 1 ml of the standard methanol solution in the above manner, beginning with the oxidation step. Read the absorbance of the samples and standard methanol at 570 mμ versus the reference ethanol blank.

The absorbance is directly proportional to the quantity of formaldehyde, which is in turn proportional to the methoxyl and the methanol, respectively. However, the color intensity is also a function of temperature, and thus it is necessary to read the absorbance of both sample and standard methanol solution at nearly identical temperatures.

Calculation. Calculate the quantity of methoxyl by the following formula:

$$\frac{A_s}{A_m} \times F \times M \times R = \text{weight per cent of methoxyl} \tag{1}$$

where A_s = absorbance of sample,
$\quad A_m$ = absorbance of methanol standard,
$\quad F$ = dilution factor of the sample,
$\quad M$ = per cent methanol by weight in standard,
$\quad R$ = molecular weight ratio of methoxyl to methanol.

Microdetermination. Introduce a weighed sample of approximately 1 mg into a 100-ml round-bottom flask and attach to a reflux condenser. Add 10 ml of concentrated sulfuric acid and heat to fumes of sulfur trioxide for 5 minutes. Then cool and add about 40 ml of water through the reflux condenser. Remove the condenser, attach the flask to the distilling column, and place the solution under total reflux for 20 minutes. Maintaining a reflux ratio of 30 to 1, collect 2 ml of distillate. Pipet 1 ml of the distillate into a 10-ml volumetric flask and add 1 drop of 95% ethanol. Cool the flask in an ice bath, add 2 ml of chilled permanganate solution, and allow the oxidation to proceed for 30 minutes. Destroy the excess permanganate with sodium bisulfite, add 1 ml of chromotropic acid solution and 6 ml of concentrated sulfuric acid, and heat the mixture for 30 minutes in 55 to 65°C water. Then cool the flask, fill to the mark with sulfuric acid, and read the absorbance at 570 mμ versus a reference blank prepared with 1 drop of ethanol. Treat 0.2 ml of standard methanol solution in the same manner as the sample. Calculate the methoxyl content of the sample according to equation 1.

The oxidation of methanol by acid permanganate solution is an equilibrium-type reaction in which less than a 50% yield of formaldehyde is present under equilibrium conditions. Methanol and ethanol are oxidized to the respective aldehydes and acids, with the possibility of the former giving some carbon dioxide. The acetaldehyde produced by the ethanol interferes to some extent with the determination of formaldehyde, giving a yellow solution, which has some absorbance at 570 mμ. Interference due to acetaldehyde is minimized by use of the reference blank, in which an identical quantity of alcohol is treated in the same manner as the sample. Results for the determination of methoxyl, both with and without the addition of ethanol to some of the samples, are shown in Table 4-1.

Beroza (3) demonstrated that methylenedioxyl or other labile methylene groups can be converted to formaldehyde by hydrolytic cleavage of the molecule; therefore, compounds containing this group interfere with this methoxyl determination. Glycolic acid and methyl iodide interfere. No interference is offered by other alkoxy or alkimide groups in the determination of methoxyl by this method.

2. Cleavage of Polyoxyethylene Compounds.

When anhydrous samples of polyoxyethylene compounds are heated in acidic solution of alcoholic 2,4-dinitrophenylhydrazine and the products are treated with alcoholic potassium hydroxide, an intense purple color with an absorption maximum at 560 mμ results. The hydrazone formed is not that of formaldehyde or acetaldehyde because these compounds give red or wine-red colors with absorption maxima at 440 and 425 mμ, respectively. Gatewood and Graham (4), who developed a method based on the reaction for polyoxyethylene compounds, suggested that possibly bisphenylhydrazones are formed in the process. These substances can then react further with the hydrazine to form 2- and 3-carbon fragments. Surface-active agents which do no

Table 4-1. Methoxyl Determination Results (2)

Compound	Methoxyl, weight %	
	Calculated	Found
Anisaldehyde (*p*-methoxybenzaldehyde)	22.8	22.8
α-Chloroisobutyraldehyde dimethyl acetal	40.6	40.2
α-Methoxyisobutyraldehyde, 2,4-dinitrophenylhydrazone	11.0	10.8
Methyl Cellosolve (2-methoxyethanol)	40.7	40.4, 40.6
Cocaine hydrochloride (benzoyl methyl ecogonine)	9.1	8.9, 8.8
Codeine sulfate hydrated (methyl morphine)	7.9	7.5,[a] 7.8
Codeine sulfate hydrated (microdetermination)	7.9	7.5,[a] 8.0, 7.8
Methyl α-chloroisobutyrate	22.7	22.9
Methyl cellulose (monomethyl ether)	17.6	17.5
Methyl methacrylate	31.0	30.6
Pectin	. . .	3.5,[a] 3.5
Vanillin (4-hydroxy-3-methoxybenzaldehyde)	20.4	20.2, 20.2
Glycine	0	0
Glycerol	0	0
Ethylene glycol	0	0
p-Dimethylaminobenzaldehyde	0	0
Lactic acid	0	0
Tartaric acid	0	0
Methyl iodide	21.8	7.6
Glycolic acid (hydroxyacetic acid)	40.8	24.7

[a] Oxidized in the absence of ethyl alcohol.

contain the ethylene oxide or propylene oxide unit did not respond. Ethylene glycol does not react, but di- and triethylene glycols, as well as propylene glycol, do.

METHOD FROM GATEWOOD AND GRAHAM (4)

Reagents.   Carbonyl-free methanol. Reflux 500 ml of methanol with 5 grams of 2,4-dinitrophenylhydrazine and 1 ml of concentrated hydrochloric acid for 3 hours and distill. Collect the fraction boiling at 64.7 at 65°C and redistill.

10% Alcoholic potassium hydroxide. Dissolve 10 grams of potassium hydroxide in 20 ml of double-distilled water and make to 100 ml with car-bonyl-free alcohol. Keep the flask in an ice bath during all additions.

2,4-Dinitrophenylhydrazine solution. Add 5 ml of concentrated hydro-chloric acid to 100 mg of 2,4-dinitrophenylhydrazine and 50 ml of carbonyl-free alcohol. Shake and warm gently, if necessary, until all the 2,4-dinitro-phenylhydrazine dissolves. Make up to 100 ml with carbonyl-free alcohol. Prepare a new batch every 3 days.

Procedure.   Prepare a sample containing 2.5 to 25 μg/ml. Pipet 1 ml into an oven-dried borosilicate glass-stoppered test tube. If less than a 1-ml sample is used, make up to 1 ml with carbonyl-free alcohol. Add 1 ml of 2,4-dinitro-phenylhydrazine solution and heat the mixture on a water bath for exactly 15 minutes. Cool in an ice-water bath and add 5 ml of distilled water. Measure the absorbance at 560 mμ against a blank and obtain the concen-tration from a calibration curve prepared from data obtained by subjecting pure compounds to the procedure.

3. Cobaltothiocyanate Complex of Polyoxyalkylene Compounds.

Brown and Hayes (5) developed quantitative procedures for poly(ethylene oxide) condensates based on the blue color formed with ammonium cobalt-othiocyanate. They extracted the complex into chloroform and measured the absorbance of the chloroform solution at 620 or 318 mμ; greater sensi-tivity was found with the latter wavelength.

Morgan (6) found that the complexes formed by several poly(ethylene oxide) surfactants were insoluble in chloroform but could be readily ex-tracted into benzene. After evaporation of the benzene, the complex was decomposed with water and the cobalt in the aqueous solution was deter-mined as its complex with nitroso-R salt. The absorbance measurements were made at 500 mμ. The sensitivity achieved by this procedure was about as high as that attained with direct measurements of the blue complex at 318 mμ.

METHOD FROM MORGAN (6)

Reagents.   Ammonium cobalt thiocyanate solution. Dissolve 174 grams of ammonium thiocyanate and 2.8 grams of cobalt nitrate hexahydrate in water and dilute to 1 liter.

Hydrochloric acid-nitric acid mixture. Dilute a mixture of 25 ml of con-centrated hydrochloric acid and 5 ml of concentrated nitric acid to 100 ml with water.

Nitroso-R solution, 0.05% (w/v), aqueous.

Sodium acetate solution, 50% (w/v), aqueous.

Procedure.   Measure approximately 5 ml of the thiocyanate solution into a stoppered 25-ml cylinder. If the surfactant is soluble in benzene, add an aliquot of benzene solution containing about 0.5 mg; if the surfactant is water soluble, add not more than 0.5 ml of its aqueous solution. Dilute to about 10 ml with benzene, shake the cylinder vigorously for 2 minutes, and allow the two layers to separate completely. Blow the benzene layer into a 100-ml beaker, and add a further 5 to 10 ml of benzene to the aqueous layer. Invert the cylinder a few times, allow the layers to separate, and again blow the benzene layer into the beaker.

Cover the beaker with a watch glass, cautiously evaporate its contents to dryness, and add 5 ml of water and 0.5 ml of the acid mixture. Add exactly 1 ml of nitroso-R salt solution and then 2 ml of sodium acetate solution. Cover the beaker with a watch glass and boil for 1 minute. Add 1 ml of concentrated nitric acid, replace the watch glass, and boil for another minute. Allow the beaker to cool in the dark, transfer its contents quantitatively to a 10-ml calibrated flask, and dilute to the mark with water. Measure the absorbance of this solution at 500 mμ against a reagent blank solution prepared from 5 ml of water, 0.5 ml of the acid mixture, etc. Calculate the concentration of nonionic surfactant present from a calibration graph plotted from results for the particular surfactant being determined.

II.   GAS CHROMATOGRAPHIC METHODS*

In most methods for determining alkoxyl groups, which are found in a variety of structures, such as the following:

$$R\text{—}O\text{—}R',\quad R\text{—}CH(OR')_2,\quad R\overset{\displaystyle OH}{\overset{|}{C}}HOR',\ R\overset{\displaystyle O}{\overset{\|}{C}}\text{—}OR'$$

the compounds being analyzed are heated with hydriodic acid to convert alkoxyl groups to alkyl iodides (—OR + HI → —OH + RI); the iodides are then volatilized out of the reaction vessel, and the individual alkyl iodides, as well as the various kinds (e.g., C_1 to C_4), are readily determined quantitatively by gas chromatography. Thus GC porvides a rapid method for multiple alkoxyl determinations.

This general procedure is used by Mitsui and Kitamura (7), who collect the alkyl iodides (C_1 to C_4) on silica gel and then heat the silica gel in a nitrogen stream for introduction of the iodides into a gas chromatograph. Ehrenberger (8) determines C_1 to C_4 alkoxyl groups and oxyalkylene groups (—OC_2H_4— and —OC_3H_6—) by essentially the same procedure in an apparatus of his own design. In another procedure, the sample is heated with

* Written by Morton Beroza and May N. Inscoe.

KI—H_3PO_4 reagent and methylcyclohexane is included as an internal standard; the reaction vessel is connected to the gas chromatograph for determination of the reaction products. A short precolumn containing soda lime and $CaCl_2$ is used to absorb acid vapors (9). Several methods utilize the same general procedure for determining alkoxy groups in methyloxypropyl cellulose (10), in acrylate and maleate polymeric esters (11), and in glyceryl ethers (12).

Alcohols, since they react with HI and form alkyl iodides, will interfere in the foregoing analyses.

A means of determining ethoxyl groups quantitatively in O-ethyl cellulose and in ethyl-hydroxyethyl cellulose is to oxidize the sample with 30% (w/v) chromic acid in water and then analyze an aliquot of the reaction mixture for acetic acid by direct GC, using a 60/80 mesh Porapak G-S column at 190°C (13).

Klesment and Kasberg (14) determined the amount of alkoxy groups in phenols by passing the compounds through a microreactor containing a platinum catalyst at 300 to 350°C and measuring by GC the hydrocarbons formed in the hydrogenolytic reaction.

III. NUCLEAR MAGNETIC RESONANCE METHODS*

Vinyl ethers are readily determined by nuclear magnetic resonance spectroscopy because of the characteristic three-spin system and the resonance position of the hydrogens involved. The ethylenic protons of vinyl ethers appear at slightly higher field than other vinyl derivatives and thus establish whether a vinyl ether is present. A variety of vinyl ethers have been studied (15), such as methyl, ethyl, *tert*-butyl, *n*-butyl, and 2-chloroethyl.

The analysis of alkylene oxide is reported by Mathias and Mellor (16); the oxyethylene and oxypropylene contents were determined by nmr and compared with the results of a gas chromatography method. The advantage of the nmr method is that a chemical change is not required. Poly(ethylene glycol)400, poly(propylene glycol)1000, and mixtures were studied, and the results obtained by the two methods were in general agreement. A synthetic mixture containing 4.8% poly(ethylene oxide) gave 4.6% by nmr and 4.6% by gas chromatography.

The stereochemistry of isopropylidene derivatives of some polyhydric alcohols was also studied (17). The chemical shift data, however, are valuable in identifying the isomers present and following the rates of hydrolysis.

* Written by Harry Agahigian.

REFERENCES

Absorption Spectrophotometric

1. T. Pavolini and A. Malatesta, *Ann. Chim. Appl.*, **37**, 495 (1947).
2. A. P. Mathers and M. J. Pro, *Anal. Chem.*, **27**, 1662 (1955).
3. M. Beroza, *Anal. Chem.*, **26**, 1970 (1954).
4. L. Gatewood, Jr., and H. D. Graham, *Anal. Chem.*, **33**, 1393 (1961).
5. E. G. Brown and T. J. Hayes, *Analyst*, **80**, 755 (1955).
6. D. J. Morgan, *Analyst*, **87**, 233 (1962).

Gas Chromatographic

7. T. Mitsui and Y. Kitamura, *Microchem. J.*, **7**, 141 (1963).
8. F. Ehrenberger, *Z. Anal. Chem.*, **210**, 424 (1965).
9. S. Araki, S. Suzuki, and M. Kitano, *Bunseki Kagaku*, **18**, 608 (1969).
10. A. A. Karnishin, *Zh. Analit. Khim.*, **23**, 1072 (1968).
11. D. L. Miller, E. P. Samsel, and J. G. Cobler, *Anal. Chem.*, **33**, 677 (1961).
12. S. Ramachandran, H. W. Sprecher, and D. G. Cornwell, *Lipids*, **3**, 511 (1968).
13. H. Jacin and J. M. Slanski, *Anal. Chem.*, **42**, 801 (1970).
14. I. Klesment and A. Kasberg, *Mikrochim. Acta*, **1967**, 1136.

Nuclear Magnetic Resonance

15. J. Feeney, A. Ledwith, and L. H. Sutcliffe, *J. Chem. Soc.*, **1962**, 2021.
16. A. Mathias and N. Mellor, *Anal. Chem.*, **38**, 473 (1966).
17. N. Bagget, K. W. Buck, A. B. Fuster, R. Jefferis, B. H. Rees, and J. M. Webber, *J. Chem. Soc.*, **1965**, 3382.

5 Epoxide groups

I. ABSORPTION SPECTROPHOTOMETRIC METHODS*

A. Direct Infrared Measurement

The infrared absorbance spectra of eleven oxirane ring compounds have been recorded in the 2- to 15-μ region by Bomstein (1). They are chiefly nonterminal epoxides. There does not seem to be any general application of these spectra to the quantitative determination of the epoxide group.

Goddu and Delker (2) based a quantitative procedure on the absorbance of terminal epoxides in the near-infrared region. These groups have sharp bands at 1.65 and 2.20 μ, the latter absorptions being several times more intense than the former. These bands are useful for the rapid determination of epoxides in a variety of mixtures. In particular, both terminal epoxides and terminal olefins may be determined simultaneously in the same sample. Other oxygen rings do not interfere. The range covered by the method is from 10 γ per milliliter of

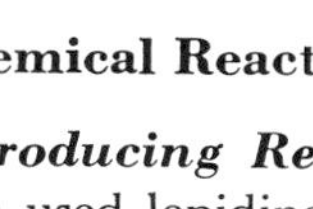

$$\begin{array}{c} \diagdown \\ C\text{-----}CH_2 \\ \diagup \diagdown \diagup \\ O \end{array}$$

to pure epoxides. The accuracy and precision over most of the range was reported to be ± 1 to 2% of the amount present.

B. Measurement after Chemical Reaction

1. Lepidine as a Color-Producing Reagent for Ethylene Oxide. Gunther and his coworkers (3) used lepidine as a colorimetric reagent for the determination of ethylene oxide, which reacts in the presence of diethyl-

---

* Written by J. Gordon Hanna.

161

ene glycol at 170°C to form an intense Hortense blue dye, wavelength maximum at 610 mμ and minimum at 490 mμ. Their method was recommended specifically for the determination of the pesticide Aramite, 2-(*p-tert*-butylphenoxy)isopropyl-2-chloroethyl sulfite. Aramite refluxed with sodium isopropoxide solution quantitatively liberates ethylene oxide. A disadvantage of the method is that careful removal of the ethylene oxide from interfering substances is required in a special evolution apparatus. Any molecule capable of liberating gaseous cyclic oxides, imines, or sulfides will produce the blue color. Ammonia and simple amines such as triethylamine prevent the color formation even in the presence of large amounts of ethylene oxide. Also, careful exclusion of oxygen is necessary. Small amounts of water and isopropyl alcohol tend to interfere with the color development.

The portion of the procedure that deals specifically with ethylene oxide determination is as follows.

METHOD FROM GUNTHER ET AL. (3)

Apparatus. Inulin tubes, Kimble Brand pressure reaction tube No. 46300. Replace the rubber cap liner with a close-fitting cork liner cut from 3-mm sheet cork.

Sand bath. Keep the bath, containing approximately 3 inches of sand, heated in an oven between runs.

Oven. This should be capable of maintaining the sand-bath temperature at 170 ± 5°C.

Aluminum foil. Wear-Ever aluminum foil No. 3383 or equivalent household grade. Cut in circles that will fit snugly into the caps of the inulin tubes.

Colorimeter or spectrophotometer. Beckman Model DU or Lumetron Model 402 EM with micro test tube adapter.

Reagents. Lepidine. Mix commercial lepidine with one-half its volume of benzene and distill at atmospheric pressure. Discard the benzene-water forerun and the first 10% of the lepidine distillate. The lepidine cut (258 to 261°C at 735 mm) should be water-white (n_D^{22}, 1.6162) and must be stored in a brown bottle. Redistill aged yellow specimens before use.

Diethylene glycol. Redistill and store the fraction boiling between 243 and 245°C at 735 mm in a brown bottle.

Reagent solution. Dilute 6.0 ml of the purified lepidine to 120 ml with purified diethylene glycol. Prepare this 5% solution daily and store it under nitrogen in a brown bottle.

Calibration Curve. To inulin pressure tubes, each containing 6.0 ml of the lepidine reagent solution, add ethylene oxides ranging from 20 to 60 μg. Top the inulin tubes with liners consisting of four disks of aluminum foil before the cork-gasketed caps are screwed on tightly. Push the set of capped tubes (including a reagent blank) vertically into the sand bath, preheated in the oven to 170 ± 5°C. After exactly 90 minutes, remove the tubes and allow them to cool to room temperature for 3 minutes. Shake the tubes and then immerse them in running water for 10 minutes. After gently inverting the cooled tubes several times, determine the absorbance of the resulting blue solution at 610 mμ. Observe that the reagent blank should always be light yellow in color.

Procedure. Treat the samples in the same manner as the standards used for preparation of the calibration curve.

2. Colorimetric Determination of Ethylene Oxide by Conversion to Formaldehyde.

Ethylene oxide is hydrolyzed to ethylene glycol in the presence of mineral acid at elevated temperature. Critchfield and Johnson (4) used this reaction as the basis for a colorimetric method for the determination of ethylene oxide. They reacted the glycol produced and sodium periodate to form formaldehyde, which they determined with sodium chromotropate.

METHOD FROM CRITCHFIELD AND JOHNSON (4)

Calibration Curve. Put approximately 50 ml of distilled water in a 100-ml volumetric flask. Tare the flask, introduce approximately 1.5 grams of ethylene oxide, and swirl to mix the contents thoroughly. When the ethylene oxide is dissolved, record the total weight of the flask and contents and calculate the net weight of the ethylene oxide. Dilute the contents of the flask to the mark with additional distilled water and mix thoroughly. Transfer a 10-ml aliquot of this dilution to a 1-liter volumetric flask containing approximately 200 ml of distilled water. Dilute to the mark and mix. Transfer aliquots to pressure bottles containing 20 ml of distilled water. Measure the absorbance of each standard after following the steps of the procedure described below.

Procedure. Into a pressure bottle containing 20 ml of distilled water introduce the sample containing no more than 0.7 mg of ethylene oxide. Pipet 1.0 ml of 0.5 N sulfuric acid into the bottle. Protect the rubber gasket on the bottle stopper with polyethylene film. Stopper the bottle, enclose it securely in a fabric bag, and heat in a steam bath for 60 minutes at 98 ± 2°C. Remove the bottle from the bath and allow it to cool to room temperature. When the bottle has cooled, loosen the bag, uncap the bottle carefully to release any pressure, and remove the bag. Quantitatively transfer the contents to a 100-ml glass-stoppered graduated cylinder. Pipet 1.0 ml of 0.5 N sodium hydroxide to the graduate, stopper, and mix. Pipet 2 ml of 0.1 N sodium periodate into the graduate, stopper, mix, and allow to react for 15 minutes at room temperature. Pipet 2.0 ml of 5.5% sodium sulfite into the graduate, dilute to the 100-ml mark with distilled water, stopper, and mix.

Transfer a 10-ml aliquot to a 100-ml glass-stoppered graduate. Add approximately 0.05 gram of sodium chromotropate and shake to effect solution. Dilute the contents to the 50-ml mark with concentrated sulfuric acid and allow the normal heat-of-reaction rise to occur. By means of a 20-ml pipet or capillary glass tubing connected to a nitrogen source and immersed in the acid solution, ebulliate vigorously with nitrogen for approximately 10 minutes. Allow the cylinder contents to come to room temperature. Measure the abosrbance of the sample against a blank at 570 mμ in 1-cm cells. Read the concentration of ethylene oxide from the calibration curve.

The method was used by Critchfield and Johnson to determine ethylene oxide in certain spices fumigated with this substance. Before the ethylene oxide determination they used an evolution procedure to remove interfering substances that react with periodate. They claimed that the method should be directly applicable to the determination of most 1,2-epoxides. When a 20-gram sample was used, the lower limit of detection was approximately 1 ppm. This limit can be extended with a larger sample size.

3. Alpha Epoxides—Isomerization to Aldehydes.

The reactivity and other properties of an oxirane oxygen containing a tertiary carbon atom deviate from those of other epoxy groups, and the usual methods for the other epoxides are not generally applicable to these types. Durbetaki (5) presented a method for this type of compound whereby the epoxide is converted to its corresponding aldehyde by catalytic isomerization with zinc bromide in benzene. Although he determined the aldehyde gravimetrically as the 2,4-dinitrophenyhydrazone, it appears possible that the final determination could be done just as well spectrometrically. Durbetaki illustrated the method by using it to determine α-pinene oxide, camphene oxide, 1,2-diisobutylene oxide, and α-methylstyrene oxide. As the reaction is carried out in the presence of air, the aldehydes are oxidized to their corresponding acids to an extent of 5 to 10% unless about 10% by weight of the corresponding olefin is present.

METHOD FROM DURBETAKI (5)

Reagents. Benzene, dried and stored over sodium.

Zinc bromide, fused. Place 0.1 gram of analytical reagent-grade zinc bromide in a dry clean tube (4 mm i.d., 6 mm o.d., 17 cm long), sealed at one end. Heat the tube to about 200°C to avoid condensation of moisture distilled from the zinc bromide during fusion. Hold the zinc bromide over a Bunsen flame and heat gently to melt (m.p., 394°C) and to remove all the water. Avoid excessive heating as it decomposes the zinc bromide. Seal the tube immediately at about 3 cm from the liquid zinc bromide with an oxygen gas torch and allow to cool in a slanted position to prevent the flow of the zinc bromide to the sealed end of the vial. A large number of these vials can be prepared in a short time and stored in a desiccated jar for use as needed.

Isopropyl alcohol. Treat with and distill over 2,4-dinitrophenylhydrazine.

Procedure.   Weigh 2.0 to 3.5 grams of the oxide in a 50-ml ground-glass Erlenmeyer flask with a spout. Add 5 ml of dry benzene by means of a dry pipet, followed by anhydrous olefin at least 10% by weight of the α-epoxide taken for analysis, and stopper the flask. The olefin can be the corresponding olefin or any liquid olefin (i.e., α-pinene, α-methylstyrene, or diisobutylene) which can be delivered readily by pipet. Do not add olefin, however, if it is already in the sample to be analyzed. Break a vial containing the fused zinc bromide into two parts and immediately add the half containing the catalyst to the benzene solution. Attach the flask to an air-cooled condenser equipped with a drying tube containing calcium chloride. Lower the flask into an oil bath at $98 \pm 0.1°$ C and heat for 10 minutes with occasional stirring. The reaction is vigorous and proceeds almost instantaneously as soon as the solution attains the desired temperature.

After 10 minutes remove the flask from the oil bath and cool to room temperature by immersing it into a beaker of cold water. Wash the condenser with carbonyl-free isopropyl alcohol. Transfer the contents of the Erlenmeyer flask to a 100-ml glass-stoppered volumetric flask by successively washing both the Erlenmeyer flask and the zinc bromide vial with small portions of isopropyl alcohol and make up to volume.

Determine the aldehyde concentration of this solution by the colorimetric 2,4-dinitrophenylhydrazine method described in Chapter 2, Section I.C.1.

II.   GAS CHROMATOGRAPHIC METHODS*

Epoxy groups are found in many natural and synthetic compounds, and their detection and determination are frequently required. Although compounds containing these groups (and no polar ones) chromatograph well, acidic conditions should be avoided because they tend to break open the oxirane ring.

Epoxyglycerides were determined, but not quantitatively, by reacting them with ketones in the presence of BF_3 to give 1,3-dioxolane derivatives; cyclopentanone was the most suitable ketone (6). Epoxides are "subtracted"

* Written by Morton Beroza and May N. Inscoe.

(>80% and usually >95%) by a reaction loop containing 100 to 200 mg of 5% phosphoric acid on Chromosorb W. (See Chapter 1, Section II.F.) The reaction has limited utility in quantification, however, because other compounds (e.g., those with reactive ethers such as methyl benzyl ether or 1,2-dimethoxyethane) are partially subtracted (7). Nevertheless, since few good methods are available for detecting and determining epoxides, especially in small amounts, the reaction loop may be worth trying.

The position of epoxide groups in microgram amounts of compound is located by reacting the compound in a halogenated solvent with dry powdered periodic acid for 5 minutes (8). The aldehyde and perone fragments produced by cleavage between the carbon atoms of the epoxide are then determined by gas chromatography of the reaction solution.

REFERENCES

Absorption Spectrophotometric

1. J. Bomstein, *Anal. Chem.*, **30**, 544 (1958).
2. R. F. Goddu and D. A. Delker, *Anal. Chem.*, **30**, 2013 (1958).
3. F. A. Gunther, R. C. Blinn, M. J. Kolbezen, J. H. Barkley, W. D. Harris, and H. S. Simon, *Anal. Chem.*, **23**, 1835 (1951).
4. F. E. Critchfield and J. B. Johnson, *Anal. Chem.*, **29**, 797 (1957).
5. A. J. Durbetaki, *Anal. Chem.*, **29**, 1666 (1957).

Gas Chromatographic

6. J. A. Fioriti, M. J. Kanuk, and R. J. Sims, *J. Chromatog. Sci.*, **7**, 448 (1969).
7. B. A. Bierl, M. Beroza, and W. T. Ashton, *Mikrochim. Acta*, **1969**, 637.
8. B. A. Bierl, M. Beroza, and M. H. Aldridge, *Anal. Chem.*, **43**, 636 (1971).

6 Organic peroxides

I. ABSORPTION SPECTROPHOTOMETRIC METHODS*

A. Direct Infrared Measurement

Hydroperoxides exhibit maximum absorptions at 1.46 to 2.07 μ, which are not shown by dialkyl peroxides. The intensity of the absorption is proportional to the peroxide content of the sample (1).

The chief distinguishing bands of peroxy acids appear to be at 3.05, 6.9, and 8.5 μ in the vapor phase (2, 3). The peroxy acids were observed to decompose while their spectra were being recorded. In fact, peroxybutyric acid vapor decomposed too quickly to be run entirely on one sample.

B. Measurement after Chemical Reaction

1. Benzoyl Leuco Methylene Blue Method. Peroxides or hydroperoxides and benzoyl leuco methylene blue react in a benzene-trichloroacetic acid solution to form the characteristic methylene blue color. This reaction was used as a basis for a method to determine organic peroxides by Eiss and Giesecke (4). Although reaction is sensitive to ultraviolet light and to a lesser extent to artificial light and heat, the color is stable for several days if kept in the dark at 24°C. Zirconium naphthenate was used to accelerate the peroxide decomposition and thereby increase the leuco dye reaction rate. Of the five peroxides tested, benzoyl peroxide, lauroyl peroxide, *p*-methane hydroperoxide, and cumene hydroperoxide followed Beer's law. *tert*-Butyl hydroperoxide deviated somewhat, and a calibration curve of concentration versus absorbance was necessary for this compound. As little as 0.5 mg of active oxygen could be detected. An estimate of the precision of

* Written by J. Gordon Hanna.

167

the method was obtained by running replicate samples for each compound. The 95% confidence limits ranged from ± 1.7 to $\pm 2.6\%$.

The reaction can be formulated as shown in equation 1:

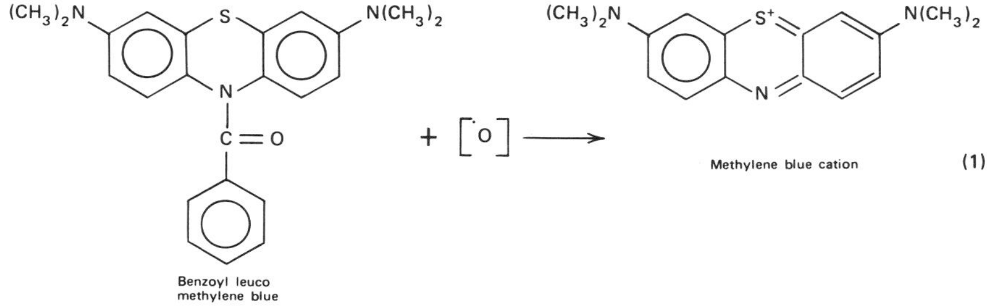

$$+ \; [\dot{\text{o}}] \longrightarrow$$

Methylene blue cation   (1)

Benzoyl leuco
methylene blue

METHOD FROM EISS AND GIESECKE (4)

Reagents. Zirconium naphthenate, 0.24%. Dilute 1 ml of commercial 6% zirconium naphthenate to 25 ml with benzene.

Trichloroacetic acid, 0.5% in benzene.

Benzoyl leuco methylene blue solution. Dissolve 0.005 gram in 100 ml of benzene. Store in a dark bottle.

Procedure. Prepare standard peroxide and hydroperoxide solutions of pure materials in benzene. To obtain a calibration curve, pipet aliquot portions of the standard solutions into 25-ml volumetric flasks containing 15 to 20 ml of 0.5% trichloroacetic acid solution and 0.3 ml of 0.24% zirconium naph-

Table 6-1.   Reaction Times and Absorbances for Peroxides Tested (4)

Compound	Active oxygen, %	Time for complete color development at 25°C	Absorbance, 1 mg/100 ml, 1-cm cell
tert-Butyl hydroperoxide	17.8	30 min (36 hr[a])	Approx. 16
Cumene hydroperoxide	10.5	40 min (38 hr[a])	9.7 ± 0.25[b]
p-Methane hydroperoxide	9.3	2 hr	(8.8 ± 0.15)[b]
Lauroyl peroxide	4.02	5 hr	(4.6 ± 0.10)[b]
Benzoyl peroxide	6.6	30 hr (120 hr[a])	(6.2 ± 0.12)[b]

[a] Time required for complete color development without use of zirconium naphthenate.

[b] 95% confidence limit.

thenate. Add 1 ml of the leuco dye solution. Fill the flask to the mark with benzene, mix thoroughly, and protect immediately from light. Allow the flasks to stand in the dark for the designated time (Table 6-1) at $24 \pm 1°C$. Measure the absorbance versus water at 662 mμ in 1-cm cells. Run a reagent blank with each set of standards. Use the same procedure for sample analysis.

If the peroxide species is unknown, check the time for the peroxide decomposition to reach completion experimentally, that is, measure the absorbance until it no longer increases. Calculate the result as percentage of active oxygen.

The results of Eiss and Giesecke for the analyses of peroxide standards by this method are shown in Table 6-2. Dialkyl peroxides such as di-*tert*-butyl peroxide do not react.

2. Ferric Thiocyanate Complex. Wagner, Clever, and Peters (5) evaluated the ferrous thiocyanate colorimetric method for the determination of peroxides. The method, originally proposed by Young, Vogt, and Nieuwland (6), consists of the reduction of the sample in an acidified solution of ferrous thiocyanate in methanol, followed by measurement of the depth of

Table 6-2. Analysis of Peroxide Standards (4)

	Peroxide, μg	
Compound	Present	Found
Lauroyl peroxide	11.0	10.9
	17.0	16.8
	22.0	21.6
Benzoyl peroxide	9.9	10.2
	19.8	20.0
	36.0	35.8
p-Menthane hydroperoxide	13.1	13.1
	26.3	25.9
	32.8	33.5
Cumene hydroperoxide	3.8	3.6
	7.6	7.3
	15.2	14.9
tert-Butyl hydroperoxide	5.9	5.7
	11.8	11.9
	23.6	24.3

color of the ferric thiocyanate produced. The method was found to be fundamentally accurate for hydrocarbon material containing no conjugated diolefins; for diolefins, however, the results varied considerably with time. The precision obtained was approximately ±5 to 10% of the absolute value.

METHOD FROM WAGNER, CLEVER, AND PETERS (5)

Apparatus. A Spekker photoelectric absorptiometer equipped with blue No. 6 and green No. 5 filters and 1-cm absorption cells. Any equivalent colorimeter or spectrophotometer probably can be used.

Reagent. Ferrous thiocyanate solution. Dissolve 1 gram of ammonium thiocyanate and 1 ml of 25% by weight of sulfuric acid in 200 ml of deaerated methanol and shake the resulting solution with 0.2 gram of finely pulverized ferrous ammonium sulfate. Place the decanted reagent (prepared fresh daily) in a brown glass-stoppered bottle.

Procedure. Introduce 1 ml of sample, preferably containing between 0.0001 and 0.0007 meq of reactive peroxide, in a 25-ml volumetric flask. If necessary, use 1 ml of a suitable methanolic dilution of the sample. Fill to the mark with ferrous thiocyanate reagent, mix well, and fill the colorimeter cell with the mixture. At 10 minutes from the time of mixing, read the absorbance and convert the value, corrected for a blank determination value, to concentration terms by comparison with a calibration curve obtained with corresponding known amounts of standard ferric chloride solution.

The method was recommended as best for frequent analyses of materials containing only small amounts of peroxide. The iodometric method (7), with which this method was compared, was alleged to be more precise and accurate for large amounts of materials. The accuracy of the method is indicated by the data shown in Table 6-3.

3. Iodine Liberated from Potassium Iodide.   Methods based on the oxidation of potassium iodide to produce iodine have long been favorite approaches for the determinations of peroxides (7–18). Heaton and Uri (19) developed a spectrophotometric procedure for the iodometric determination of traces of lipide peroxides. A 2:1 mixture of acetic acid and chloroform was used as the solvent under continuous deaeration. The ionic species determined was the triodide complex. Although an absorbance maximum was noted at 362 mμ, measurement was made at 400 mμ, where fewer compounds interfered. Calibration curves with linoleic acid peroxide and pure iodine were identical, and Beer's law was obeyed by concentrations of peroxide below 5×10^{-4} M. The application of this procedure to other comercially available peroxides was not demonstrated.

Table 6-3. Accuracy of Colorimetric Thiocyanate Method in Comparison with Sodium Iodide-Isopropanol Method (5)

Compound	Sodium iodide-isopropyl alcohol method (6)	Colorimetric method (5)
	Apparent peroxide content, %	
tert-Butyl hydroperoxide	99.8	105
Cumene peroxide	94.2	112
Tetralin peroxide	95.4	95
Hydrogen peroxide, 30%	29.2	24
	Peroxide number, meq/liter	
Cyclohexane	121	120
2-Pentene	51	48
Diisobutylene	19.7	19
Diethyl ether	12.2	7.4
Methylpentadiene	61	266

Banerjee and Budke (20) modified the method of Heaton and Uri and applied it to a wide range of peroxides of varying reactivity. Quantitative results were obtained with seventeen commercially available peroxides. No color formation was observed with di-tert-butyl peroxide and dicumyl peroxide, which are among the least reactive of all peroxides.

The sample is diluted with a mixture of acetic acid and chloroform and treated with potassium iodide after deaeration. The iodine liberated is measured spectrometrically at 470 mμ in 1-cm cells. The method was satisfactory for the determination of peroxides in benzene, chloroform, 2-propanol, methanol, pentane, hexane, toluene, ethyl ether, acetone, vinyl acetate, and ethyl acetate. The same authors (21) later showed that the method was applicable also to the determination of peroxides in unsaturates —octene-1, mixed octenes, heptenes, cyclohexene, isoprene, acrolein, and sorbic acid. No iodine absorption by the olefins was observed, even when the solutions were allowed to stand for 2 hours before the absorbance measurements were made.

METHOD FROM BANERJEE AND BUDKE (20).

Apparatus. Beckman Model DU spectrophotometer. For the range of 0 to 400 μg of active oxygen per 25 ml, use matched 1-cm cells.

Because of the relatively high blank values and other experimental difficulties experienced with conventional Beakman 5-cm and 10-cm cells, a special cell was designed for the range covering 0 to 40 μg of active oxygen per 25 ml. The cell body consisted of a Coleman precision absorption tube of approximately 1.5-cm path length. The tube was fitted with a standard taper joint from which a glass capillary extended to the bottom of the absorption tube for purging with nitrogen *in situ*. The glass capillary was so positioned that it was out of the light path and had no influence on the absorption measurements. All transfers of solutions between color development and measurement were completely eliminated. Whenever the special cell was used, the regular Beckman cell carriage was replaced with the attachment provided for the measurement of absorbance in test tubes. The absorption cell is illustrated in Figure 6-1.

Procedure.   For the range 0 to 400 μg of active oxygen per 25 ml, prepare a calibration curve as follows, using pure iodine. Dissolve 0.1270 gram of iodine in 2:1 acetic acid-chloroform and dilute to 100 ml in a volumetric flask. This solution contains 1.27 mg of iodine per milliliter, which is equivalent to 80.0 μg of active oxygen per milliliter. Pipet 0-, 1-, 2-, 3-, 4-, and 5-ml aliquots of this solution into 25-ml volumetric flasks and dilute to volumes with the acetic acid-chloroform mixture. Purge the solutions with nitrogen for 1 to $1\frac{1}{2}$ minutes, add 1 ml of freshly prepared 50% potassium iodide solution, and continue purging for 1 minute. Measure the absorbance at 470 mμ, using 1-cm matched cells and water reference. Plot absorbance versus micrograms of active oxygen per 25 ml.

Determine the peroxide content of solvent samples as follows. Pipet a 5-ml aliquot of the solvent into a 25-ml volumetric flask and dilute to volume with the acetic acid-chloroform mixture. Insert a hypodermic needle or a glass capillary to the bottom of the flask and purge with a fine stream of nitrogen for $1\frac{1}{2}$ minutes. Add 1 ml of fresh 50% potassium iodide solution and continue to purge for an additional minute. Remove the tube, stopper the flask, shake, and let stand in the dark for 1 hour. At the end of this time, measure the absorbance of the solution at 470 mμ, using covered 1-cm cells and water as a reference. Measure the absorbance as rapidly as possible to minimize any air oxidation. Subtract the absorbance of a blank run through the entire procedure. Determine the micrograms of active oxygen in the sample by reference to the calibration curve, and convert them to parts per million of active oxygen in the solvent.

Peroxides in solid samples can be determined also if they are soluble in the acetic acid-chloroform solvent. If a specific peroxide is known to be present, convert parts per million of active oxygen to parts per million of peroxide with the appropriate factor.

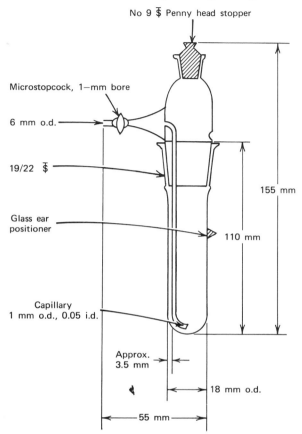

No 9 $ Penny head stopper

Microstopcock, 1—mm bore

6 mm o.d.

19/22 $

155 mm

Glass ear positioner

110 mm

Capillary
1 mm o.d., 0.05 i.d.

Approx.
3.5 mm

18 mm o.d.

55 mm

Figure 6–1. Absorption cell for low active oxygen (20).

Procedure for Low Active Oxygen Range. Prepare a calibration curve covering the range 0 to 40 μg of active oxygen as follows. Prepare an iodine solution in 2:1 acetic acid-chloroform containing 63.4 μg of iodine per milliliter, which is equivalent to 4.0 μg of active oxygen per milliliter. Pipet 0-, 1-, 3-, 5-, and 8-ml aliquots of this solution into 25-ml volumetric flasks and dilute to volumes with the acetic acid-chloroform solvent containing 4% water. Transfer a portion of each standard to the special absorption cell and purge slowly with nitrogen for 3 minutes. Add 5 drops of freshly prepared, deaerated 50% potassium iodide solution through the standard taper joint and replace the stopper loosely. Continue the purging with nitrogen for an additional 3 minutes. Then tighten the stopper and close the nitrogen inlet so that the solution is under a slightly positive nitrogen pressure. Measure the absorbance immediately at 410 mμ, using water in another matched Colemen absorption tube as reference. Subtract the absorbance of the blank and plot absorbance versus micrograms of active oxygen per 25 ml.

Determine the peroxide content of a solvent sample by diluting a 5-ml sample to 25 ml with 2:1 acetic acid-chloroform containing 4% water. Develop the color as described for the calibration curve and allow the solution to stand in the dark for 1 hour before the absorbance is read. After correcting for the absorbance of a blank carried through the entire procedure, obtain the micrograms of active oxygen present in the sample from the calibration curve and convert to parts per million in the solvent.

Results of the analyses for peroxides obtained by Banerjee and Budke (20)

Table 6-4. Analysis of Peroxides (20)

Compound	Sample size, ml	Active oxygen, ppm	
		Present	Found
tert-Butyl peroxide	2	154.0	152.0
	4	154.0	152.5
p-Methane hydroperoxide	2	105.1	99.0
	4	105.1	97.0
2,5-Dimethylhexane-2,5-	2	94.0	98.0
dihydroperoxide	4	94.0	96.0
Cumene hydroperoxide	3	67.3	74.0
	5	67.3	73.2
Lauroyl peroxide	3	42.7	40.0
	5	42.7	40.8
Myistoyl peroxide	3	61.0	59.3
	5	61.0	58.8
Benzoyl peroxide	3	76.4	73.3
	5	76.4	73.2
	4	34.8	34.7
	4	66.0	64.6
	4	86.5	85.1
2,4-Dichlorobenzoyl peroxide	3	84.3	86.0
	5	84.3	83.8
tert-Butyl perbenzoate	3	72.8	78.0
	5	72.8	78.0
	5	45.0	48.8
	5	66.0	70.8
	5	84.4	90.0

are shown in Table 6-4. Table 6-5 gives the results obtained by the same authors for peroxides in the presence of unsaturates.

4. Yellow Titanium Complex. Strohecker, Vaubol, and Tenner (22) and Furmanek and Manikowski (23) used dilute acid to hydrolyze certain peroxides to hydrogen peroxide and determined the hydrogen peroxide colorimetrically as a titanium complex. Pobiner (24) extended the titanium method to include organic hydroperoxides. Under controlled conditions different peroxide compounds can be acid-degraded to hydrogen peroxide and measured by the titanium method.

Hydroperoxides are acidic and can be removed quantitatively from hydrocarbons by alcoholic caustic extraction. The alcoholic caustic contain-

Table 6-4. (cont.)

Compound	Sample size, ml	Active oxygen, ppm	
		Present	Found
tert-Butyl peracetate	3	69.4	79.3
	5	69.4	74.0
Di-*tert*-butyl diperphthalate	2	138.5	141.0
	3	138.5	142.7
tert-Butyl peroxyisobutyrate	2	88.0	107.0
	4	88.0	109.0
tert-Butyl peroxypivalate	2	51.5	61.0
	4	51.5	62.5
Cyclohexanone peroxide	2	161.5	157.0
	3	161.5	161.6
Methyl ethyl ketone peroxide	3	87.8	94.0
	5	87.8	94.4
Succinic acid peroxide	3	79.1	73.3
	5	79.1	75.2
	4	40.7	39.2
	4	79.3	76.2
	4	78.3	73.5
Hydroxyheptyl peroxide	5	54.0	54.0
	5	84.6	84.6
	5	120.0	119.6
Di-*tert*-butyl peroxide	5	91.0	None recovered after 24 hr
Dicumyl peroxide	5	54.6	None recovered after 24 hr

Table 6-5. Recovery of Peroxides in the Presence of Unsaturates (20)

Sample	Time of absorbance measurement after KI addition, minutes	Active oxygen, μg Added	Found
Octene-1	60	0	0
	30	336	318
	60	336	314
	120	336	320
Mixed octenes	60	0	0
	30	336	324
	60	336	332
	120	336	334
Heptenes	60	0	0
	30	336	332
	60	336	342
	120	336	346
Cyclohexene	60	0	132
	60	200	332
	60	200	344
	60	200	346
Isoprene	60	0	0
	30	332	346
	60	332	340
	120	332	348
Acrolein	60	0	48
	60	258	318
Sorbic acid	60	0	0
	60	258	250

ing the hydroperoxide is treated with sulfuric acid at elevated temperatures:

$$ROOH + H_2SO_4 \xrightarrow{57-63°C} ROSO_2OH + H_2O_2 \tag{2}$$

Hydrogen peroxide in turn is dissociated in the acid solution:

$$H_2O_2 \rightarrow 2H^+ + O_2{}^{2-} \tag{3}$$

Reaction with a titanium salt in sulfuric acid produces the titanium-peroxysulfate complex:

$$O_2^{2-} + Ti^{4+} + 2H_2SO_4 \rightarrow [TiO_2(SO_4)_2]^{2-} + 4H^+ \tag{4}$$

The conversion to hydrogen peroxide is quantitative for the hydroperoxides tested. Reaction 4 is dependent on heat and acid treatment. The complex is stable indefinitely, and 1 ppm hydroperoxide oxygen extracted from hydrocarbon can be measured.

METHOD FROM POBINER (24)

Reagent. Titanium-sulfuric acid solution. Weigh 10.00 ± 0.01 grams of potassium titanium oxalate into a 1-liter beaker. Add 20.0 ± 0.1 grams of ammonium sulfate. Add 100 ml of concentrated sulfuric acid. Bring to a boil and continue to boil for 10 minutes to eliminate oxalic acid. Allow to cool. Pour the solution slowly with constant stirring into 350 ml of distilled water. Finally dilute to 500 ml with distilled water in a volumetric flask and mix well.

General Calibration with Hydrogen Peroxide. Prepare a master solution of 1 gram, weighed to 0.1 mg, of standardized hydrogen peroxide in 350 ml of distilled water. Dilute this 50 ml to 250 ml in distilled water and pipet the following aliquots into 100-ml volumetric flasks: 0 (reagent blank), 1, 2, 5, 10, 15, 20, 25, and 35 ml. Pipet in 10 ml of the titanium-sulfuric acid solution and 25 ml of 1:1 sulfuric acid. Heat at 57 to 63°C for 10 minutes in a constant-temperature water bath. Cool to room temperature and dilute to 100 ml with distilled water. Run the visible spectrum of each solution versus the reagent blank from 700 to 360 mμ, using 1-cm cells and a recording spectrophotometer, such as the Beckman DK-2. Determine the absorbance at the maximum, 407 mμ, and take the absorbance at 700 mμ as a reference zero reading. Calculate the grams of peroxide oxygen (O_2^{2-}) in each aliquot. Use the hydrogen peroxide standardization value, the peroxide oxygen-hydrogen peroxide conversion factor, and the dilution factor to determine the grams of peroxide oxygen. Plot the grams cf peroxide oxygen versus absorbance at 407 mμ. A tenfold increase in sensitivity is realized by the use of 10-cm absorption cells.

Calibration with Hydroperoxide. Prepare a blend of 0.1 gram weighed to 0.1 mg of a standardized hydroperoxide in 100 ml of 1% sodium hydroxide in methanol. Pipet aliquots of 5, 10, 20, and 30 ml into 100-ml volumetric flasks. Proceed as in the hydrogen peroxide calibration, beginning with the addition of the titanium-sulfuric acid solutions.

Analysis of Hydrocarbon Samples. Extract 100 ml of the hydrocarbon sample with four 10-ml volumes of 1% sodium hydroxide in methanol. If the hydrocarbon and alcohol phases are miscible, as they should be in xylene solution, obtain the required phase separation by adding 10 ml more of water with each 10 ml of alcoholic caustic. Collect the extracts together in a 100-ml volumetric flask and dilute to the mark with methanol.

Pipet 20 ml of this solution into each of two 100-ml volumetric flasks. To one flask add 10 ml of titanium-sulfuric acid solution and 25 ml of 1:1 sulfuric acid, heat at 57 to 63°C for 10 minutes, and dilute to the mark with distilled water. The first flask is the complexed hydroperoxide solution; the second, the hydrocarbon blank solution.

Run each of the solutions in the visible spectrum versus the reagent blank prepared in the peroxide calibration. For all absorbance curves measure the absorbance at the maxima (405 to 407 mμ), in relation to a zero reading at 700 mμ. Correct the complexed hydroperoxide solution for any absorbance in the hydrocarbon blank solution. Determine the grams of hydroperoxide oxygen by referring to the calibration curve.

Analytical data obtained by Pobiner are shown in Table 6-6.

Table 6-6. Recovery of Hydroperoxides (24)

Compound	Weight % in heating oil	Recovery, % of theoretical value
tert-Butyl hydroperoxide	0.22	100
Cumene hydroperoxide	0.34	95
	1.27[a]	96
	6.78[a]	101
p-Menthane hydroperoxide	0.25	91
tert-Butyl perbenzoate	0.98	97
Di-*tert*-butyl peroxide	0.36	0
	0.06	0
Benzoyl peroxide	0.06	0
Peracetic acid	1.01	0

[a] In xylene solution.

5. N,N-Dimethyl-p-phenylenediamine Reaction.   Dugan (25, 26) and Dugan and O'Neill (27) studied peroxides and presented a colorimetric method for microgram quantities of lauroyl and benzoyl peroxides, based on the catalytic action of the peroxides on the reaction of *N,N*-dimethyl-*p*-phenylenediamine sulfate and methanol to produce a colored complex.

METHOD FROM DUGAN (25, 26) AND DUGAN AND O'NEILL (27)

Apparatus. Beckman Model DU or DK-1 spectrophotometer with matched 1-cm silica cells.

Reagent. N,N-Dimethyl-*p*-phenylenediamine solution. Dissolve 0.3 gram in 10 ml of distilled water in a 100-ml volumetric flask and dilute to the mark with absolute methanol.

Procedure. Mix 2 ml of the N,N-dimethyl-*p*-phenylenediamine sulfate reagent with 2 ml of benzene containing either lauroyl or benzoyl peroxide in a concentration of 5 to 100 μg/ml. Allow the mixture to react at 25°C for 30 minutes. Determine the absorbance of the solution at 560 mμ versus that of a suitable reagent blank without peroxide.

II. GAS CHROMATOGRAPHIC METHODS*

Although most organic peroxides decompose in the gas chromatograph, the lower hydroperoxides have been chromatographed successfully. Methyl and ethyl hydroperoxides were separated by GC and the peroxides individually determined colorimetrically, using a flowing liquid interface (28). The hydroperoxides eluting from the column were picked up by a flowing liquid containing ferrous thiocyanate (colorimetric reagent) and passed through a colorimetric-recorder arrangement that registered transmittance automatically. Quantitative measurements were made from the chromatogram.

III. ELECTROANALYTICAL METHODS†

A. Polarographic Determination of Peroxides

Peroxides of various types are utilized as oxidizing agents in a wide variety of applications. Their oxidizing power ranges from extremely strong to very weak. Since almost all peroxides are oxidizing agents, it is safe to say that reduction of the peroxide group can be carried out quite readily; electrochemical reduction of peroxides can be produced very easily. In fact, the need for deoxygenation of polarographic test solutions is based on the reduction of a peroxide. Dissolved oxygen is reduced to form hydrogen peroxide, which, in turn, is reduced.

---

* Written by Morton Beroza and May N. Inscoe.
† Written by Alan F. Krivis.

The polarographic wave for the reduction of peroxides usually appears to be irreversible, that is, the electrochemical reaction is not thermodynamically reversible and the wave does not have the desirable S-shaped form, with an almost vertical center segment. Instead, the wave is commonly spread out and unsymmetrical. This makes determination of the half-wave potential more difficult (if not impossible), but a quantitative analysis still can be excellent.

The solvent system which seems to be most universally applicable to the analysis of peroxides is a 1:1 mixture of benzene and methanol containing 0.3 M lithium chloride (29–34). Although other solvent systems have been investigated, a combination of both peroxide solubility and total sample solubility is best served by the benzene-methanol solvent.

Alternative calibration curves are in use for the polarographic determination of peroxides. One may prepare a calibration curve of current versus concentration for the specific peroxide compound of interest or use *tert*-butyl peroxide and cumene peroxide to prepare "peroxide number" samples (30, 34).

EXPERIMENTAL METHOD [ADAPTED FROM THE STUDY OF WHISMAN AND ECCLESTON (34)]*

Apparatus.   Any polarograph, recording or manual, may be used for this analysis. An H-cell with a calomel reference electrode should be thermostated in a constant-temperature bath (25°C). For convenience in cleaning, a two-piece type of cell is recommended; several variants are available commercially.

Solvent.   A solution of 0.3 M lithium chloride in equal volumes of absolute methanol and benzene should be prepared. All solvents should be the purest grade available.

Procedure.   Bubble a stock solution of electrolyte-solvent with helium to remove oxygen. Transfer a weighed sample containing about 0.01 mM of peroxide to a 10-ml volumetric flask, and use the oxygen-free solvent to fill the flask to the mark. Transfer the mixed sample solution to the cell and de-gas for 4 minutes with helium. (Nitrogen may be used instead but may require longer bubbling periods for complete oxygen removal.) The helium must be presaturated with the benzene-methanol solution to avoid evapora·tion in the test cell. Mount the dropping mercury electrode and bubble gas for another minute.

Obtain the polarogram of the sample and measure the diffusion current

---

* Reprinted in part with the permission of the copyright owner, the American Chemical Society.

for the peroxide wave. Compare this current to a calibration curve for the compound under study in order to find the peroxide content of the sample.

Alternatively, for samples in which no interfering materials are present and for which the units are pertinent, measure the current at -0.3 V and -1.3 V. Relate the difference in current between these points to the peroxide number by use of a calibration curve of current versus peroxide number.

B. Amperometric Titration of Peroxides

A method for the determination of very small quantities of peroxides is based on reaction of the peroxide with iodide to form iodine and the amperometric titration of the latter with thiosulfate (35–37). A modification for trace levels requires the formation of iodine, reaction with excess thiosulfate, and titration of the excess with standard iodate solution (37).

The method described below utilizes the dead-stop end-point detection technique. The current in the detecting cell is high at the beginning of the titration and falls to a very low level at the end point. Although the apparatus and the procedure are quite simple, the precision claimed is excellent (0.5%), even with the dilute titrant (0.001 N) employed.

EXPERIMENTAL METHOD [ADAPTED FROM THE PROCEDURE OF ABRAHAMSON AND LINSCHITZ (35)]*

Apparatus. A flat-bottomed cylindrical tube of about 100-ml capacity, carrying a 45/50 standard taper outer joint, is used for the titration. An inner joint, fitting the titration vessel, is equipped with a mechanical stirrer, two platinum electrodes, a nitrogen bubbler to provide an inert atmosphere, and an opening for a buret delivery tip.

A variable voltage is supplied to the platinum electrodes by a 100-ohm potentiometer, connected to a 1.5-V dry cell. A galvanometer with a sensitivity of 0.005 μA/mm is connected in series with the electrodes. The circuit is similar to that shown in Chapter 1, Section III, for used in the bromination of phenols.

Reagents and Chemicals. Sodium thiosulfate solution (0.001 N) was prepared from reagent-grade materials and standardized against potassium iodate. The solution was saturated with nitrogen as a precaution against possible oxygen errors.

Solutions of peroxide samples and standards were made up in various peroxide-free hydrocarbon solvents.

---

* Reprinted in part with the permission of the copyright owner, the American Chemical Society.

Procedure. Thirty milliliters of anhydrous isopropyl alcohol and 2 ml of glacial acetic acid are placed in the titration vessel. The solution is refluxed under a Soxhlet condenser for 1 minute, and 0.2 grams of anhydrous sodium or potassium iodide is added slowly. After another 30 seconds of refluxing, an aliquot of 1 to 2 ml of the peroxide stock solution is added by means of a calibrated pipet. The solution is then refluxed for 10 minutes, 5 ml of water is added, and the solution is titrated with thiosulfate titrant. The volume of titrant should not exceed 25 ml; otherwise, an emulsion may result.

At the beginning of the titration both platinum electrodes are depolarized so that only a small voltage need be applied. (A voltage sufficient to keep the galvanometer on scale is adequate.) as the titration proceeds, the applied potential is slowly increased to a maximum of about 10 mV near the end point. As the end point is approached, the galvanometer deflection falls rapidly to a final, very small value. The last few drops of titrant should be added slowly (especially if an emulsion develops) to allow for equilibration of the system. Moderate stirring and continuous flushing of the vessel with nitrogen are maintained during the titration.

The chemical reactions involved in the procedure are as follows:

$$R—OO—R' + 2I^- + 2H_2O \rightarrow I_2 + ROH + R'OH + 2OH^-$$

$$I_2 + 2S_2O_3^{2-} \rightarrow 2I^- + S_4O_6^{2-}$$

IV. NUCLEAR MAGNETIC RESONANCE METHODS*

Quantitative analyses of mixtures of organic peroxides, hydroperoxides, and alcohols have been reported (38). The authors used either the alpha or the beta protons adjacent to the functional group as a means of obtaining the quantitative data. Mixtures such at *tert*-butyl alcohol in *tert*-butyl hydroperoxide were studied. Benzyl hydroperoxides and aliphatic hydroperoxides were also analyzed by this technique. For example, the methyl resonances of β,β-dimethyl benzyl peroxides, hydroperoxides, and alcohols were measured, and the differences were sufficient to yield quantitative data as to the peroxide, hydroperoxide, and alcohol contents.

Another comprehensive study (39) provided spectral data on more than twenty peroxides, hydroperoxides and peroxy acids, dialkyl peroxides, and diacyl peroxides. The study consisted of observing the chemical shifts of the labile proton and the substituents alpha to the functional group. Although mixtures were not analyzed, spectral data were reported.

* Written by Harry Agahigian.

REFERENCES

Absorption Spectrophotometric

1. R. T. Holman, C. Nickell, O. S. Privett, and P. R. Edmondson, *J. Am. Oil Chemists' Soc.*, **35**, 422 (1958).
2. P. A. Giguere and A. W. Olmos, *Can. J. Chem.*, **30**, 821 (1952).
3. E. R. Stephens, P. L. Hanst, and R. C. Doerr, *Anal. Chem.*, **29**, 776 (1957).
4. M. I. Eiss and P. Giesecke, *Anal. Chem.*, **31**, 1558 (1959).
5. C. D. Wagner, H. L. Clever, and E. D. Peters, *Anal. Chem.*, **19**, 980 (1947).
6. C. A. Young, R. R. Vogt, and J. A. Nieuwland, *Ind. Eng. Chem. Anal. Ed.*, **8**, 198 (1936).
7. C. D. Wagner, R. H. Smith, and E. D. Peters, *Anal. Chem.*, **19**, 976 (1947).
8. D. H. Wheeler, *Oil & Soap*, **9**, 89 (1932).
9. S. Marks and R. S. Morrell, *Analyst*, **54**, 503 (1929).
10. A. Taffel and C. Revis, *J. Soc. Chem. Ind.*, **50**, 87T (1931).
11. C. H. Lea, *Proc. Roy. Soc. (London)*, **108B**, 175 (1931).
12. P. S. Panyutin and L. G. Gindin, *Bull. Acad. Sci. USSR*, 841 (1938).
13. H. A. Liebhafsky and W. H. Sharkey, *J. Am. Chem. Soc.*, **62**, 190 (1940).
14. M. E. Stansby, *Ind. Eng. Chem., Anal. Ed.*, **13**, 627 (1941).
15. V. R. Kokatnur and M. Jelling, *J. Am. Chem. Soc.*, **63**, 1432 (1941).
16. A. Lips, R. A. Chapman, and W. D. McFarlane, *Oil & Soap*, **20**, 240 (1943).
17. P. D. Bartlett and R. Atschul, *J. Am. Chem. Soc.*, **67**, 816 (1945).
18. K. Nosaki, *Ind. Eng. Chem., Anal. Ed.*, **18**, 583 (1946).
19. F. W. Heaton and N. Uri, *J. Sci. Food Agr.*, **9**, 781 (1958).
20. D. K. Banerjee and C. C. Budke, *Anal. Chem.*, **36**, 792 (1964).
21. D. K. Banerjee and C. C. Bukde, *Anal. Chem.*, **36**, 2367 (1964).
22. R. Strohecker, R. V. Vaubol, and A. Tenner, *Fette Seifen*, **44**, 246 (1937).
23. C. Furmanek and K. Manikowski, *Roczniki Panstwowego Zakladu Hig.*, **4**, 447 (1953).
24. H. Pobiner, *Anal. Chem.*, **33**, 1423 (1961).
25. P. R. Dugan, *Anal. Chem.*, **33**, 696 (1961).
26. P. R. Dugan, *Anal. Chem.*, **33**, 1630 (1961)
27. P. R. Dugan and R. D. O'Neill, *Anal. Chem.*, **35**, 414 (1963).

Gas Chromatographic

28. T. E. Healy and P. Urone, *Anal. Chem.*, **41**, 1777 (1969).

Electroanalytical

29. W. R. Lewis and F. W. Quackenbush, *J. Am. Oil Chemists' Soc.*, **26**, 53 (1949).
30. W. R. Lewis, F. W. Quackenbush, and T. DeVries, *Anal. Chem.*, **21**, 762 (1949).

31. C. O. Willits, C. Riccuti, H. B. Knight, and D. Swern, *Anal. Chem.*, **24**, 785 (1952).

32. C. O. Willits, C. Riccuti, D. L. Ogg, S. G. Morris, and R. W. Riemenschneider, *J. Am. Oil Chemists' Soc.*, **30**, 420 (1953).

33. S. S. Kalbag, K. A. Narayan, S. S. Chang, and F. A. Kummerow, *J. Am. Oil Chemists' Soc.*, **32**, 271 (1955).

34. M. L. Whisman and B. H. Eccleston, *Anal. Chem.*, **30**, 1638 (1958).

35. E. W. Abrahamson and H. Linschitz, *Anal. Chem.*, **24**, 1355 (1952).

36. L. Tsuk and G. Zöllner, *Magy. Kem. Lapja*, **14**, 417 (1959).

37. K. Oette, M. L. Peterson, and R. L. McAuley, *J. Lipid Res.*, **4**, 212 (1963).

Nuclear Magnetic Resonance

38. G. A. Ward and R. D. Mair, *Anal. Chem.*, **41**, 538 (1969).

39. D. Swern, A. H. Clements, and T. M. Luong, *Anal. Chem.*, **41**, 412 (1969).

7 Unsaturation

I. ABSORPTION SPECTROPHOTOMETRIC METHODS*

A. Direct Infrared Measurement

Infrared group-type analyses were used by Johnston, Appleby, and Baker (1) for several olefinic classes in gasolines. Anderson and Seyfried (2) extended the method to include five olefinic classes and oxygenated compounds. Saier, Pozeksky, and Coggeshall (3) used the carbon-hydrogen bending vibration absorptions to discriminate among the types listed in Table 7-1. The exact position of the absorption band for the trisubstituted olefins, R_2C=CHR, depends essentially on the nature of the R groups, and the analysis is unsatisfactory. The full substituted olefin, R_2C=CR_2, has no ethylenic hydrogen and therefore no infrared bands of this type.

Table 7-1. Infrared Absorptions of Olefinic Classes

Class	Wavelength maximum, μ
RCH=CH_2	11
RR'=CH_2	11.2
RCH=CHR' (*trans*)	10.4
$RR'C$=CHR	11.9–12.7
RCH=CHR' (*cis*)	14.4

Goddu (4) reported that near-infrared spectrometry is well suited for the determination of many types of unsaturation. Both terminal methylene and cis double bonds may be determined selectively. Other centers of unsatura-

* Written by J. Gordon Hanna.

tion do not interfere with the terminal methylene determinations. Mixtures of cis, trans and terminal unsaturation may be analyzed for their cis double bonds and terminal methylene contents. Bands at about 1.62 and 2.10 μ may be used for determining terminal unsaturation, and a band at 2.14 μ for cis unsaturation. No method was developed for trans unsaturation.

B.   Direct Ultraviolet Measurement

Conjugated double bonds give rise to intense, well-defined absorption maxima which permit quantitative determinations of conjugated unsaturation in mixtures (5), for example, in fats and fatty acids. Examples are 9,11,13-octadecatrienoic acid, with a major absorption peak at 270 mμ, and 9,11-octadecadienoic acid, with a strong absorption band at about 233 mμ.

In the case of nonconjugated materials, the absorptions are weak and general and cannot serve for analytical purposes.

C.   Measurement after Chemical Reaction

1. Ultraviolet Measurement of Iodine Complexes. Long and Neuzil (6) described a procedure for distinguishing types of alkyl-substituted olefins by means of the ultraviolet spectra of reversible complexes formed with iodine. The procedure is alleged to be useful for all olefin types. It is of particular interest for the determination of the tri- and tetrasubstituted olefins because these types are difficult to determine by infrared spectrometry.

Iodine and hydrocarbons undergo a reversible reaction to form addition complexes of an acid-base type (7, 8). The iodine functions as an electron acceptor (Lewis acid), and the hydrocarbon as an electron donor (Lewis base). The iodine complexes of olefins have intense absorptions in the ultraviolet region.

METHOD FROM LONG AND NEUZIL (6)

Apparatus.   Cary Model 11 spectrophotometer with cells 1 cm in path length.

Reagents.   Isooctane. Shake material of 99% purity with concentrated sulfuric acid. Percolate a liter of the acid-washed isooctane through a 3-foot column of 1-inch diameter filled with 200-mesh silica gel (Davison Chemical Code 950). The solvent should have an absorbance of less than 0.05 down to 230 mμ when compared with distilled water in a cell of 1-cm path length.

Iodine reagent. Weigh approximately 0.1 gram of iodine to the nearest 0.1 mg, dissolve in purified isooctane, and dilute to 100 ml with the isooctane.

Calibration and Procedure.   Dilute olefin standards to contain known concentrations of approximately 0.1 mole/liter in purified isooctane. Add 5 ml of the prepared olefin solution to 5 ml of the iodine reagent. Measure the spectrum of the resulting solution from 400 to 250 mμ in a silica absorption cell of 1-cm path length. Measure the solution immediately after mixing to avoid the catalytic effect of light on the formation of undesired diiodo compounds.

Obtain a correction for background due to uncomplexed iodine by adding 5 ml of iodine reagent to 5 ml of isooctane and then measuring the solution as described above.

Table 7-2 shows the characteristic maximum absorption wavelengths of the complexes of different olefin types. Cyclopentene and cyclohexene behave like the cis forms of the corresponding open-chain olefin, RCH=CHR. Table 7-3 gives the results reported by Long and Neuzil for a two-component olefin blend, and Table 7-4 those for a five-component blend.

2. Brominations. Sweetser and Bricker (9) developed a spectrophotometric method for the titration of olefins and other substances with tribromide ion at wavelengths ranging from 270 to 360 mμ. Miller and DeFord (10) adapted this method for electrically generated bromine; they carried out the titrations in a 3:1 mixture of glacial acetic acid and methanol

Table 7-2. Maximum Absorption Wavelength of the Iodine Complexes Characteristic of Each Olefin Type (6)

Olefin type	Wavelength maximum, mμ
RCH=CH$_2$	275
R$_2$C=CH$_2$	290–295
RCH=CHR	295–300
R$_2$C=CHR	317
R$_2$C=CR$_2$	337

Table 7-3. Analysis of Two-Component Olefin Blend (6)

Olefin type	Compound	Known value, %	Found, %
R$_2$C=CHR	2-Methyl-2-hexene	72.1	71.5
R$_2$C=CR$_2$	2,3-Dimethyl-2-hexene	27.9	28.5

Table 7-4.   Analysis of a Five-Component Olefin Blend (6)

Olefin type	Compound	Known value, %	Found, %
R_2C=CR_2	2,3-Dimethyl-2-hexene	16.9	18.3
R_2C=CHR	3-Methyl-*cis*-3-hexene	25.4	26.1
R_2C=CH_2	2,3-Dimethyl-1-butene	19.1	17.3
cis-RCH=CHR	2,2-Dimethyl-*cis*-3-hexene	20.9	22.5
RCH=CH_2	1-Tetradecene	17.7	19.5
		100.0	103.7

with a small amount of potassium bromide and hydrochloric acid added. They found it necessary for mercuric chloride to be present as a catalyst to permit direct titrations. A wavelength of 360 mμ was selected when this catalyst was employed. This was the shortest wavelength that could be used in the presence of mercuric chloride, which has a high absorbance in the ultraviolet region.

A major difficulty associated with the usual bromination methods for the determination of unsaturation is that in the presence of excess bromine many compounds undergo substitution side reactions that may cause high results. Direct titration procedures offer the advantage that no excess bromine is present until after the end point is reached; therefore side reactions are less likely to occur.

Two procedures were presented by Fritz and Wood (11) for the determination of unsaturation with the bromination reaction. Macro amounts of olefins are determined by rapid direct titration with a spectrophotometric end point in the visible spectral region. Minor quantities of olefinic unsaturation are measured by a quick spectrophotometric procedure, based on the decrease in absorbance resulting from the reaction of bromine and the C—C double bond in a bromine-hydrobromic acid solution in acetic acid-water.

METHOD FROM FRITZ AND WOOD (11)

Apparatus.   A Beckman Model B spectrophotometer was used by Fritz and Wood. A hole was drilled through the floor of the cell compartment for the shaft of a small motor, a magnetic stirrer was screwed to the shaft by a set screw after removal of the cell carriage. A specially built aluminum compartment painted a dull black fitted over the magnetic stirrer, which was air or water driven. Two holes were drilled near the bottom of the front side of the titration compartment to accommodate the tubing used for water or air to run the stirrer. A 10-ml automatic buret, which extended through the cover and dipped into the solution, served to deliver the titrant. The titration cells used were 180-ml tall form beakers with a light path of 4.5 cm.

Direct Titration Procedure. Add 85 ml of glacial acetic acid and 10 ml of water to a 180-ml tall form beaker. Pipet into this mixture a 5-ml sample containing 0.3 to 1.0 mM of unsaturated compound dissolved in carbon tetrachloride or other suitable solvent. Place the beaker in the titration compartment, adjust for efficient stirring, place the cover on the compartment, and insert the buret through the hole. Adjust the spectrophotometer to zero absorbance at 400 mμ and titrate with 0.12 M bromine in glacial acetic acid. Titrate rapidly until the absorbance readings indicate the presence of some excess bromine. When the absorbance is between 0.1 and 0.3, record the amount of titrant added and the absorbance. Add three or four more 0.2-ml increments of titrant and record the absorbance within a few seconds after each addition. If the substance being titrated reacts more slowly than simple olefins, additional time may be required to reach equilibrium after the addition of each increment of titrant. Construct a plot of absorbance versus volume of titrant. The end point of the titration is at the intersection of the straight-line plot with the zero absorbance line.

Table 7-5 contains data obtained by Fritz and Wood for the direct titration of unsaturated compounds. They also found that direct titrations of allyl ether, allyl acetate, and 3-bromopropene were possible with mercury(II) catalyst. However, in the presence of mercury(II) the reaction of conjugated unsaturated compounds such as *trans*-cinnamaldehyde, *trans*-cinnamic acid, and ethyl acrylate was too slow for a practical direct titration but was fast enough to interfere with the determination of simple olefins. Also, when brominated in the presence of a mercury(II) catalyst, isoprene apparently begins to add a second mole of bromine too slowly for a direct titration to be of practical value. Aniline, phenols, alkyl sulfides and disulfides, and thiols were found to interfere.

Quite small amounts of simple organic unsaturation can be measured by the indirect method of Fritz and Wood, which is based on the decrease in absorbance resulting from the reaction of C—C double bonds and excess bromine. As a small total volume of bromine is used, a small consumption of the element causes a proportionately large change in bromine concentration. The change in bromine concentration is measured spectrometrically at a specific wavelength such as 410 mμ, and the amount of olefinic unsaturation is calculated from a Beer's law plot.

Hydrogen bromide formed during bromination of the olefin combines with the bromine present to form the tribromide complex, which has a

Table 7-5. Direct Titration of Unsaturation with Bromine (11)

Compound[a]	Found, %[b]	Number of determinations
Allyl acetate	Reacts slowly[c]	. . .
Allyl alcohol	96.0 ± 0.3	4
Allyl ether	93.9 ± 0.8[d,e]	2
2-Bromopropene	Does not react[f]	. . .
2-Butenenitrile	Does not react[f]	. . .
3-Butyn-1-ol	Does not react[f]	. . .
trans-Cinnamaldehyde	Does not react[f]	. . .
trans-Cinnamic acid	Does not react[f]	. . .
Cyclohexene	100.0 ± 0.4	4
Cyclohexylacetylene	0.2 ± 0.1	4
2,3-Dimethyl-2-butene	100.0 ± 0.2	2
2,5-Dimethyl-2-hexene	98.6 ± 0.4	2
trans-2,5-Dimethyl-3-hexene	110.3 ± 0.0	2
Ethyl acrylate	Does not react[f]	. . .
3-Ethyl-2-pentene	113.8 ± 0.5	4
Ethynylbenzene	Reacts slowly[c]	. . .
trans-3-Hexene	99.8 ± 0.2	4
Isoprene	100.6 ± 0.19	4
d-Limonene	99.2 ± 5[d]	2
1,2-Octene	100.4 ± 0.3	4
1,4-Octadiene	Interferes[h]	. . .
Oleic acid	97.2 ± 0.5	4
1,3-Pentadiene	99.6 ± 0.9[g]	2
Pyrrole	Interferes[h]	. . .
Styrene	100.0 ± 0.1	4

[a] Amounts taken range from 0.2 to 1.0 mM.

[b] Results based on the assumption that the samples are 100% pure.

[c] Reacts too slowly to be determined.

[d] Results based on the assumption that both double bonds add bromine.

[e] 20 to 25 minutes required per determination.

[f] No detectable reaction in 5 minutes.

[g] Results based on the assumption that only one double bound adds bromine.

[h] Initial rapid reaction followed by much slower reaction. No end point observed.

higher molar absorptivity than the bromine itself. The final absorbance is higher, therefore, than it should be, and it appears that less bromine has been consumed than is actually the case. This problem can be avoided by adding sufficient hydrogen bromide to the bromine solution used so that any additional hydrogen bromide produced in the reaction has a negligible effect on the absorbance. A plot of absorbance of bromine in acetic acid-water as a function of added hydrogen bromide shows a leveling off around a mole ratio of approximately 6 hydrogen bromide to 1 bromine. A brominating solution of this composition was found to be quite successful for the indirect determination of small amounts of unsaturated compounds. Linear Beer's law plots were obtained at several different wavelengths, so that the sensitivity of the method can be varied by changing the wavelength.

Spectrophotometric Determination of Unsaturation

Apparatus. Bausch and Lomb Spectronic 600 spectrophotometer with a set of matched quartz 1 × 1 cuvettes as cells.
   Micrometer buret.

Procedure. Measure accurately into a 1 × 1 cuvette 2.60 ml of 0.001 *M* tribromide solution in 90% acetic acid-10% water. To prepare the tribromide solution add 6.5 ml of bromine to 1 liter of a solvent mixture prepared by mixing 86 ml of approximately 47% hydrobromic acid (sp. gr., 1.49), 32 ml of distilled water, and 882 ml of glacial acetic acid. This gives a solution with a mole ratio of bromide to bromine of approximately 6:1. For a more dilute solution, dilute this solution 120-fold with 90% acetic acid-10% water. Place the cuvette in the sample compartment of the spectrophotometer opposite a matched cuvette containing the same amount of hydrobromic acid in the same solvent. When the absorbance at 410 mμ (which should be approximately 0.5) is stable for 1 minute, remove the cuvette and add a 50- to 150-μl sample containing 0.2 to 1.6 μeq of reactive unsaturation. Shake the cuvette to mix thoroughly and replace it in the sample compartment. Record the absorbance automatically until it is constant for 1 minute. From a previously prepared plot of absorbance versus tribromide concentration and from a knowledge of the initial and final volumes of solution present, calculate the microequivalents of bromine consumed, which is the same as the microequivalents of unsaturation present.
   Amount of unsaturation ranging from 0.25 to 0.2 μeq may be determined by using more dilute tribromide solution (1.7 × 10^{-5} *M*) and a lower wavelength (down to 270 mμ) at which the absorbance of tribromide is greater. At these low concentrations, a blank should be determined to correct for probable impurities in the sample solvent.

Results obtained by Fritz and Wood using this method are shown in Table 7-6.

Table 7-6. Indirect Determination of Unsaturation (11)

Compound[a]	Found, %[b]	Number of determinations
Allyl alcohol in CCl_4	93.7 ± 3.0	14
Allyl ether in CCl_4	94.9 ± 3.1[c]	7
trans-Cinnamaldehyde in CCl_4	Does not react[d]	...
Cyclohexene in benzene	97.8 ± 6.0	9
Cyclohexene in CCl_4	99.0 ± 2.0	8
Cyclohexene in CCl_4[e]	100.9 ± 0.5	3
Cyclohexene in methanol	100.4 ± 0.6	3
2,3-Dimethyl-2-butene in CCl_4	96.6 ± 4.0	7
trans-2,5-Dimethyl-3-hexene in CCl_4	101.8 ± 3.9	3
Ethyl acrylate in CCl_4	Does not react[d]	...
3-Ethyl-2-pentene in CCl_4	99.7 ± 5.0	3
Isoprene in CCl_4	101.3 ± 1.5[f]	4
d-Limonene in CCl_4	96.3 ± 5.0[c]	9
1,2-Octene in CCl_4	96.0 ± 1.1	3
1,2-Octene in methanol	97.0 ± 4.5	5

[a] Amounts taken range from 0.3 to 1.1 μM.
[b] Results based on the assumption that samples are 100% pure.
[c] Results based on the assumption that both double bonds add bromine.
[d] No detectable reaction in 10 minutes.
[e] Equimolar amount of allyl acetate also present.
[f] Results based on the assumption that only one double bond adds bromine.

II. GAS CHROMATOGRAPHIC METHODS*

Because compounds may contain cis and trans double bonds, triple bonds, conjugated structures, branched or cyclic structures, epoxides, and other groupings, the application of GC retention times to identify unsaturates is not considered reliable; a common recourse is the use of hydrogenation to determine the presence and amount of unsaturation and to help establish structure assignments. For example, fatty acid methyl esters differing in carbon number and unsaturation have almost identical retention times in

* Written by Morton Beroza and May N. Inscoe.

many instances; most often these mixtures may be resolved and quantified by means of hydrogenation.

Some typical procedures used to hydrogenate unsaturated compounds are listed in Table 7-7. Hydrogenation may be carried out before, or during chromatography (on-line) by using a catalyst in the GC system and hydrogen as the carrier gas or in the carrier gas (see description of analysis in Section A below). From chromatograms made before and after hydrogenation, the amounts of saturated and unsaturated compounds may be determined. A polar liquid phase is usually employed in analyses with and without hydrogenation because differences in the t_R's of olefins and their saturated analogs are large. The chain lengths of compounds can usually be determined from chromatograms after hydrogenation.

Table 7-7. Some Typical Hydrogenation Procedures for Determining Unsaturation

Type of compound	Method	Reference
Fatty acid methyl esters	Hydrogenation, then GC	[1,2]
	On-line hydrogenation, then GC	[3-6]
α-Olefins	Hydrogenation, then GC	[7]
Long-chain hydrocarbons	Hydrogenation, then GC	[8]
Wide variety of compounds	On-line hydrogenation, then GC	[9]

[1] J. L. Iverson, *J. Assoc. Offic. Anal Chemists*, **50**, 1118 (1967).
[2] J. Adda, *Rev. Franc. Corps Gras*, **11**, 527 (1964).
[3] T. L. Mounts and H. J. Dutton, *Anal. Chem.*, **37**, 641 (1965).
[4] H. J. Dutton and T. L. Mounts, *J. Catal.*, **3**, 363 (1964).
[5] I. Hornstein, P. F. Crowe, and J. B. Ruck, *J. Gas Chromatog.*, **5**, 319 (1967).
[6] I. Hornstein, P. F. Crowe, and R. Hiner, *J. Food Sci.*, **32**, 650 (1967).
[7] H. Hachenberg and J. Gutberlet, *Brennstoff-Chem.*, **45**, 132 (1964).
[8] B. Hallgren and S. Larsson, *Acta Chem. Scand.*, **17**, 1822 (1963).
[9] M. Beroza and R. Sarmiento, *Anal. Chem.*, **38**, 1042 (1966).

An analysis of mixtures of unsaturated compounds applicable to a wide variety of compounds is described in detail in Section A.

A. Analysis of Mixtures of Unsaturated Compounds

METHOD FROM BEROZA AND SARMIENTO (12)

Unsaturated compounds passed through a hot hydrogenation catalyst in a hydrogen carrier gas are instantaneously and quantitatively hydrogenated in the gas chromatographic pathway; the saturated compounds separate in the chromatographic column, and detector response is recorded as the compounds emerge.

Apparatus.   A gas chromatograph equipped with a flame ionization detector is used, although an instrument with a different type of detector responsive to the compounds being analyzed (e.g., a thermal conductivity detector) may serve equally well. The chromatographic column is a 6-foot-$\frac{1}{4}$-inch o.d. aluminum column containing 15% stabilized DEGS (diethylene glycol succinate: Analabs, North Haven, Conn.) on 60/80 mesh Gas Chrom P (Applied Science Lab., State College, Pa.). The flow rate of the hydrogen carrier gas is 60 ml/minute. Nitrogen at 60 ml/minute is led into the port at the flame head, where hydrogen enters in ordinary GC.

One of the two hydrogenators shown in Figure 7-1 is used. The oven hydrogenator is quickly assembled by attaching a $\frac{1}{4}$-inch stainless steel union (Swagelok No. 400-6-316, Crawford Fitting Co., Cleveland, Ohio) to a 1-inch length of $\frac{1}{4}$-inch o.d. stainless steel tubing and attaching the free nut of the union to the other end of the stainless steel tube (see Figure 7-1A). A $\frac{1}{4}$-inch length of catalyst (ca. 25 mg) is held between two glass wool plugs (previously rinsed with solvent and dried) in the $\frac{1}{4}$-inch stainless steel tube. The oven hydrogenator is connected between the exit end of the injection port and the head of the chromatographic column. Since the unit is in the oven, the temperature of the catalyst within it is that of the oven. The oven hydrogenator can be used only when the oven temperature is in the 140 to 250°C range; hydrogenation is incomplete when the catalyst temperature is below 140°C.

The injection port hydrogenator (Figure 7-1B) is identical to the oven hydrogenator except that a length of stainless steel needle stock tubing (Vita Needle Co., Needham, Mass.), which fits into the injection port of the gas chromatograph and extends almost to the septum, is silver-soldered into the 1-inch length of $\frac{1}{4}$-inch o.d. tubing, two $\frac{1}{8}$-inch back ferrules (Swagelok No. 204-1) being used to hold the needle stock tubing in place. (For an F & M Model 810 gas chromatograph a $5\frac{7}{8}$-in. length of 11-gauge (0.120

Figure 7-1.   Hydrogenators.   A: Oven hydrogenator; B: Injection-port hydrogenator.

inch o.d. $\times$ 0.094 inch i.d.) stainless steel needle stock tube was used.) The catalyst (ca. 25 mg) is held in the needle stock tubing between glass wool plugs and thus is located in the injection port and assumes its temperature. The injection port hydrogenator is more versatile than the oven hydrogenator because the catalyst can be held at one temperature (140 to 250°C) while the oven is held at another; for example, the oven may be kept at 100°C to secure adequate separation of low-molecular-weight compounds, while the catalyst is held at 140°C to assure complete hydrogenation.

Reagents. The catalyst is a so-called neutral 1% palladium on 60/80 mesh Gas Chrom P. It is prepared by dissolving palladium chloride in 5% aqueous acetic acid and adding enough nonvolatile alkali to neutralize the HCl that forms on activation ($PdCl_2 + H_2 \rightarrow Pd + 2HCl$; typically 88.8 mg $PdCl_2$ will require 53 mg Na_2CO_3). The solution is evaporated to dryness in contact with the Gas Chrom P in a rotating evaporator and dried at 110°C. The catalyst is activated by heating it to 200°C for 1 hour in the chromatographic pathway (H_2 carrier gas). Several hundred milligrams of catalyst may be activated each week and stored in a vial. The catalyst activity is checked with a known compound twice a day and replaced if evidence of incomplete hydrogenation is found. The activated catalyst is installed and used after 5 to 10 minutes at the operating temperature. The neutral catalyst (not activated) is available from National Instruments Laboratory, Rockville, Md.

Procedure. Mixtures of compounds in the 10- to 500-μg range or pure compounds are injected with or without solvent into the gas chromatograph with and without the hydrogenator in place. Because of the slight difference in t_R of the compounds in the two runs, the t_R of one of the saturated compounds is adjusted to be the same in the two runs by making a small change in flow rate. Each peak area is determined, and the amount of compound for each peak is calculated by using appropriate response factors.

Results. Chromatograms of a mixture of fatty acid methyl esters obtained with and without the hydrogenator are shown in Figure 7-2. The formation of saturates from the unsaturates is apparent. The hydrogenation is quantitative.

Discussion. The foregoing procedure was used to hydrogenate fifty unsaturated compounds, which included alcohols, amides, amines, carbonyl compounds, esters, ethers, halides, epoxides, and other types. Of these, aldehydes, halides, and sulfides underwent some hydrogenolysis of the functional group. The analysis of aldehydes, halides, and sulfides will therefore not be quantitative by this technique.

Polymerized samples should be avoided because they contain nonvolatile material that will coat and thereby weaken the catalyst.

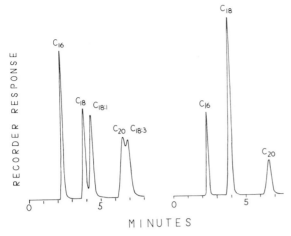

Figure 7–2. Chromatograms of a mixture of fatty acid methyl esters (C_{16}, palmitate; C_{28}, stearate, $C_{18:1}$, oleate; $C_{18:3}$, linolenate; C_{20}, arachidate). Left, without hydrogenator; right, with hydrogenator.

In an elegant modification of the hydrogenation technique used to identify flavor chemicals, Issenberg, Kobayashi, and Mysliwy (13) employed a post-column reactor, which could be included or excluded from the system, with a valve that led to a mass spectrometer. The saturated compounds were readily identified, and the number of double bonds was determined from the difference in parent masses of the olefin and its corresponding saturated analog. The combination of GC and mass spectrometry is one of the finest ways to establish the quantitative validity of a GC peak.

B. Hydrogenation Detectors for Gas Chromatography

Several clever devices that respond to hydrogenatable compounds, especially olefins, have been reported. In one, a thermal conductivity cell is followed by another identical to it except that a hydrogenation catalyst has been deposited on its sensing elements. When hydrogenation occurs, the thermal effect due to the catalytic reaction is superimposed on the thermal conductivity effect, and the discrepancy in the responses of the two cells is quantitatively related to the heat of hydrogenation(14).

In another device, gas from the GC column mixes with a gas containing electrolytically generated hydrogen and passes through a reactor containing a palladium catalyst to an electrochemical cell sensitive to hydrogen but not to the hydrogenation products. When an olefin enters the detector, hydrogen is consumed and a servomechanism causes extra hydrogen to be generated. A recording of the amount of hydrogen generated gives a quantitative measure of the amount of unsaturate without being affected by saturated compounds that may be present (15).

In still another device, unsaturated compounds or those that may be dehydrogenated are measured in an argon-hydrogen stream by passage through a postcolumn platinum-kieselguhr catalyst. The results of the reaction are registered as changes of hydrogen concentration in the gas stream by a thermal conductivity detector (16).

C. Location of Double-Bond Position Using Gas Chromatography

Procedures for determining the positions of double bonds in compounds have been most extensively investigated by those interested in the analysis of fatty acids and oils, and generally the compounds are analyzed in gas chromatographs as fatty acid methyl esters. Privett (17) has reviewed the methods used to determine the positions of unsaturation in fatty acids employing permanganate oxidation and ozonolysis. He prefers ozonolysis methods and concludes that they are fast, simple, and capable of determining the structures of compounds unambiguously. Privett also notes that ozonolysis may be carried out on an ultramicro scale and can be employed for quantitative analysis of mixtures of fatty acids.

Apparatus for the ozonolysis of the small amounts of compound used in GC analysis is simple to construct (18, 19). The ozonized compound is usually cleaved to produce an aldehyde or a ketone, depending on whether or not the doubly bound carbon atom is substituted, in accordance with the following equation:

$$R\text{—}CH\text{=}\overset{\overset{\displaystyle R''}{\displaystyle |}}{C}\text{—}R' \xrightarrow[\substack{2:\text{cleavage}\\ \text{reagent}}]{1:O_3} R\text{—}CHO + R''\text{—}\overset{\overset{\displaystyle O}{\displaystyle ||}}{C}\text{—}R' \qquad (1)$$

The aldehydes and ketones are usually identified from their t_R's in GC analysis.

Davison and Dutton (20) devised a microreactor apparatus consisting of a soldering gun and a reactor U-tube terminating in a needle. Fatty acid methyl esters were successively reacted with ozone, thermally cleaved, and injected into the gas chromatograph with this device. Zinc oxide granules were used to trap carboxylic acids formed in the pyrolysis. The results are said to be quantitative but are not in accordance with equation 1.

Nickell and Privett (21) use a system of "controlled" pyrolysis. The compound is added to a solution of ozone in pentane at −65°C. Pyrolysis at

225°C in contact with Lindlar catalyst produces aldehydes and ketones, which are determined by GC.

The use of ozonolysis with GC on a wide variety of compounds has been demonstrated by Beroza and Bierl (22). Although they do not claim that their procedure is quantitative, quantification is usually not necessary to locate double-bond positions in a molecule. Their procedure, which is described in detail below, is simple, rapid, and easily carried out. It may be applied to as little as 1 to 5 μg of chemical.

1. Alkylidene Analysis

METHOD ADAPTED FROM BEROZA AND BIERL (22)

Ozone from an easily constructed generator (18) is led into a solution of the compound(s) at − 70°C to form the ozonide(s). The ozonides are cleaved to give aldehyde and/or ketone fragments for analysis by GC.

Apparatus. A flame ionization gas chromatograph equipped with a 12-foot-$\frac{1}{4}$-inch o.d. copper column containing 5% Carbowax 20M on 60/80 mesh Gas Chrom P is used with a nitrogen carrier gas flow rate of 30 to 60 ml/minute. Injection port and detector temperatures are set at 225°C. Temperature programming is useful for determining the identity of large and small molecules in a single analysis. Other types of chromatographs that will detect the compounds being analyzed may be used, as well as other GC columns providing adequate separation of products.

A micro-ozonizer, shown in Figure 7-3, is made by wrapping aluminum foil, *D*, around the single leg, *C*, of a three-way stopcock, covering it with insulation, *E*, and grounding it, *F* to *G*, for example, to a water tap. A $5\frac{1}{2}$-inch length of 21-gauge stainless steel needle stock tubing, *A* (Vita Needle Co., Needham, Mass.), is pushed through *B*, an injection port septum (W-10 septum, Applied Science Lab., State College, Pa.), which is fitted into the end of the glass tube, *C*. Sources of O_2 and N_2 are attached by rubber tubing to the inlets indicated above the stopcock, allowing either gas to be led into the reaction tube by rotating the stopcock. The reaction tube, *J*, may be made from a medicine dropper by sealing the glass at its small end. Teflon tubing, *L* ($\frac{1}{16}$ inch o.d.), conducts the gases emerging from the tube, *J*, through another injection port septum, *I*, into a 10-ml Erlenmeyer flask, *M*, containing 4 ml of an indicating solution, *N*, which turns blue when ozone enters. A modified form of this apparatus and all reagents for the analysis are available from Supelco Co., Bellefonte, Pa.

A vacuum tester, *H* (Tesla coil), such as Fisher Scientific Co. 15-340-75V3, serves as a source of high voltage.

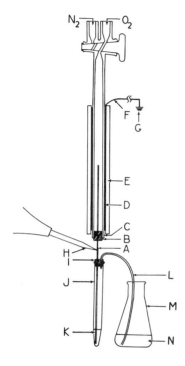

Figure 7-3. Micro-ozonizer.

Reagents. Carbon disulfide and pentyl acetate are "chromatoquality" reagents, supplied by Matheson, Coleman, and Bell, Norwood, Ohio. The CS_2 is used as received. The pentyl acetate is slowly distilled in an all-glass apparatus, and a heart cut that is checked at high sensitivity for absence of interference is taken. Triphenylphosphine is used as received from Distillation Products Industries, Rochester, N.Y. The indicating solution is 5% KI in 5% aqueous sulfuric acid with added starch.

Procedure. In a typical analysis, a solution, K, of 25 µg of compound in 100 µl of CS_2 (or pentyl acetate) is placed in the reaction tube, J, of the assembly shown in Figure 7-3. Ten milliliters per minute of O_2 is passed into the solution while the tube is held at ca. $-70°C$ in a xylene-Dry Ice bath (not shown in Figure 7-3). The vacuum tester, H, is applied to the inner electrode, A, to generate ozone until the indicator solution turns blue (ca. 15 seconds). The stopcock is then reversed to sweep out the ozone and O_2 from the reaction tube (ca. 15 seconds). The cold bath is removed, the reaction tube is slipped off its rubber stopper, *I*, and 1 mg of powdered triphenylphosphine is dropped into the solution. The tube is then stoppered and swirled. When the solution reaches room temperature (ca. 5 minutes), 20 μl of the solution (equivalent to 5 μg of compound) is injected into the gas chromatograph for analysis. Identifications are based on t_R.

Carbon disulfide is used as the solvent to observe C_5 and larger fragments, and pentyl acetate to observe C_5 and smaller fragments. Temperature programming was used with the CS_2 solvent as follows: Hold at 50°C for 6 minutes, program at 6.4°C/minute to 200°C, and hold. (Any similar program that will separate products will do.) With pentyl acetate as the solvent, hold the column temperature at 35°C until the pentyl acetate appears; then raise the temperature to 200°C to clear the column.

Results. A total of thirty-three compounds of different classes (alcohols, amides, esters, halogenated compounds, a phosphate, ring compounds, aliphatic, aromatic, heterocyclic, and other compounds) gave the expected products, as judged by their t_R's against those of authentic compounds.

Discussion. Although compounds with isolated double bonds were readily cleaved under the conditions described, certain double bonds resisted ozonolysis; for example, several α,β-unsaturated nitriles remained uncleaved. Ozonization of triple bonds also did not occur. (Presumably, ozonization can therefore be used to distinguish double and triple bonds.) End-methylene groups will produce formaldehyde, which is not detected by a flame ionization detector. (The other end of the molecule, however, should be detectable.) Malonaldehyde, which would be the normal product from the methylene-interrupted double-bond structure of natural fatty acids, does not appear in the chromatograms, probably because of its instability and low response factor.

Since no evaporations are required, low-molecular-weight compounds may be analyzed.

The procedure has been used for determining alkylidene groups (fragments from terminal methyl to first double bond) in glyceride oils. Black and Beal (23) determined quantitatively the Δ15 double bond in fatty acids of hydrogenated soybean oil by GC measurement of the amount of propanal formed by the ozonization-triphenylphosphine reaction.

2. Oxidation Followed by Gas Chromatography. In one form or another the periodate-permanganate method of von Rudloff (24) has been adapted to determining the location of double bonds by oxidative cleavage at the double bond followed by GC of the fragments. The fragments are carboxylic acids, and they are usually determined as methyl esters. Kuemmel (25) utilized continuous extraction of the fatty acids, conversion to soaps during solvent removal, and programmed-temperature GC of the methyl esters formed from the soaps (without concentration) for the quantitative recovery of C_4 and higher mono- and dicarboxylic acid esters. His procedure is said to overcome difficulties in recovering the short-chain acids, which tend to have a high volatility and considerable solubility in water. Several similar procedures also give quantitative recovery of all but the short-chain acid fragments (26, 27). In another modification of the von Rudloff method (28), tetramethylammonium hydroxide is added to the aqueous solution of the acid fragments and an aliquot is placed in a probe, dried at 100°C, and injected into the gas chromatograph; methyl esters form when the salts contact the hot (>250°C) injection port.

One difficulty with the periodate-permanganate procedure is that ambiguity can result from the production of dibasic acids from the carbons between the double bonds as well as from the carbons between the double bond and the carboxylic acid end group; that is, the position of each dibasic acid is not defined. This difficulty does not occur with ozonolysis because the cleavage products are aldehydes or ketones.

Several methods that employ combined GC and mass spectrometry to locate double-bond position have been described. Double bonds are first oxidized to the dihydroxy compounds (—CH=CH— → —CHOH— CHOH—) with osmium tetroxide, and then the mass spectra are determined of the O-isopropylidene derivatives (29), methyl ethers (30), or trimethylsilyl ethers (31).

At this time ozonolysis appears to be more rapid, more efficient, and far simpler than oxidation, derivatization, and GC-mass spectrometry as a method for locating double bonds.

D. Removal of Olefins for Analysis by Gas Chromatography

Hydrogenating an olefin, followed by GC on a polar liquid phase, removes its peak to another part of the chromatogram, thereby indicating which peak arises from the olefin. It is also possible to "subtract" double-bonded compounds completely. For example, peaks of fatty acid (C_{10} to C_{20}) methyl esters containing double bonds were subtracted from the chromatogram of a mixture of methyl esters by prior treatment with bromine (32). The bromine adds to the double bond, and the compound becomes relatively involatile.

Most subtractive techniques deal with separations of hydrocarbon fractions. A chemical or chemical mixture placed in the GC pathway reacts with a given species of hydrocarbon and essentially removes it from the carrier gas stream. (See Chapter 1, Section II.F, for a discussion of subtractive loops.) With a dual-column, dual-recording instrument having the subtraction agent in one column but not in the other, data may be acquired rapidly and complex analyses simplified. Olefins are removed with packings containing sulfuric acid (33–36), silver (33, 37, 38), mercury (33, 39, 35, 40, 41), or cuprous salts (37). By selecting the proper agent it is possible to subtract olefins with or without accompanying aromatics. See Table 7-8.

Dienes have been absorbed with a column packed with 20% 1:1 maleic anhydride-stearic acid on a support (42).

E. Unsaturation (Dienes)

In a nonquantitative method for distinguishing cis and trans dienes, the diene is chromatographed with the dienophile chloromaleic anhydride as

Table 7-8.   Absorption of Hydrocarbon Classes by Chemical Absorbents (33)

				Absorbed, %		
Gas	20% HgSO$_4$-20% H$_2$SO$_4$	Saturated solution of mercuric acetate	4% Ag$_2$SO$_4$-95% H$_2$SO$_4$	H$_2$SO$_4$		
				95%	80%	60%
Methane	0	0	0	0	0	0
Ethane	0	0	0	0	0	0
Propane	0	0	0	0	0	0
n-Butane	0	0	0	0	0	0
n-Hexane	0	0	0	11	7	0
n-Octane	0	0	0	0	0	0
Cyclohexane	0	0	8	31	0	0
Ethylene	100	94	100	11	6	0
Propylene	100	100	100	100	67	0
Isobutylene	100	100	100	100	100	100
2-Pentene	100	67	100	100	88	0
3-Heptene	100	60	100	100	85	0
4-Methylcyclohexene	100	70	100	100	70	0
Benzene	5	46	100	94	32	13
Toluene[a]	0	22	100	100	33	0
p-Xylene[a]	0	0	100	100	0	0
Acetylene	100	100	100	16	11	0

[a] Accuracy is only ca. ±10% because of hangup; otherwise, estimated to be ±5%.

the stationary phase; the resulting adducts (Diels-Alder) are relatively non-volatile and remain on the colunn. The peak of the trans diene is diminished more than its cis analog because more of it reacts in passage (43).

III. ELECTROANALYTICAL METHODS*

Amperometric Titration of Olefins

The addition of bromine to olefins forms the basis for several methods of analysis of unsaturation (44). One of the difficulties with this general approach is the fact that the addition reaction is not instantaneous; on the other hand, a lengthly reaction time (or a large excess of bromine), if used to accelerate the addition reaction, fosters substitution and /or other undesirable reactions. Therefore, the success of the analysis depends on choosing the right set of conditions to maximize addition while minimizing undesirable reactions.

Dubois and Skoog (45) proposed the amperometric titration of olefins with bromine from a bromate-bromide titrant. A solvent consisting of acetic acid, carbon tetrachloride, methanol, sulfuric acid, and a mercuric chloride catalyst gave good results. Subsequently (46, 47), the omission of mercuric chloride was found to produce even better results. The pertinent reactions are as follows:

$$BrO_3^- + 5Br^- + 6H^+ \rightarrow 3Br_2 + 3H_2O \tag{1}$$

$$-\overset{|}{\underset{|}{C}}=\overset{|}{\underset{|}{C}}- + Br_2 \rightarrow -\overset{|}{\underset{\underset{Br}{|}}{C}}-\overset{|}{\underset{\underset{Br}{|}}{C}}- \tag{2}$$

EXPERIMENTAL METHOD [ADAPTED FROM THE STUDY OF UNGER (47)]†

Apparatus. Several types of apparatus may be used for this analysis. The electrodes may be two small platinum foils, rods, coils, or rings. The monitoring device may be similar to the biamperometric system described for coulometric bromination (cf. Section III of Chapters 1 and 12). Instead, it may be the type of circuit used for Karl Fischer titrations; this type of circuit will be found on most modern pH meters. A commercial automatic Karl Fischer titrator, the Beckman Aquameter, was used by Unger (47), and part of the procedure is based on this.

---

* Written by Alan F. Krivis.
† Reprinted in part with the permission of the copyright owner, the American Chemical Society.

Reagents. The bromate-bromide titrant (0.25 N) is prepared by dissolving 51 grams of potassium bromide and 13.92 grams of potassium bromate in 2 liters of water. This solution can be standardized against 0.1 N sodium thiosulfate solution.

The titration solvent is prepared by mixing 712 ml of glacial acetic acid, 134 ml of carbon tetrachloride, 134 ml of methanol, and 18 ml of sulfuric acid (1:5, v/v).

Procedure. The titration must be carried out at 0 to 5°C; this requires that the titration beaker be placed in an ice bath.

If an automatic titrator is being used, transfer 110 ml of solvent to the titration beaker and chill to 0 to 5°C. Pretitrate the blank solvent, using a polarizing current of 5 μA and a time delay of 60 seconds. When the blank titration is completed, add an appropriate aliquot of the sample and titrate again.

If an automatic titrator is not used, separately titrate a blank as above and subtract the volume for the blank from that of the unknown titration.

From the volume of titrant used for the sample, calculate the percentage of olefinic compound present.

IV. NUCLEAR MAGNETIC RESONANCE METHODS*

The determination of unsaturation by nmr spectroscopy is facilitated by the fact that the chemical shifts of the vinyl protons are sufficiently different from those of the other hydrogens so that there is no difficulty in identification. The conversion of the olefin to the saturated compound by either hydrogenation or bromination can also be followed with relative ease by the same technique. It is just a question of the rate of disappearance of the vinyl protons and the appearance of the protons of a saturated system.

One interesting article related to the determination of unsaturation was written by Stehling and Bartz (48). These authors discuss the utilization of nmr spectroscopy to obtain characteristic patterns for the identification of sixty known olefins. Correlations are made whereby the structures can be determined by the chemical shift differences and also the coupling constants. The only purpose here is to show that this particular publication is an excellent reference for the determination of unsaturation in which the double-bond geometry is easily established. The series of unsaturated hydrocarbons that were studied ranged from a simple monoolefin such as $CH_3(CH_2)_n CH_2CH{=}$ CH_2 to complex olefins in which there are both beta substitution, $R{-}C{=}CH_2$,

$$\underset{R}{\overset{}{|}}$$

and alpha substitution, $R_2C{=}CHCH_3$.

* Written by Harry Agahigian.

The correlation study is geared primarily toward the determination of structure but is also extremely valuable in that the nature of an olefin present in the system can be determined as well. For example, the dimer of 3-methylbutene-1, the dimer of 2-methylpentene-1, copolymers of isoprene and isobutylene, polypentadiene-1,3, and polyisoprenes were all studied and reported by Stehling and Bartz. The authors' approach was to look at the various substituent groups and to determine the coupling constants and the chemical shifts; hence they were able to do some interesting determinations. The experimental data were obtained in the usual manner, using 1% TMS as the internal standard. The significance of this article is that it enables one to tell the substituents on a particular double bond. In addition, one can determine that unsaturation is present from the chemical shift data. Thus, this publication makes it possible to obtain the kind of information that enables one to determine the presence and also the type of unsaturation.

Fatty acids and their triglycerides have been reported, and a method of determining the degree of unsaturation has been described by Johnson and Shoolery (49). The assignments are straightforward, and the authors are concerned with the olefinic protons in the triglycerides. The procedure is to establish the integral area per proton. In this case, the methylenes of the glycerol portion of the natural fat are used; by subtracting the methine proton of the glycerol portion, which overlaps the olefinic protons, the percentages of unsaturation are easily determined.

Comparing with the iodine number values of various fats the results shown in Table 7-9 were obtained.

Table 7-9.

Oil	Nuclear magnetic resonance values	WIJS iodine #
Cocoanut	10.5 ± 1.3	8.0–8.7
Olive	80.8 ± 0.9	83.0–85.3
Peanut	94.5 ± 0.6	95.0–97.2
Soybean	127.1 ± 1.6	125.0–126.1
Sunflower	135.0 ± 0.9	136.0–137.7
Safflower	141.2 ± 1.0	140.0–143.5
Whale	150.2 ± 1.0	149.0–151.6
Linseed	176.2 ± 1.2	179.0–181.0
Tung	225.2 ± 1.2	146.0–163.5[a]

[a] The discrepancy in the case of tung oil is probably due to the conjugated double bond of oleostearic acid.

Comparison of nmr values with saponification values yielded the results shown in Table 7-10.

Table 7-10. Average Molecular Weights of Fats from Saponification Values and Nuclear Magnetic Resonance

Oil	Saponification value	Molecular weight	Nuclear magnetic resonance molecular weight
Olive	189.3	887.1	873.8 ± 5.3
Peanut	188.8	891.5	882.3 ± 7.4
Safflower	191.5	879.0	874.9 ± 9.3

The advantage of the nmr approach for the determination is the relatively short time required for the analysis. There is generally good agreement with the results of other methods.

The analysis of α-olefins by the method of Flanagan and Smith (50) offers another example of the advantage of nmr with these systems. The α-olefins studied to establish the method were 1-hexane, 1-octene, 2-methyl-1-pentene, and 2-ethyl-1-dodecene. The presence of *trans*-2-hexene in 1-hexene was determined.

Another method reported (51) made use of the addition of sulfenyl chlorides across double bonds. The sulfenyl chlorides used were methane sulfenyl chloride and benzenes sulfenyl chloride. Adducts were obtained for olefins such as ethylene, propylene, isobutylene, 3-methylbutene, styrene, acenaphthylene, norbornene, and vinyl chlorides. Depending on the substituents on the double bond, Markovnikof and anti-Markovnikof addition can be obtained. Here the advantage of the nmr approach is that not only can the disappearance of the unsaturation be followed *in situ* but information concerning the structure can also be obtained.

The reaction of iodonium nitrate to alkenes at room temperature has been reported by Diner and Lown (52). The yields are in the range of 50 to 80% and offer a convenient way of studying alkenes when the adducts have a strong deshielding effect and the proton chemical shifts appear downfield. The reactions are carried out in chloroform and pyridine. The olefins reported were hexene-1, *cis*-pentene-2, ethyl vinyl ether, stryene, 3,3-dimethylbutene-1, and 2-methylbutene-2.

V. RADIOCHEMICAL METHODS*

Radioreagent methods for unsaturation have been developed principally for determining unsaturated fats and fatty acids on the micro level and for analyzing synthetic polymers containing low concentrations of double bonds. Radioisotopes of the halogens are especially advantageous in these applications. Adduct formation in the presence of methanol-^{14}C and mercuric acetate is useful for determining double bonds in many monomeric compounds, as well as in poly(oxypropylene) glycols and in cyclic ether elastomers containing antioxidants.

A. Conversion to Adducts of Iodine Monobromide

The reaction of unsaturated fats and fatty acids with IBr to yield derivatives containing one atom of iodine and bromine is the basis of the well-established titrimetric method of Hanus (53). The ^{131}I-labeled compound has been adapted to the determination of micro amounts of these substances by using a technique in which a specimen is treated on filter paper with an anhydrous methanol solution of the radioreagent saturated with sodium bromide (54, 55). The reactivity of deposited compounds is considerably enhanced by the relatively large surface area exposed to the reagent on the capillary substrate. The solution of ^{131}IBr is prepared by evaporating an aliquot of aqueous Na^{131}I solution, containing 1 mM of the salt and 0.5 mCi of activity, in a 10-ml volumetric flask and treating the thoroughly dried compound with the theoretical volume of 0.2 N bromine in dry methanol; completion of reaction is indicated by the color change from brown to orange. The NaBr arising in the reaction serves to saturate the reagent, which must not contain free bromine.

The specimen is deposited on the paper from a volatile hydrocarbon solvent in amounts of from 20 to 60 μg. If unsaturated fatty acids are to be determined, the dried paper is dipped first in 5% copper acetate solution to fix the acid, thus preventing its diffusion on addition of the reagent; the spots of the copper soaps are easily located by their blue color. Before analysis, the paper is washed with water to remove most of the copper acetate and dried. The spot is then treated dropwise with the reagent solution until no further color change occurs, and additional reagent is added to provide an excess of at least 100% (56). Unreacted ^{131}IBr is removed by washing the dried paper for 3 to 5 minutes with 1:9 ethanol-water. The spots are cut out, and the activity in each one is assayed by G-M counting.

* Written by D. Campbell.

The calibration is made by treating, in an identical manner, pure oleic or linoleic acid in quantities of from 20 to 60 μg and preparing a graph relating net count rate to amount of acid taken. Other calibration standards are similarly useful, for example, acids or fats the unsaturation of which has been determined titrimetrically may be employed. The technique is straightforward and appears well suited to the rapid, routine determination of total unsaturated fats and higher fatty acids in the absence of other unsaturated substances. As in all quantitative methods using ^{131}I, a standard prepared with the same lot of reagent used to derivatize the sample must be assayed concurrently to obtain accurate decay corrections for the 8-day half-life. For oleic and linoleic acids, lard, and various vegetable oils the radioreagent method gave results which were in very good agreement with those obtained by means of a standard bromination method.

B. Conversion to Iodinated Derivatives

Elemental iodine adds only slowly to most unsaturates in solution. This fact, together with the availability of the versatile IBr and ICl reagents, has limited the use of iodine in conventional methods. However, deposition or separation of a water-insoluble unsaturate on a suitable supporting medium, such as chromatographic paper impregnated with silicone oil or certain alkanes, furnishes the substance in a form highly reactive toward even elemental iodine in aqueous solution. Free iodine is easily generated as needed by acidification of a solution of iodide and iodate ions:

$$5^{131}\text{I}^- + \text{IO}_3^- + 6\text{H}^+ \rightarrow 3^{131}\text{I}_2 + 3\text{H}_2\text{O}$$

Uncombined iodine is readily removed from the paper by volatilization. Radioreagent methods based on these principles have been developed for determining unsaturation in a variety of fats and fatty acids.

Unsaturation in oleic and linoleic acids was determined initially (57, 58) after their separation by ascending chromatography on paper strips impregnated with a low-boiling (b.p. 180 to 190°C) mixture of synthetic alkanes. A 9:1 acetic acid-water solution saturated with the same hydrocarbon mixture was used for development. After chromatography, the strip was dried for 1 hour at 60°C in nitrogen. It was then immersed in 6.5 ml of a solution 0.0022 M in Na^{131}I and 0.0044 M in NaIO$_3$ and containing ca. 5 μCi/ml of activity. Iodine was generated by making the solution 0.025 M in H$_2$SO$_4$. The strip was immersed for 2.5 hours and dried overnight at room temperature to remove excess reagent. The activity on the strip was determined using a G-M tube and a chromatogram scanning technique. Calibration graphs prepared from 7 or 8 spots of the pure acids in amounts up to 300 to 400 μg exhibited good linearity between integrated spot activity and amount of acid deposited. A satisfactorily linear relationship more convenient for routine analysis of these acids separated on chromatograms, however, was that observed between maximum spot activity and logarithm of spot content. Conventional reagents are useful for making the spots visible and thereby facilitating location of maximum spot density.

An essentially identical technique was used to determine 5 to 10% rape seed oil in other edible oils with an accuracy of $\pm 8\%$ (59). After hydrolysis, the erucic acid was separated and the activity of its treated spot was compared with the result obtained using a known quantity of the pure acid.

Oleic, linoleic, and linolenic acids obtained by the saponification of linseed oil were determined in a similar manner after their chromatographic separation on paper impregnated with silicone oil (60). In this case a linear relationship observed between maximum spot activity and spot content on the chromatograms was useful for rapidly determining 20 to 400 μg of total unsaturated fatty acids, with 0.0022 M ^{131}NaI having a specific activity of 1 μCi/ml. Reduction of the specific activity of the NaI solution to 0.05 μCi/ml gave improved peak resolution and permitted the same relationship to be used for determining the three acids when their total amount was within the range of 50 to 150 μg.

Direct iodination on paper has also been employed for determining unsaturation in edible and industrial oils without preliminary hydrolysis. In the method developed originally for linseed oil (61), several 10-μg spots of the oil were deposited from hexane onto a strip of Whatman No. 1 filter paper that had been treated with silicone oil. The strip was then immersed in the acidified solution of sodium iodide-^{131}I and sodium iodate for 2 hours at 60°C. The specific activity of the reagent solution was 10 μCi/ml. After removal of excess iodine with warm air, the activity in the excised spots was determined by end-window G-M counting. In this method, the specific activity of the iodine was determined directly by assaying a weighed amount of the sodium iodide-^{131}I deposited by evaporation on an aluminum planchet. Since the elemental iodine combining with the unsaturate is derived partly from the sodium iodate, its specific activity, as calculated from the observed specific activity of the sodium iodide-^{131}I, must be corrected for this dilution. The double-bond content of the sample is obtained from the expression

$$U = \frac{A}{2SW} \tag{1}$$

where U = unsaturation (mM/gram),
$\quad A$ = activity in the treated spot (μCi),
$\quad S$ = specific activity of the elemental iodine (μCi/meq),
$\quad W$ = weight of sampled deposited on the paper (grams).

The technique was studied extensively in its application to 22 oils that had iodine numbers ranging from 7 to 155 (62). Significantly higher iodine numbers obtained for several oils by the radioreagent method in comparison with those observed using the established titrimetric procedures of Hanus and Hübl were attributed to more nearly complete halogen addition in the radiochemical method. The accuracy of the technique is $\pm 0.5\%$.

Elemental ^{131}I is useful also for determining sitosterol on paper chromatograms (63) and for locating and determining certain steroid hydrazones after their separation by electrophoresis (64). In the latter work, a linear relationship was observed on the electropherograms between maximum count rate and amount of steroid derivative deposited.

Although direct assay of combined ^{131}I activity by G-M counting has been employed exclusively in the methods described, the excellent efficiency with which this isotope can be determined by liquid scintillation counting (65) suggests that elution of iodinated derivatives is a useful modification for enhancing sensitivity. Iodine-125, with a half-life of approximately 60 days, can also be determined with reasonable efficiency by liquid scintillation counting (65). The longer-lived isotope necessitates less frequent preparation of reagents, and its much weaker gamma radiation requires substantially less shielding.

C.   Conversion to Chlorinated Derivatives

Although free chlorine does not have a history of significant use for chemically measuring olefinic unsaturation, [36]Cl has proved useful for determining very low concentrations of double bonds in certain synthetic hydrocarbon polymers under carefully controlled reaction conditions. This radioisotope is a pure beta emitter ($E_{max} = 0.714$ MeV), with a half-life of 3.1×10^5 years. Chlorine is readily prepared by the decomposition of palladous chloride (66) and can be transferred quantitatively on a high-vacuum line equipped with greaseless stopcocks. Radioassays of the gas are conveniently made by absorbing a known weight into a solution of excess styrene in carbon tetrachloride. The chlorine reacts with the styrene immediately by addition, and the radioactivity in the solution can be counted with a G-M tube designed for use with liquids. Chemical determinations of chlorine are easily performed by thiosulfate titration of the iodine liberated on absorption of the gas into aqueous potassium iodide.

The use of [36]Cl$_2$ as a quantitative reagent for unsaturation requires that reactions be performed under rigorously controlled, reproducible conditions for several reasons. First, such control is necessary to regulate the mechanism, since this may be ionic as well as free radical. Moreover, the extent to which addition or substitution takes place is determined not only by the structure of the olefin but also by the conditions of chlorination. Furthermore, the reactive nature of chlorine often leads to side reactions, necessitating a series of carefully controlled analyses in which the amount of reagent is varied. To suppress free radical reactions, chlorination is performed in sealed tubes at 20° or 25°C in the absence of light and air. Under these conditions, branched olefins exemplified by $-CH_2-C(CH_3)=CH-CH_2-$ and $-CH_2-C(CH_3)=CH_2$ react by substitution on the methyl group.

The methodology used with $^{36}Cl_2$ is illustrated by its initial application to the determination of combined isoprene in butyl rubber (67). A 20-mg specimen of elastomer is dissolved in 10 ml of CCl_4, and the solution is transferred quantitatively by syringe to a two-arm reaction bulb attached to a high-vacuum line. After adjustment of the volume to 15 ± 0.5 ml, the entrance arm is sealed, and the solution is thoroughly de-gassed. Sufficient $^{36}Cl_2$ to provide a $1.4 \pm 0.4\%$ by weight concentration is condensed into the bulb, which is then sealed and thawed. After 10 minutes at 20°C in the dark, the tube is opened and the solution is washed with aqueous 0.1 N $Na_2S_2O_3$ to remove $^{36}Cl_2$ and $H^{36}Cl$. After further washing, the solution is made up to 25 ml for radioassay. The process is repeated 5 to 7 times, using $^{36}Cl_2$ in concentrations up to $17 \pm 2\%$ by weight. A plot is prepared of weight per cent chlorine in the polymer as ordinate versus weight per cent chlorine in the reaction mixture as abscissa, and the two clearly defined linear portions of the curve are extrapolated to an intersection corresponding to completion of the analytical reaction. Determination of the chlorine contents of polymers of known unsaturation showed the reaction product to be $[-CH_2-C(CHCl_2)=CH-CH_2-]_n$.

In applying the method to terminal unsaturation in polyisobutene (68), model compounds, such as 2,4,4-trimethyl-1-pentene, 2,4,4-trimethyl-2-pentene, and 2,4,4,6,6-pentamethyl-1-heptene, as well as polymer of molecular weight <1000, were used to demonstrate that a monochloro derivative, $-CH_2-C(CH_2Cl)=CH_2$, was obtained, rather than the dichloro product originally assumed (69).

Radiochlorine has similarly been used to determine unsaturation in polystyrene prepared by both cationic (70) and free radical (71) polymerization. In these studies, the assumption was made that, by analogy with styrene monomer, the double bonds present react by addition to incorporate two chlorine atoms for each unsaturated center.

D. Conversion to Brominated Derivatives

Trace concentrations (>0.064 $\mu M/ml$) of unsaturated hydrocarbons in alkanes have been determined rapidly and accurately using $^{82}Br_2$ (72). An aliquot of 50 λ of elemental ^{82}Br having an optimum specific activity of 8 mCi/gram is added to 10 ml of alkane, and the solution is maintained in the dark at 0°C for 30 minutes. Excess bromine is then removed by extraction with 10 ml of aqueous sulfite solution. The concentration of dibromo derivatives is determined by counting the radioactivity in an aliquot of the dried alkane; the specific activity of the bromine is determined by similarly assaying an aliquot of the aqueous phase. Although the method is straightforward and sensitive, the 35-hour half-life of ^{82}Br is somewhat disadvantageous for routine analyses. Qualitative identification of brominated derivatives at concentrations >5 $\mu M/ml$ was possible by gas radiochromatography.

E. Conversion to *vic*-Methoxyacetoxymercuri Derivatives

Many compounds having terminal double bonds, as well as substances possessing certain forms of cis-type unsaturation, react quantitatively in the presence of an excess of both mercuric acetate and methanol to yield vicinal adducts of the olefinic linkage. If methanol-^{14}C is used, methoxy-^{14}C-acetoxymercuri derivatives are obtained, as indicated for styrene:

The adducts, in general, are sufficiently stable to be isolated by evaporation of the methanol, and use has been made of this stability in a radioreagent method for a number of reactive unsaturates (73). In practice, a 2-ml aliquot of ethylene dichloride containing 75 to 250 μM of double bond, but not more than 20 mg of total nonvolatile organic material in the case of an unknown, is transferred to a 10-ml round-bottom flask. The flask is fitted with a ball joint permitting its attachment to an apparatus designed for the determination of ^{14}C by wet combustion and quantitative collection of the $^{14}CO_2$ in an ionization chamber (74–77). A 1-ml aliquot of a solution of mercuric acetate in methanol-^{14}C (150 mg/ml) is added, and the flask is fitted with a cap. After a 1-hour reaction period at 40°C, ethylene dichloride and excess methanol are removed with a current of inert gas and collected in an efficient cold trap; final removal of uncombined methanol is performed under vacuum at 30 to 40°C. A 1.5-gram portion of an intimate 2:1 $K_2Cr_2O_7$-KIO_3 mixture is added to the residue, which is then decomposed by heating with 5 ml of an anhydrous mixture of phosphoric and fuming sulfuric acids. The $^{14}CO_2$ is collected in a 250-ml ionization chamber, and the activity present is assayed using a dynamic condenser, or vibrating reed, electrometer. The residual activity, when corrected for a blank, is proportional to the amount of unsaturation present. The specific activity of the methanol-^{14}C is determined in the same manner after first converting the reagent to the p-nitrobenzoate. Analyses of representative compounds to which the method is applicable are given in Table 7-11.

Table 7-11. Determination of Unsaturates by Methoxy-^{14}C-mercuration

Compound analyzed	Unsaturation, mM/gram		Number of analyses	Average deviation, mM/gram
	Theoretical	Experimental		
Allyl alcohol	17.22	16.93	2	0.06
Allylcyclohexane	8.05	8.17	2	0.06
3-Cyclohexene-1-carboxaldehyde	9.08	9.25	2	0.03
Diallyl ether	20.38	20.07	3	0.01
4,5-Epoxy-1-hexene	10.19	9.99	2	0.01
5-Hexene-2-one	10.19	10.23	2	0.05
1-Octene	8.91	8.80	2	0.01
Oleic acid	3.54	3.55	4	0.06
Squalene	14.32[a]	14.31	2	0.03
Styrene	9.60	9.59	4	0.01

Copyright 1968, by Academic Press, Inc. Reproduced in part by permission of the coypright owner.

[a] Based on 98% purity.

The method is widely applicable to substances containing terminal double bonds, as well as to unsaturates of the cis type, such as oleic acid and cyclohexene derivatives. No interference from the carboxyl, hydroxyl, carbonyl, or oxirane groups was observed with the unsaturates analyzed. Although acrylic and methacrylic acids may be determined, compounds containing double bonds alpha-beta to carbonyl groups do not usually react quantitatively. The ionization chamber-electrometer radioassay technique has an accuracy of $\pm 0.5\%$ (77) and has been shown to be capable of a precision of $\pm 0.25\%$ in a study using dry combustion (78).

The sensitivity of this analytical method has made it useful for determining low concentrations of double bonds in poly(oxypropylene) glycols (73). The substantial oxygen content of these polymers permits use of up to 100 mg of sample, thus making possible the quantitative determination of ca. 1 μM of double bond when methanol having a specific activity of 1 μCi/mM is employed. Radioreagent and titrimetric methods for determining unsaturation in such glycols are compared in Table 7-12.

Table 7-12. Comparison of Radiochemical and Macrotitrimetric Techniques for Unsaturation in Commercial Poly(oxypropylene) Glycols

Nominal molecular weight of polyol	Unsaturation, mM/gram	
	Tritimetric	Radiochemical
400	0.0046	0.0048
	0.0047	0.0032
1000	0.0193	0.0192
	0.0191	0.0184
2000	0.0447	0.0448
	0.0448	0.0437
3000	0.0417	0.0400
	0.0419	0.0428

The technique has also been applied to elastomers prepared from propylene oxide and an unsaturated epoxide (79). The availability of the procedure obviates the need for the difficult separation of these polymers from antioxidants, which interfere significantly in the usual halogenation methods.

Adaptation of the method to liquid scintillation counting (80), using the counting vial as reaction container, greatly increases the speed and convenience of the analysis. After removal of ethylene dichloride and unconsumed methanol, the residue is dissolved for radioassay in 15 ml of a dioxane-naphthalene-methanol scintillator containing 5 ml/liter of glacial acetic acid. A toluene-based scintillator is required for analysis of higher cyclic oxide polymers. The methanol-^{14}C is assayed directly.

F. Conversion to Adducts of Maleic Anhydride

Although definite analytical procedures have been not established, the potential use of the ^{14}C-labeled dieneophile, maleic anhydride, as a radioreagent has been indicated (81) on the basis of its high reactivity toward cyclic and linear dienes. Cyclopentadiene, for example, has been converted to the maleic anhydride adduct in 97% yield (82).

REFERENCES

Absorption Spectrometric

1. R. W. B. Johnston, W. G. Appleby and M. O. Baker, *Anal. Chem.*, **20**, 805 (1948).
2. J. A. Anderson, Jr., and W. D. Seyfried, *Anal. Chem.*, **20**, 998 (1948).
3. E. L. Saier, A. Pozefsky, and N. D. Coggeshall, *Anal. Chem.*, **26**, 1258 (1954).
4. R. F. Goddu, *Anal. Chem.*, **29**, 1790 (1957).
5. R. T. O'Connor, D. C. Heinzelman, A. F. Freeman, and F. C. Pack, *Ind. Eng. Chem., Anal. Ed.*, **17**, 467 (1945).
6. D. R. Long and R. W. Neuzil, *Anal. Chem.*, **27**, 1110 (1955).
7. H. A. Benesi and J. H. Hildebrand, *J. Am. Chem. Soc.*, **71**, 2703 (1949).
8. S. H. Hastings, J. L. Franklin, J. C. Schiller, and F. A. Matsen, *J. Am. Chem. Soc.*, **75**, 2900 (1953).
9. P. B. Sweetser and C. E. Bricker, *Anal. Chem.*, **24**, 1107 (1952).
10. J. W. Miller and D. D. DeFord, *Anal. Chem.*, **29**, 475 (1957).
11. J. S. Fritz and G. E. Wood, *Anal. Chem.*, **40**, 134 (1968).

Gas Chromatographic

12. M. Beroza and R. Sarmiento, *Anal. Chem.*, **38**, 1042 (1966).
13. P. Issenberg, A. Kobayashi, and T. J. Mysliwy, *J. Agr. Food Chem.*, **17**, 1377 (1969).
14. J. Guillot, H. Bottazzi, A. Guyot, and Y. Trambouze, *J. Gas Chromatog.*, **6**, 605 (1968).
15. A. B. Littlewood and W. A. Wiseman, *J. Gas Chromatog.*, **5**, 334 (1967).
16. I. Klesment, *J. Chromatog.*, **31**, 28 (1967).
17. O. S. Privett, *Progress in Chemistry of Fats and Other Lipids*, Vol. IX: *Polyunsaturated Acids*, R. T. Holman, ed., Part I, Pergamon, New York, 1966, pp. 91–117.
18. M. Beroza and B. A. Bierl, *Anal. Chem.*, **38**, 1976 (1966).
19. M. Beroza and B. A. Bierl, *Mikrochim. Acta*, **1969**, 720.
20. V. L. Davison and H. J. Dutton, *Anal. Chem.*, **38**, 1302 (1966).
21. E. C. Nickell and O. S. Privett, *Lipids*, **1**, 166 (1966).
22. M. Beroza and B. A. Bierl, *Anal. Chem.*, **39**, 1131 (1967).
23. L. T. Black and R. E. Beal, *J. Am. Oil Chemists' Soc.*, **46**, 312 (1969).
24. E. von Rudloff, *Can. J. Chem.*, **34**, 1413 (1956).
25. D. F. Kuemmel, *Anal. Chem.*, **36**, 426 (1964).
26. A. D. Tulloch and B. M. Craig, *J. Am. Oil Chemists' Soc.*, **41**, 322 (1964).
27. E. P. Jones and V. L. Davison, *J. Am. Oil Chemists' Soc.*, **42**, 122 (1965).
28. D. T. Downing and R. S. Greene, *Lipids*, **3**, 96 (1968).

29. J. A. McCloskey and M. J. McClelland, *J. Am. Chem. Soc.*, **87**, 5090 (1965).
30. W. G. Niehaus, Jr., and R. Ryhage, *Tetrahedron Letters*, **49**, 5021 (1967).
31. C. J. Argoudelis and E. G. Perkins, *Lipids*, **3**, 379 (1968).
32. P. W. O'Keefe, G. H. Wellington, L. R. Mattick, and J. R. Stouffer, *J. Food Sci.*, **33**, 188 (1968).
33. W. B. Innes, W. E. Bambrick, and A. J. Andreatch, *Anal. Chem.*, **35**, 1198 (1963).
34. T. P. Maher, *J. Gas Chromatog.*, **4**, 355 (1966).
35. D. L. Klosterman and J. E. Sigsby, Jr., *Environ. Sci. Technol.*, **1**, 309 (1967).
36. R. L. Martin, *Anal. Chem.*, **32**, 336 (1960).
37. A. B. Littlewood, *J. Gas Chromatog.*, **1**(11), 34 (1963).
38. R. L. Hoffman, G. R. List, and C. D. Evans, *J. Am. Oil Chemists' Soc.*, **43**, 675 (1966).
39. W. B. Innes and W. E. Bambrick, *J. Gas Chromatog.*, **2**, 309 (1964).
40. R. J. Gordon, H. Mayrsohn, and R. M. Ingels, *Environ. Sci. Technol.*, **2**, 1117 (1968).
41. D. J. McEwen, *Anal. Chem.*, **38**, 1047 (1966).
42. K. V. Alekseeva and L. S. Solomatina, *Zavodsk. Lab.*, **33**, 1376 (1967).
43. E. Gil-Av and Y. Herzberg-Minzly, *J. Chromatog.*, **13**, 1 (1964).

Electroanalytical

44. S. Siggia, *Quantitative Organic Analysis via Functional Groups*, 3rd Ed., John Wiley, New York, 1963, pp. 297–371.
45. H. D. Dubois and D. A. Skoog, *Anal. Chem.*, **20**, 624 (1948).
46. J. C. S. Wood, *Anal. Chem.*, **30**, 372 (1958).
47. E. H. Unger, *Anal. Chem.*, **30**, 375 (1958).

Nuclear Magnetic Resonance

48. F. C. Stehling and K. W. Bartz, *Anal. Chem.*, **38**, 1467 (1966).
49. L. F. Johnson and J. N. Shoolery, *Anal. Chem.*, **34**, 1136 (1962).
50. P. W. Flanagan and H. F. Smith, *Anal. Chem.*, **37**, 1699 (1965).
51. W. Mueller and P. E. Butler, *J. Am. Chem. Soc.*, **90**, 2075 (1968).
52. U. E. Diner and J. W. Lown, *Chem. Commun.*, 333 (1970).

Radiochemical

53. S. Siggia, *Quantitative Organic Analysis via Functional Groups*, 3rd Ed., John Wiley, New York, 1963, pp. 316–318.
54. H. P. Kaufmann and J. Budwig, *Fette und Seifen*, **53**, 253 (1951).
55. H. P. Kaufmann, *Fette, Seifen, Anstrichmittel*, **58**, 492 (1956).
56. M. Jaky and K. Kaffka, *Fette, Seifen, Anstrichmittel*, **62**, 682 (1960).
57. A. Z. Budzynski, Z. J. Zubrzycki, and I. G. Campbell, *Nature*, **182**, 178 (1958).
58. A. Z. Budzynski, Z. J. Zubrzycki, and I. G. Campbell, *Proc. 2nd Intern. Conf. Peaceful Uses At. Energy*, Geneva, 1958, Vol. 24, p. 274.

59. J. Budzynska and A. Z. Budzynski, *Roczniki Panstwowego Zakladu Hig.*, **13**, 39 (1962); *Chem. Abstr.*, **57**:2637c.

60. M. F. Abdel-Wahab and S. A. El-Kinawi, *Z. Anal. Chem.*, **184**, 40 (1961).

61. M. F. Abdel-Wahab and S. A. El-Kinawi, *Z. Anal. Chem.*, **180**, 420 (1961).

62. M. F. Abdel-Wahab and S. A. El-Kinawi, *Z. Anal. Chem.*, **186**, 364 (1962).

63. M. F. Abdel-Wahab and N. F. Selim, *Z. Anal. Chem.*, **211**, 195 (1965).

64. M. F. Abdel-Wahab and R. H. Bishara, *Z. Anal. Chem.*, **219**, 183 (1966).

65. B. A. Rhodes, *Anal. Chem.*, **37**, 995 (1965).

66. I. C. McNeill, *J. Chem. Soc.*, **1961**, 639.

67. I. C. McNeill, *Polymer*, **4**, 15 (1963).

68. R. McGuchan and I. C. McNeill, *J. Polymer Sci.*, **A5**, 1425 (1967).

69. R. McGuchan and I. C. McNeill, *J. Polymer Sci.*, **A4**, 2051 (1966).

70. I. C. McNeill and S. I. Haider, *European Polymer J.*, **3**, 551 (1967).

71. I. C. McNeill and T. M. Makhdumi, *European Polymer J.*, **3**, 637 (1967).

72. O. C. Gadeken, R. L. Ayres, and E. P. Rack, *Anal. Chem.*, **42**, 1105 (1970).

73. D. R. Campbell, *Microchem. J.*, **13**, 630 (1968).

74. O. K. Neville, *J. Am. Chem. Soc.*, **70**, 3501 (1948).

75. V. F. Raaen and G. A. Ropp, *Anal. Chem.*, **25**, 174 (1953).

76. W. A. Bonner and C. J. Collins, *J. Am. Chem. Soc.*, **75**, 3994 (1953).

77. C. J. Collins and G. A. Ropp, *J. Am. Chem. Soc.*, **77**, 4160 (1955).

78. R. H. Wiley and B. Davis, *J. Polymer Sci.*, **A1**, 2819 (1963).

79. D. R. Campbell, *J. Appl. Polymer Sci.*, **14**, 847 (1970).

80. D. R. Campbell, Unpublished work.

81. V. F. Raaen, G. A. Ropp, and H. P. Raaen, *Carbon-14*, McGraw-Hill, New York, 1968, p. 18.

82. J. A. Berson and W. M. Jones, *J. Am. Chem. Soc.*, **78**, 6045 (1956).

8 Active hydrogen

I. ABSORPTION SPECTROPHOTOMETRIC METHODS*

Measurement after Chemical Reaction

Harp and Eiffert (1) presented a generalized infrared method for the determination of active hydrogen, involving exchange of the active hydrogen in the sample with deuterium of D_2O. The use of a large amount of D_2O shifts the equilibrium toward the products. The amount of active hydrogen is then calculated from the intensity of the 2.97-μ OH band generated in the D_2O. Chemically reactive hydrogens, such as those bonded to oxygen, nitrogen, sulfur, or phosphorus, rapidly equilibrate with the deuterium of D_2O to establish a statistical distribution of the active hydrogen and the deuterium between the exchange compound and D_2O.

The sample containing active hydrogen is dissolved in or placed in contact with D_2O, and the mixture is shaken thoroughly for periods up to $\frac{1}{2}$ hour. The ratio of deuterium to active hydrogen should be greater than 30 to 1, so that the exchange can be considered essentially complete. All the exchanged hydrogens, regardless of origin, in the equilibrium mixture become bonded to oxygen atoms, and the OH bonds so formed show the typical hydrogen-bonded OH stretching vibration absorptions at 2.79 μ. Thus, after suitable calibration, the active hydrogen content of the unknown can be determined from this OH absorption of the D_2O phase. In general a determination is accurate to $\pm 2\%$ of the amount of the active hydrogen present, while repeatability is somewhat better. As little as 0.005 wt $\%$ active hydrogen can be determined with no modification of the method.

METHOD FROM HARP AND EIFFERT (1)

Apparatus. Any commercial infrared spectrophotometer with fused quartz cells 1 mm thick and with a light path of 0.025 to 0.50 mm.

* Written by J. Gordon Hanna.

218

Reagent. D_2O of 99.8% purity or better. Because D_2O is extremely hygro-scopic, prepare and transfer samples in a dry box. Prepare the samples in volumetric flasks stoppered with rubber serum stoppers and make sample transfers with hypodermic syringes. Equilibrate all sample-handling equip-ment with D_2O. When the sample size permits, rinse several volumes of the sample through the absorption cell before making absorbance measure-ments.

Instrument Calibration.   Use water, a convenient source of active hydrogen, to prepare calibration samples covering the concentration range of 0 to 25 grams of water per liter for cells of 0.025- or 0.050-mm thickness, respectively. Weigh a known amount of water into a 5-ml volumetric flask, stopper the flask with a rubber serum stopper, and then in a dry box add D_2O by means of a hypodermic syringe. Remove the flask from the dry box and reweigh to determine the exact amount of D_2O added. Express the concentration of the calibration solution as moles of active hydrogen per liter, using as the densities of water and D_2O at 25°C values of 0.9971 and 1.023, respectively.

Determine the absorbance equal to log T_0/T for the calibration solutions, where T_0 equals the transmittance at 2.30 μ and T equals the transmittance at 2.97 μ, both relative to air.

The transmittance at 2.30 μ was selected as the base-line reference point because D_2O has maximum transmittance at this wavelength, as do most samples to be analyzed. Correct the sample absorbance for the blank absorb-ance of the cell filled with D_2O. Construct the usual calibration curve of corrected absorbance versus concentration for use in the analysis of un-known samples.

Procedure for Samples Soluble in D_2O.   Weigh a portion of the sample into a 5-ml volumetric flask and fill to volume with D_2O. Provide good mixing to achieve equilibrium. Determine the absorbance of the sample-D_2O solution at 2.97 μ as described for the calibration and correct the absorbance for the cell blank and residual D_2O background. Use this corrected absorbance to find the moles of OH groups per liter of solution from the calibration curve of corrected absorbance versus concentration.

Calculate the average number of active hydrogens per molecule as follows:

$$\text{Average number of hydrogens/molecule} = \frac{\text{moles of OH/liter of solution}}{\text{moles of sample/liter of solution}}$$

Procedure for Samples Insoluble in D_2O.   For liquids weigh a portion of the sample into a 5-ml flask and fill to volume with D_2O as for the soluble sam-ples. Shake long enough to ensure complete exchange. If the sample is very viscous, dilute it with a dry nonreactive solvent (e.g., hexane, benzene, amyl acetate, or carbon tetrachloride). If an emulsion forms, use gentle shaking or centrifugation to obtain a clear D_2O phase. After the exchange is complete and the two phases have separated, measure the absorbance at 2.97 μ and make corrections as described previously. Calculate the active hydrogen as described but take the volume of the D_2O phase as the volume of the sample being analyzed.

Results obtained by Harp and Eiffert are shown in Table 8-1. These analyses were performed on a Perkin-Elmer Model 21 instrument, using 0.025-mm cells for the insoluble samples.

Table 8-1. Active Hydrogen Content Determinations of Known Samples (1)

	Active hydrogens per molecule	
Compound	Theoretical	Found
Water-soluble		
Urea	4.0	4.12, 3.95
2-Amino-2-hydroxymethyl-1,3-propanediol	5.0	4.80
Potassium acid phthalate	1.0	0.95
Water insoluble		
Aniline	2.0	2.01, 1.98
Cetyl alcohol	1.0	0.99
Fluorene	0	0
4-Methyl-2-pentene	0	0

II. GAS CHROMATOGRAPHIC METHODS*

MODIFICATION OF METHOD OF NORTON, TURNER, AND SALMON (2)

A solution of lithium aluminum hydride in tetrahydrofuran is injected through a septum into a closed reaction vessel containing the sample. The hydrogen evolved is swept into the gas chromatograph by the nitrogen carrier gas and is determined quantitatively by a thermal conductivity detector.

---

* Written by Morton Beroza and May N. Inscoe.

Apparatus. A gas chromatograph equipped with a thermal conductivity detector and a copper column, 900 mm × 4 mm i.d. packed with 60/80 mesh molecular sieve 5A was used with nitrogen as the carrier gas. The carrier gas was dried with molecular sieves, and its flow rate was 60 ml/minute. The temperatures of the column and the detector were ambient and 100°C, respectively.

The construction and details of the reaction vessel are apparent from Figure 8-1, and its position in the system is shown schematically in Figure 8-2. Two reaction vessels are desirable; one can be cleaned (with water, dil. HCl, water, and acetone in this order) and dried (105°C) while the other is in use.

The syringes used are 100 μl, 3 ml, and 10 ml in capacity, the last two having wide-bore needles.

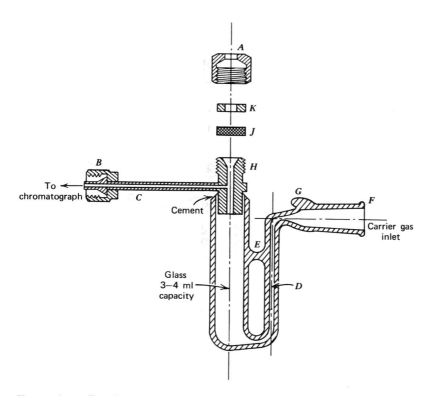

Figure 8–1. Details of the reaction vessel. A = Injection port; B = Stainless-steel Swagelock coupling; C = ⅛-inch-o.d. stainless steel side arm brazed on; D = Capillary; E = Brace; F = B.7 joint; G = Glass hook for retaining spring; H = ⅛-inch Wade stainless steel double-ended union, drilled to ⅛-inch bore; J = Septum; K = Spacer washer.

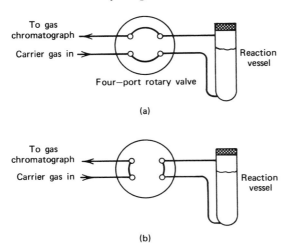

Figure 8–2. Details of the flow system: (a) sampling position, and (b) isolated position.

The sample vials are of about 15-ml capacity and are fitted with plastic caps. Rubber septums that fit over the mouths of these vials are needed.

Reagents. Tetrahydrofuran is technical grade and is dried over molecular sieves for at least 1 week.

Lithium aluminum hydride (2 grams) is refluxed gently in 100 ml of dried tetrahydrofuran for about 1 hour under a water-cooled condenser fitted with a drying tube. Then the condenser is replaced with the drying tube and allowed to cool. The clear solution above the gray sediment (undissolved reagent) is transferred to a clean dry 50-ml flask with a 10-ml syringe and stoppered immediately.

Molecular sieves 5A (60/80 mesh) are activated by heating at 250°C for 2 hours.

Benzoic acid, the standard, is analytical-reagent grade.

Procedure. Weigh the sample accurately into a sample vial and stopper with a plastic cap. Transfer by pipet 5 ml of dried tetrahydrofuran into the vial and shake until the sample dissolves. (For resins or other materials containing water, add 10 to 15 particles of activated molecular sieves and allow vials to stand for $\frac{1}{2}$ hour in desiccator to remove traces of water.) Then connect the clean dry reaction vessel into the GC system and inject 2 ml of lithium aluminum hydride solution into the reaction vessel. Turn the sampling valve to bubble carrier gas through the reagent, thus flushing hydrogen into the gas chromatograph. Turn the valve to admit carrier gas directly into the gas chromatograph. Repeat the valve movements until no hydrogen is evolved. Quickly replace the plastic cap with a rubber septum. Fill a 100-μl syringe with sample solution via the rubber septum and inject 50 μl into the reaction vessel through its septum, taking care to pass the contents directly into the hydride solution without touching the walls of the vessel. The evolution of hydrogen is instantaneous. After 1 minute, switch the valve to the sampling position to sweep the evolved hydrogen into the gas chromatograph. Measure the area of the hydrogen peak.

Additional determinations may be carried out on the same sample, on the benzoic acid standard, and on a blank (50 μl of tetrahydrofuran). Three or four determinations can be carried out before the lithium aluminum hydride reagent becomes gelatinous, at which point quantitative recovery is no longer obtained and fresh reagent is prepared. After about 30 injections raise the GC column temperature to 150°C for an hour (preferably overnight) to flush out the tetrahydrofuran that accumulates in the column.

Calculation. Calculate R, the milliliters of hydrogen per unit area for the benzoic acid standard, as follows:

$$R = \frac{11200 \times w \times v}{122 \times (a - b) \times t} \tag{1}$$

where w is the weight in grams of the benzoic acid taken, a is the number of area units of the hydrogen peak from the benzoic acid, b is the number of area units of the hydrogen peak from the blank, v is the volume (μl) of benzoic acid solution injected, and t is the total volume (μl) of benzoic acid solution (5000 μl).

For hydroxy compounds calculate the hydroxyl content as follows:

$$\text{OH content as OH/gram} = \frac{R \times (a_1 - b) \times t_1}{w_1 \times v_1 \times 11200} \tag{2}$$

where R is the value calculated in equation 1, and the values with the subscript (a_1, t_1, w_1, v_1) are identical to those described for benzoic acid except that they apply to the unknown. Multiplying the OH content by the molecular weight of the unknown will give the OH content per mole, and dividing this value by 17 will give the number of OH groups per molecule.

Similar calculations are used for active hydrogen in other types of compound.

Discussion. The method described, although used by Norton et al. to determine the OH content of epoxy resins, has been modified here to determine active hydrogen in general. The procedure is a modification of an earlier one (3, 4), which was said to be accurate within 0.03% absolute.

The same general procedure has been used also to determine hydridic hydrogen in borane compounds (5) by measuring the hydrogen liberated when dilute HCl is added to these compounds, as well as to determine Si—H in silicone monomers and intermediates by measuring the hydrogen liberated when the silicon compound is treated with KOH (6).

III. ELECTROANALYTICAL METHODS*

Conductometric Titration for Active Hydrogen

The method described in Chapter 1, Section III.A, for the titration of hydroxyl groups can be applied directly to the determination of active hydrogen.

IV. RADIOCHEMICAL METHODS†

Radioisotope methods for determining active hydrogen are particularly valuable because of the excellent sensitivity which they offer. Such highly sensitive techniques permit analysis of micro amounts of organic compounds and determination of end groups in certain functionally terminated polymers. The procedures available make use either of the isotope exchange of active hydrogen in the sample with that in hydroxyl-tritiated alcohols and tritiated water, or of the reaction of active hydrogen with tritiated lithium aluminum hydride.

A. Conversion to the Tritiated Form via Isotope Exchange

Dissolution of an organic compound containing active hydrogen in an aliphatic alcohol is accompanied by an exchange of the active hydrogens of the two substances, which attain equilibrium virtually instantaneously (7, 8). Similarly, addition of tritiated water to an aliphatic alcohol results in immediate equilibrium distribution of the isotope between the water and the hydroxyl group of the alcohol; no exchange, however, involving hydrogen attached to carbon in methanol, ethanol, and *tert*-butanol has been detected (9). Removal of the water by a suitable drying agent, or the use initially of only a trace amount of water having high specific activity, yields the analytical reagent. After the sample has dissolved in the alcohol, the latter is

---

* Written by Alan F. Krivis.
† Written by D. Campbell.

quantitatively removed as rapidly as possible. The molar specific activity of the substance being analyzed is then determined and compared with that of the monofunctional substrate. A large excess of alcohol is commonly used to avoid significant dilution of its active hydrogen by that of the sample.

Although isopropanol-^{3}H was employed initially in a macro method (10), the technique using ethanol-^{3}H given here requires only 5 to 30 mg and may be performed in 30 minutes exclusive of counting time (11).

METHOD OF J. A. GILES (11)*

Reagents. Hydroxyl-tritiated ethyl alcohol. Dilute tritiated water (New England Nuclear Corp., 1.8 Ci/mole) with sufficient absolute ethyl alcohol to obtain an activity of about 0.6 μCi/mM of alcohol. Standardize the tritiated ethyl alcohol by adding 1.00 ml to a mixture of 4.00 ml of methanol and 5.00 ml of ethyl alcohol, and diluting to 50.0 ml with the toluene scintillation reagent. Count the radioactivity in four 10-ml aliquots of three such dilutions and calculate the activity of the tritiated ethyl alcohol.

Alcoholic scintillation reagent. Dissolve 3.0 grams of 2,5-diphenyloxazole (PPO) and 0.1 gram of 1,4-bis-2-(5-phenyloxazolyl)benzene (POPOP) in 1 liter of methanol-ethyl alcohol-toluene (10:15:75).

Toluene scintillation reagent. Dissolve 3.0 grams of PPO and 0.1 gram of POPOP in 1 liter of toluene.

Apparatus. Liquid scintillation spectrometer. Packard Tri-Carb, Model 314X (or equivalent).

Counting vials. Wheaton Glass Co., 20-ml screw cap vials made of low-potassium glass.

Distillation column. Use a 15 × 1 cm, vacuum-jacketed, straight-through column with no dead spaces and a total take-off head. The column must facilitate rapid distillation and minimize transfer or fractionation of the hydrogen isotopes.

Procedure. Accurately weigh 0.1 meq of the sample into a 25-ml, 19/38 T, round-bottomed flask and dissolve in 2.00 ml of standard tritiated ethyl alcohol. Add 10 ml of dry toluene. Connect the vacuum-jacketed, microdistillation column and immerse the flask in a wax bath preheated to 170°C. Collect 4 to 5 ml of distillate in 5 minutes; then remove the distillation column, allowing 15 to 30 seconds to vaporize any residual tritiated ethyl alcohol out of the T joint. Quickly cool the distillation residue, dilute to approximately 10 ml with toluene, transfer this solution to a 50-ml volumetric flask, and dilute to volume with the alcoholic scintillation reagent. Determine the activity of the sample by counting the radioactivity in four 10-ml aliquots of the dilution with the liquid scintillation spectrometer. Correct for quenching, if present, by the internal standard method:

$$\text{Number of active H atoms} = \frac{\text{molar specific activity of compound}}{\text{molar specific activity of ethanol}} \quad (1)$$

Since the ratio of the milliequivalents of alcohol to that of sample is well over 300, exchangeable sample hydrogen does not significantly dilute the hydrogen of the alcohol. Azeotropic distillation is a quick and convenient means for removing the ethanol, which must be separated rapidly to avoid fractionation of the hydrogen isotopes and loss of tritium to the glass of the distilling apparatus. Use of a scintillator solution containing alcohols minimizes such losses during radioassay.

Values for the number of active hydrogens in twenty-six chemically diverse compounds analyzed are given in Table 8-2. The greatest deviations of experimental values for the number of active hydrogens from the theoretical values in the compounds analyzed are observed for benzoic and nicotinic acids. The method is not applicable to substances which are volatile under the conditions used to remove the ethanol. Although nitro groups do not interfere with the exchange reaction, they have a severe quenching effect in liquid scintillation counting.

In conjunction with the application of isopropanol-^{3}H to the determination of active hydrogen (10) the use of a distribution coefficient, α (12), was proposed as a measure of the uniformity of distribution of tritium between sample and alcohol. The value of α for exchange involving a monofunctional substrate is determined by dividing the apparent number of active hydrogens in the compound analyzed, obtained using equation 1, by the true number of active hydrogens. Procedures giving values for α close to unity for diverse known compounds would then be directly applicable to a wide variety of unknowns.

In the case of isopropanol-^{3}H, values for α closest to unity were obtained by means of several successive exchanges. Initially, 200 mg of sample was refluxed for 20 minutes with 25 ml of isopropanol-^{3}H, the solvent was removed at reduced pressure, and the sample was dried. The recovered sample was then heated with another 10 ml of labeled isopropanol, the latter was removed, and the process was repeated twice. A group of twenty-three diverse compounds gave values for the distribution coefficient close enough to unity to make the method reliable with unknowns. The greatest deviations of α from unity were observed with aromatic carboxylic acids; however, treatment of these acids with tritiated water in the presence of the sodium salt of the acid gave results for α very close to 1. Although the time-consuming

Table 8-2. Determination of Active Hydrogen Using Tritiated Ethanol

Compound	Equivalent weight Theoretical	Equivalent weight Experimental	Number of active H
Tetradecanol	214	206 ± 7.3	1.04
Hexadecanol	242	233 ± 4.2	1.04
Octadecanol	270	271 ± 7.7	1.00
Farnesol	222	237 ± 8.7	0.94
Phytol	297	305 ± 3.0	0.97
Solanesol	631	630 ±16.9	1.00
Cholesterol	387	391 ± 2.3	0.97
β-Sitosterol	415	408 ± 3.3	1.03
Stigmasterol	413	416 ± 2.0	0.99
Hydroquinone[a]	55	60 ± 0.8	1.80
Phenyl salicylate[b]	214	216 ± 2.0	0.99
Solanachromene[b]	749	741 ±12.5	1.01
Thymol[a]	150	169 ± 7.7	0.89
Palmitic acid	256	259 ± 0	0.99
Stearic acid	284	279 ±11.5	1.02
Benzoic acid[a]	122	155 ± 4.7	0.79
Nicotinic acid	123	148 ± 6.0	0.85
p-Hydroxybenzoic acid	69	73 ± 0	1.89
3-Indoleacetic acid	88	91 ± 3.5	1.93
Benzamide	61	66 ± 1.9	1.82
Cyclohexylbenzamide	203	197 ± 4.5	1.03
Nicotinamide	61	63 ± 1.0	1.94
Cyclohexanone oxime	113	113 ± 2.5	1.00
1-Menthylphenylurethane	275	258 ± 0.5	1.07
n-Butylphenylurethane	193	182 ± 7.5	1.06
Butyrophenone semicarbazone	68	70 ± 1.0	2.93

Reprinted from *Anal. Chem.*, **32,** 1717 (November 1960). Copyright 1960, by the American Chemical Society. Reprinted by permission of the copyright owner.

[a] Low results are believed to be due to slight volatility of sample and have been shown to be not due to quenching.
[b] Corrected for quenching.

reduction method (13) was used for radioassays, liquid scintillation counting is more advantageous if the method is not to be applied to nitro compounds.

The tritium exchange technique is especially valuable for determining very low concentrations of active hydrogen present as terminal hydroxyl and carboxyl groups in poly(ethylene terephthalate), which is an important and difficultly soluble condensation polymer. Tritiated water has been used exclusively in the two procedures developed. In the initial method (14), the film or fiber, after removal of any finish present, was left for several days at room temperature in a large excess of ^3H_2O having a known specific activity. The specimen was isolated by freeze drying and was then heated to 80°C to remove traces of tritiated water. No study was made, however, of the effect of drying conditions on the specific activity of treated polymers. After re-exchange of tritium by immersion of the specimen in a definite quantity of water, an aliquot of the latter was taken for assay by liquid scintillation counting.

Alternatively, tritium may be removed by immersing the specimen after exchange directly in a methanol-toluene scintillator solution in a counting vial and following the increase in activity until it reaches a maximum value (15). A constant activity corresponds to essentially complete re-exchange of tritium. In the more recent investigation, a drying study of films 0.2 mm thick revealed that water was effectively removed by drying for 4 hours at 80°C and 0.1 torr. The specific activities of substantially thinner films (0.012 to 0.050 mm) were significantly lower than those predicted from the known functional group content. This effect is attributable, at least in part, to re-exchange of sample tritium with moisture in the atmosphere.

Since the concentration of active hydrogen in the end groups of the polymer is extremely low, no correction is necessary for the dilution of the hydrogen of the water by the exchangeable hydrogen of the specimen. If the total concentration of the terminal OH and COOH groups is known, the number average molecular weight of the polymer may be calculated on the assumption that there are two and only two of these groups in each molecule.

A similar technique has been used to determine active hydrogen in soluble specimens of poly(phenylene oxide), poly(propylene oxide), and poly-(ethylene imine) and its derivatives (16). The polymer is dissolved in dry dioxane to which tritiated water has been added, and after a suitable exchange period the solvent and water are removed by rapid freeze drying. Radioassays of the finely divided polymer recovered and of the substrate are made by liquid scintillation counting.

In the exchange methods which employ tritiated water, the number of equivalents of active hydrogen per gram of polymer, E, is obtained from

$$E = \frac{S_p}{S_r} \qquad (2)$$

where S_p = specific activity of the polymer (μCi/gram) and S_r = specific activity of the water (μCi/equivalent).

When active hydrogen is determined by isotope exchange, careful protection of the treated samples is necessary before radioassay to prevent loss of tritium by re-exchange with atmospheric moisture. The solvents used in the assay of the polymers and water by liquid scintillation counting should also contain as much methanol or ethanol as is compatible with polymer solubility in order to prevent loss of tritium to the walls of containers.

B. Conversion to Elemental Hydrogen Using Tritiated LiAlH$_4$

Substances which contain active hydrogen decompose lithium aluminum hydride with the liberation of hydrogen. The reagent is reasonably soluble in various ethers, and solutions 0.3 to 0.7 M in diethyl ether were first used in a manometric method for active hydrogen (17). If the reagent is labeled with ^{3}H, active hydrogen may be determined by quantitatively collecting and assaying the gaseous tritium produced. The method is standardized by analyzing in the same way known amounts of the compound to be determined, or of one chemically similar to it. Since the reaction yields elemental tritium, the decomposition is performed most conveniently by mixing the solutions of reagent and sample in a specially constructed apparatus, or other closed system, which permits use of a stream of counting gas for transfer of the labeled hydrogen to an internal-flow proportional counter.

In the original method (18, 19), a 15-ml aliquot of 0.2 M LiAl^{3}H$_4$ (specific activity 0.5 to 2.0 μCi/mM) in highly purified diethylene glycol diethyl ether was placed in a reaction cell having a usable volume of ca. 25 ml. The cell was fitted with an entrance tube for flow gas which extended well below the surface of the liquid; an exit tube at the top led to the counter. Proportional-counting gas was passed through the solution at the rate of 40 ml/minute, and background was measured. The sample, which was dissolved in tetrahydrofuran (THF) distilled from LiAlH$_4$, was then introduced from a side-arm buret, and the radioactivity passing through the proportional counter was assayed for 10 minutes. A blank analysis was made on the THF in the same manner. Fifty or more samples could be analyzed with a single filling of the cell, although refilling was recommended after half of the reagent initially present was consumed.

Four representative compounds containing active hydrogen, namely, n-butanol, benzoic acid, aniline, and water, were analyzed. Each compound was analyzed with only a single lot of reagent, and each lot of reagent possessed a different specific activity. Graphs prepared by plotting total net counts, as ordinate, against amount of active hydrogen taken, as abscissa, were linear for the four compounds examined. The fact that the extrapolation of all plots gave an ordinate intercept below the origin was taken as an indication of fixed losses of hydrogen due to incomplete sample transfer, of volatilization of the compound before reaction, or possibly of nonquantitative removal of gas from the cell within 10 minutes. Since the purpose of this initial investigation was evaluation of the linearity of response and demonstration of the analytical utility of the procedure, no analyses were performed in which a single lot of radioreagent was used for determining different classes of compounds. The stoichiometry of the reaction had previously been studied extensively in the manometric technique (17).

Subsequent use of the procedure revealed, however, that it is not an absolute method for active hydrogen, wherein a calibration plot prepared with a suitable known compound may be applied to any unknown (20). In this later study, substances containing different forms of active hydrogen were analyzed with the same lot of labeled reagent. Analyses of benzoic acid and n-butanol gave results which were in agreement, whereas the analysis of aniline yielded values which differed significantly from those observed with the alcohol. In addition, a soluble sample of poly(ethylene adipate), prepared to contain essentially terminal hydroxyl groups, gave a linear plot but liberated less tritium than did an equivalent amount of n-butanol, Also, with the modified experimental arrangement used (and described below), fresh reagent was required for aniline in order to obtain reasonable agreement between the amounts of tritium liberated from a given quantity of the amine. No such effect of residual reaction products was observed in the analysis of n-butanol, which could be determined with a reagent solution used previously for both aniline and benzoic acid.

The production of elemental tritium thus depends on the nature of the sample and may be affected significantly by the quantity and the nature of products remaining from previous analyses. These effects apparently arise from complex and variable exchange reactions which may involve lithium aluminum hydride, elemental hydrogen, and hydrogen of by-products of the reaction. Before this method is applied to an unknown, it should be calibrated with a substance as chemically similar as possible to the sample of interest. The analysis of polymeric materials may present special difficulties, as shown by the results obtained with n-butanol and poly(ethylene adipate).

The experimental apparatus and technique also have important effects on reproducibility, and these effects may also depend on the nature of the compound analyzed. In the more recent work (20), the reaction was conducted in a three-neck 100-ml, round-bottom flask, one neck of which was fitted with a rubber septum for introducing the sample by syringe. The LiAl^{3}H$_4$ reagent was 0.4 M in diethylene glycol diethyl ether, and samples were prepared in the same solvent, rather than in THF. The counting gas, which was dried over calcium sulfate, was passed into the flask above the surface of the liquid. A magnetically driven, Teflon-covered stirring bar served to obtain the vigorous agitation desirable for quantitative liberation of hydrogen. A 20-minute counting time was employed. Although an extensive comparison was not made, results observed for aniline by both the original and the modified procedures suggested that the reproducibility of the original procedure was better. For n-butanol, the procedures appeared to be equivalent.

The method using $LiAl^3H_4$ has certain advantages in comparison with the techniques employing isotope exchange for the determination of active hydrogen both in low-molecular-weight compounds and at very low concentrations. It is applicable to liquids as well as to soluble solids, provided that the former are not appreciably volatilized during the time allowed for their decomposition by the reagent. In addition, the use of a closed system for both reaction and radioassay favors the detection of trace amounts of active hydrogen, whereas the sensitivity of the exchange methods is adversely affected by incomplete removal of the tritiated alcohol and, perhaps to a greater extent, by re-exchange of tritium in the isolated sample with atmospheric moisture. The main disadvantage of the hydride method lies in the fact that it is not absolute. This limitation severely restricts its applicability to polymeric materials. Although polymers analyzed by other techniques may be employed as standards, often only semiquantitative or relative values are possible. A minor drawback to the hydride method is the interference from nitro compounds.

REFERENCES

Absorption Spectrophotometric

1. W. R. Harp, Jr., and R. C. Eiffert, *Anal. Chem.*, **32**, 794 (1960).

Gas Chromatographic

2. E. J. Norton, L. Turner, and D. G. Salmon, *Analyst*, **95**, 80 (1970).
3. M. N. Chumachenko and L. B. Tverdyukova, *Dokl. Akad. Nauk SSSR*, **142**, 612 (1962).
4. M. N. Chumachenko, L. B. Tverdyukova, and F. G. Leenson, *Zh. Analit. Khim.*, **21**, 617 (1966).
5. I. Lysyj and R. C. Greenough, *Anal. Chem.*, **35**, 1657 (1963).
6. J. Franc and F. Mikes, *Collection Czech. Chem. Commun.*, **31**, 363 (1966).

Radiochemical

7. J. Hine and C. H. Thomas, *J. Am. Chem. Soc.*, **76**, 612 (1954).

8. K. B. Wiberg, *Chem. Rev.*, **55**, 713 (1955).

9. M. J. Pro, W. L. Martin, and A. D. Etienne, *U.S. At. Energy Comm.*, TID-13828.

10. J. F. Eastham and V. F. Raaen, *Anal. Chem.*, **31**, 555 (1959).

11. J. A. Giles, *Anal. Chem.*, **32**, 1716 (1960).

12. V. Gold and D. P. N. Stachell, *Quart. Rev. (London)*, **9**, 51 (1955).

13. K. E. Wilzbach, L. Kaplan, and W. G. Brown, *Science*, **118**, 522 (1953).

14. C.-Y. Cha, *J. Polymer Sci.*, **B2**, 1069 (1964).

15. M. Mozisek and L. Klimanek, *Plaste Kautschuk*, **15**, 99 (1968).

16. C. C. Price, Private communication, 1962.

17. J. A. Krynitsky, J. E. Johnson, and A. W. Carhart, *J. Am. Chem. Soc.*, **70**, 486 (1948).

18. D. J. Chleck, F. J. Brousaides, W. Sullivan, and C. A. Ziegler, *U.S. At. Energy Comm.* AECU-4493 (1959).

19. D. J. Chleck, F. J. Brousaides, W. Sullivan, and C. A. Ziegler, *Intern. J. Appl. Radiation Isotopes*, **7**, 182 (1960).

20. W. Seaman and D. Stewart, Jr., *Intern. J. Appl. Radiation Isotopes*, **15**, 565 (1964).

9 Acetylenic hydrogen

I. ABSORPTION SPECTROPHOTOMETRIC METHODS*

Measurement after Chemical Reaction

Reaction with Cuprous Chloride. Cuprous chloride and compounds containing acetylenic hydrogen react to form cuprous acetylide:

$$Cu_2Cl_2 + 2HC{\equiv}CR \rightarrow 2CuC{\equiv}CR + 2HCl$$

This reaction has been used for many years for the colorimetric determination of small amounts of acetylene (1–4). The cuprous reagents are usually prepared by the hydroxylamine reduction of an ammoniacal cupric solution. Colloids are normally added to retard flocculation of the cuprous acetylide precipitate. Difficulties encountered have been reagent instability, precipitation of copper metal by overreduction, and flocculation of the cuprous acetylide.

Hobart, Bjork, and Katz (5) substituted an acetic acid-acetate-buffered medium for the usual ammoniacal solution to control the reduction potential of the hydroxylamine used to reduce cupric to cuprous in the reagent preparation. The addition of gelatin and potassium stabilizes the suspensoid and thus prevents flocculation of the cuprous acetylide formed for as much as 1000 μg of acetylene in 100 ml of solution. The colored colloid is formed within 15 minutes and is stable for at least 3 days if protected from free access to atmospheric oxygen.

METHOD FROM HOBART, BJORK, AND KATZ (5)

Apparatus. Beckman Model DU spectrophotometer with 1-cm cells.

Reagent. Potassium chloride-gelatin. Add 0.40 gram of gelatin to 30 grams of potassium chloride in 300 ml of water. Heat to dissolve, cool to room temperature, and dilute to 500 ml.

---

* Written by J. Gordon Hanna.

Procedure. Add the following reagents, in the order given, to a 100-ml volumetric flask: 25 ml of 0.8 M potassium chloride-0.08% gelatin, 10 ml of 0.1 M cupric chloride solution, 10 ml of 1 M ammonium acetate, and 10 ml of 0.3 M hydroxylamine hydrochloride solution.

Allow about 5 minutes for the reduction of the copper; the solution will be a pale blue, not quite colorless. Failure to allow sufficient time for reduction of the copper will retard color development in the procedure.

Meanwhile pass the gas to be analyzed through cold acetone traps. Remove the traps, transfer their acetone contents to a 100-ml volumetric flask, and stopper. Warm the stoppered flask to room temperature; then dilute to volume with acetone. Remove a 20-ml aliquot and add to the reagent solution previously prepared. Allow about 15 minutes for complete color development. Dilute to volume with water and determine the absorbance at 552 or 575 mμ. Compare with a standard curve prepared by determining the absorbance developed by 20-ml aliquots of various dilutions of acetylene stock solutions.

It was noted that the absorbance is sensitive to the order of addition of the reagent mixtures. Therefore the acetylene-acetone addition to the mixture must be made in a single 20-ml volume.

Absorbances at the 552-mμ peak and at the 450 to 500-mμ plateau are directly proportional to the acetylene concentrations from 0.2 to at least 10 μg/ml. At 552 mμ the molar absorbance was found to be dependent on the potassium chloride-gelatin and copper chloride concentrations, while ammonium acetate or hydroxylamine variations of $\pm 20\%$ were without significant effect. At 450 to 500 mμ the absorptivity, although somewhat lower, was found to be independent of $\pm$ 20% variations of all the reagents.

The analytical recoveries obtained by Hobart, Bjork, and Katz are shown in Table 9-1.

II. NUCLEAR MAGNETIC RESONANCE METHODS*

The unconjugated acetylene protons exhibit resonances from 2.45 to 2.65 ppm from TMS, and the conjugated protons from 2.8 to 3.1 ppm. This region is in the range between the alkane and alkene proton resonances; unfortunately it is also the position for methylene and methine protons. This is a limitation of the method. If the acetylene molecule contains a series of methylenes and methines, there is a problem of overlapping. The coupling constant and solvent sensitivity, however, can be utilized for identification. The reduction of the acetylene to the alkene to saturation can be followed by nmr, and the type of acetylene determined.

* Written by Harry Agahigian.

Table 9-1.   Recovery of Known Amounts of Acetylene from Argon

Acetylene, μg		Acetylene recovered, %
Added	Found	
90	85	95.6
172	168	97.7
262	250	95.1
262	254	96.7
326	313	96.0
326	317	97.2
446	431	96.9
446	442	99.2
	Average recovery:	96.8
	Relative standard deviation:	1.4

A variety of acetylenes have been studied in this way and chemical shift data reported. The acetylenic proton of phenylacetylene appears at 2.93 ppm, and that of propargyl alcohol is at 2.33 ppm (6). The effect of halogen substitution is quite small, as shown by the following values:

$HC\equiv CCH_2Cl$:   2.24 ppm   chemical shift   relative to TMS
$HC\equiv CCH_2Br$:   2.24 ppm   chemical shift
$HC\equiv CCH_2I$:   2.19 ppm   chemical shift

The conjugated acetylenic proton appears at 2.71 ppm, about 0.2 ppm to higher field than the phenyl derivative.

A series of monosubstituted acetylenes was studied by nmr, and the chemical shifts and coupling constants were reported. These studies were accomplished on the neat samples. Some of the acetylenes studied were 3-chloropropyne-1, 3-bromopropyne-1, propargyl alcohol, 3-dimethylaminopropyne-1, 3-diethylaminopropyne-1, 3-hydroxybutyne-1, and 3-methyl-3-n-butylamino-propyne-1. The chemical shift range (7) of the acetylenic protons is quite small, thus constituting a possible limitation of the method. Other spectroscopic techniques must be used in conjunction with nmr if complex acetylenes are to be determined. Systems such as polyacetylenes have also been studied (8).

A characteristic of acetylenes is their solvent-dependent chemical shifts. If the acetylenic hydrogen resonance is measured as dilution is made with an inert solvent, a diamagnetic shift of 0.4 ppm is observed. Depending on the solvent used and the nature of the association with the acetylenic proton, differences as great as 1 ppm are observed (9).

REFERENCES

Absorption Spectrophotometric

1. L. Ilsovay, *Berichte*, **32**, 2697 (1899).
2. E. R. Weaver, *J. Am. Chem. Soc.*, **38**, 352 (1916).
3. T. Geissman, S. Kaufman, and D. Dollman, *Anal. Chem.*, **19**, 919 (1947).
4. H. Pieters, *Chem. Weekblad*, **43**, 456 (1947).
5. E. W. Hobart, R. G. Bjork, and R. Katz, *Anal. Chem.*, **39**, 224 (1967).

Nuclear Magnetic Resonance

6. E. B. Whipple, J. H. Goldstein, L. Mandell, G. S. Reddy, and G. R. McClure, *J. Am. Chem. Soc.*, **81**, 1321 (1959).
7. J. V. Hatton and R. E. Richards, *Trans. Faraday Soc.*, **56**, 315 (1960).
8. E. I. Synder and J. D. Roberts, *J. Am. Chem. Soc.*, **84**, 1582 (1962).
9. R. E. Richards and J. V. Hatton, *Trans. Faraday Soc.*, **57**, 28 (1961).

10 Acetals, ketals, and vinyl ethers

I. ABSORPTION SPECTROPHOTOMETRIC METHODS*

Measurement after Chemical Reaction

Under acidic conditions, acetals, ketals, and vinyl ethers hydrolyze to produce the parent carbonyl compounds:

$$\underset{\substack{\text{Ketal or} \\ \text{acetal}}}{RCH \begin{smallmatrix} OR' \\ \\ OR'' \end{smallmatrix}} + H_2O \xrightarrow{\ H^+\ } RCHO + R'OH + R''OH \qquad (1)$$

$$\underset{\text{Vinyl ether}}{ROCH = CH_2} + H_2O \xrightarrow{\ H^+\ } CH_3CHO + ROH \qquad (2)$$

Spectrometric procedures based on acidic reagents such as 2,4-dinitrophenylhydrazine, Chapter 2, Section I.C.1, can be applied directly. The acidity of the reagent solution will rapidly convert these compounds to the parent carbonyl compound, which will in turn react with the reagent. Nonacidic reagents can be used also if the acetal, ketal, or vinyl ether is first hydrolyzed in the presence of dilute mineral acid to produce the carbonyl compound.

Bowman, Beroza, and Acree (1) hydrolyzed compounds that give acetaldehyde with sulfuric acid, steam-distilled the acetaldehyde produced, and treated it with a solution of p-phenylphenol and cupric sulfate in concentrated sulfuric acid to give a violet color absorbing at 572 mμ. They claimed an accuracy within 3% in the microgram range and reproducible results.

* Written by J. Gordon Hanna.

METHOD FROM BOWMAN, BEROZA, AND ACREE (1)

Apparatus.   Beckman Model DU spectrophotometer with square Corex cells having a 1-cm light path.

The reaction apparatus described by Giang and Schechter (2), consisting of an all-glass system with two vertical, parallel condensers. One permits refluxing and the other distillation in the same apparatus without transfer. The reflux condenser is at least 15 cm long and is attached to the reaction flask. The second condenser, 25 cm long, is attached to the reflux condenser by a suitable adapter, so that the condensate runs into a suitable collection vessel joined to the second condenser by means of a ground-glass joint.

Reagents.   Sulfuric acid-cupric sulfate reagent. Add 5 ml of 5% $CuSO_4 \cdot 5H_2O$ to 500 ml of sulfuric acid and mix well.

p-Phenylphenol solution. Dissolve 1 gram of crystalline material in 25 ml of hot 2 N sodium hydroxide, add 75 ml of water, and store in a brown bottle.

Preparation of Standard Curve.   Attach a test tube (15 × 150 mm) containing 5 ml of 2% sodium bisulfite solution (freshly prepared) to the reaction apparatus so that the lip of the delivery tube extends to the bottom of the solution. Partly immerse the test tube in an ice bath during the reflux and distillation processes. Chill in an ice bath a 50-ml distilling flask containing 10 ml of 10% aqueous sulfuric acid (v/v) and 1 to 2 mg of Carborundum powder. Add the appropriate volume of standard metaldehyde in chloroform to the flask and attach the flask to the apparatus. Reflux the mixture for 15 minutes. During the reflux pass water maintained at 5°C through both condensers. Then drain the water from the reflux condenser and distill the mixture until 3 to 4 ml remains. Disconnect the apparatus, wash both condensers and delivery tube with small portions of cold water, and combine the washes and the distillate. Separate and extract the chloroform with three 10-ml portions of cold water. Combine the washes and the distillate and dilute to 50 ml in a volumetric flask. Extract the solution with 5 ml of re-distilled hexane to remove traces of chloroform.

Mix 1 ml of the aqueous solution by swirling with 8 ml of cold sulfuric acid-cupric sulfate reagent in a 15 × 150-mm test tube partly immersed in an ice bath. Add 0.2 ml of the *p*-phenylphenol in a similar manner, remove the tube from the ice bath, and allow it to stand in the dark at room temperature for 1 hour. Heat the tube in a water bath at 100°C for 90 seconds; then return it to the dark for about 30 minutes to adjust to room temperature. With distilled water in the reference cell, determine the absorbance at 572 mμ. Prepare a standard curve by plotting concentration versus absorbance corrected for the blank of the reagents carried through the entire process.

The color reaction conformed to Beer's law, and 1 µg of acetaldehyde produced an absorbance of about 0.150.

Procedure.   Prepare chloroform solutions of the samples for analysis by diluting 200 µl of each to 50 ml in volumetric flasks. Calculate the weight of the samples in these solutions from the volumes and specific gravities of the samples. This procedure is used to minimize any error resulting from the high volatility of some of the compounds. Then dilute each solution with chloroform to contain 100 to 200 µg of combined acetaldehyde per milliliter. Analyze 1- or 2-ml aliquots of each solution by the method used for the preparation of the standard curve.

Recoveries obtained by Bowman, Beroza, and Acree are shown in Table 10-1.

Table 10-1.   Recoveries of Acetaldehyde from Compounds Containing Combined Acetaldehyde Groups (1)

Compound	Amount added µg	Acetal-dehyde equivalent, µg	Acetaldehyde recovered	
			Micro-grams	Per cent
Ether				
2-Butoxyethyl vinyl	343	105	106	101
	686	210	207	99
Butyl vinyl	312	137	136	99
	624	274	274	100
Isobutyl vinyl	306	134	138	103
	612	268	265	99
2-Ethylhexyl vinyl	644	181	181	100
	1288	363	364	100
Ethyl vinyl	300	183	181	99
	600	366	363	99
Methoxyethyl vinyl	358	154	151	98
	716	308	301	98
Acetaldehyde diisobutyl acetal	656	166	168	101
	1312	332	331	100
2-(2-Ethoxyethoxy)ethyl-3,4-	904	134	135	101
methylenedioxyphenyl acetal (sesamex)	1808	267	275	103

II. GAS CHROMATOGRAPHIC METHODS*

Plasmalogen aldehydes, liberated from tissue lipids by methanolysis, are determined by GC as their dimethyl acetals. Under given conditions, however, the dimethyl acetals are quantitatively converted to the corresponding alk-1-enyl methyl ethers, and this procedure is suggested for the analysis of aldehydes (3). The alk-1-enyl methyl ethers are composed of almost equimolar proportions of cis and trans isomers (4).

REFERENCES

Absorption Spectrophotometric

1. M. C. Bowman, M. Beroza, and F. Acree, Jr., *Anal. Chem.*, **33**, 1053 (1961).
2. P. A. Giang and M. S. Schechter, *J. Agr. Food Chem.*, **4**, 623 (1956).

Gas Chromatographic

3. V. Mahadevan, C. V. Viswanathan, and F. Phillips, *J. Lipid Res.*, **8**, 2 (1967).
4. C. V. Viswanathan, F. Phillips, and V. Mahadevan, *J. Chromatog.*, **30**, 405 (1967).

* Written by Morton Beroza and May N. Inscoe.

11 Amino, imino, and quaternary ammonium compounds

I. ABSORPTION SPECTROPHOTOMETRIC METHODS*

A. Direct Near-Infrared Measurement of Primary and Secondary Amines

Using the N=H overtone and combination bands near 1.49 μ and 1.97 μ, respectively, Whetsel, Roberson, and Krell (1) analyzed mixtures of aniline and N-ethylaniline containing up to 99% of either constituent. The standard deviations were no greater than ±1%. They recommended the procedure also for other mixtures of primary and secondary aromatic amines. Lohman and Norteman (2) who extended the technique to aliphatic amines, based the determination of primary amines on their characteristic combination band at 2.023 μ. Secondary amines were determined by their first overtone N—H stretching absorption band at 1.538 μ after correction for primary amine absorption at that wavelength. Amides, nitriles, alcohols, and esters up to a concentration of 10% in the amine mixture do not interfere. Tertiary amines do not exhibit any specific absorption bands in these wavelength regions.

B. Measurement after Chemical Reaction

1. Schiff's Base Formation for Primary Amines. A bright yellow Schiff's base is formed when a primary amine and salicylaldehyde react:

* Written by J. Gordon Hanna.

Mulin (3) used reaction 1 to determine primary amines on the basis of the absorption of Schiff's base at 410 mμ. Interference due to secondary and tertiary amines is avoided by the presence of acetic acid in the reaction mixture.

METHOD FROM MILUN (3)

Apparatus. Beckman Model DU spectrophotometer with the cell compartment thermostated at 30°C and with 1-cm cells.

Reagents. Salicylaldehyde solution. Dilute exactly 5 ml of pure salicylaldehyde to 100 ml in a volumetric flask with chloroform. Prepare fresh every fourth day.

Acetic acid solution. Dilute exactly 4 ml of glacial acetic acid to 200 ml in a volumetric flask with chloroform.

Procedure. Weigh a sample containing up to 1.6 mM of primary amine in a weighing bottle. For samples containing over 85% primary amine use 1.3 to 1.6 mM for optimum results. Add 14 ml of chloroform from a graduate to dissolve the sample. Pipet 6 ml of the acetic acid solution, washing down the sides of the bottle during the addition, and swirl gently to mix the solution. Add 5 ml of the salicylaldehyde solution from a pipet in the same manner as the acetic acid solution and swirl gently to mix. Place the bottle in a 30°C constant-temperature bath.

One hour and 20 minutes after the salicylaldehyde solution addition, quantitatively transfer the contents of the bottle to a 500-ml volumetric flask, rinsing the bottle thoroughly with chloroform. Add chloroform to the volumetric flask to within only approximately 1.5 inches of the mark to allow for subsequent thermal expansion of the solution. Place the flask in a 30°C constant-temperature bath for 10 minutes, make up to volume with chloroform, and mix. Rinse and then fill the spectrophotometer cell with the solution. Place the cell in the spectrophotometer and allow it to sit in the compartment for 5 minutes. One hour and 40 minutes after addition of the salicylaldehyde, measure the absorbance versus chloroform at 410 mμ.

The timing should be within several minutes of that specified. With samples containing appreciable color, use a comparable concentration of the sample in chloroform as the blank.

Calculate the percentage of primary amine according to the following equation:

$$a = \frac{A}{c \times l}$$

where A = absorbance, c = concentration (grams/liter), and l = length of cell (cm); then

$$\text{Per cent of primary amine} = \frac{a \times 100 \times \text{molecular weight}}{\epsilon}$$

Calibration.   Determine the absorbance of a sample of pure primary amine as described in the procedure. Calculate the molar absorptivity as follows:

$$\epsilon = \frac{A \times \text{molecular weight}}{c \times l}$$

Results obtained by Milun for the analysis of mixtures are given in Table 11-1.

Amides and nitriles do not interfere in this determination. The 99% confidence interval for the mean of 12 determinations carried out over a period of 6 days was ±0.6% relative.

2. *Acylation of Amines with Cinnamic Anhydride.*   Hong and Connors (4) treated aliphatic amines with cinnamic anhydride, extracted the amide produced with chloroform, and measured the ultraviolet absorption of the solution:

$$(C_6H_5CH{=}CHCO)_2O + RR'NH \rightarrow$$

$$C_6H_5CH{=}CHCONRR' + C_6H_5CH{=}CHCOOH \quad (2)$$

The powerful light absorption exhibited by the *trans*-cinnamoyl group (ϵ_{max}, about 2.2×10^4 for most compounds containing this function) makes the method quite sensitive.

METHOD FROM HONG AND CONNORS (4)

Reagents.   *trans*-Cinnamic anhydride solution, 5 mg/ml in acetonitrile. First recrystallize the cinnamic anhydride three times from benzene.

Procedure.   Transfer 1.0 ml of the sample solution containing 0.7 to 4.2 μM of amine (dissolved in acetonitrile) to a 50-ml volumetric flask. Add 1.0 ml

Table 11-1.   Analysis of Mixtures of Amines (3)

Sample no.[a]	Composition of mixture, %			Determined, % primary
	Primary	Secondary	Tertiary	
1	0	100	0	0.3
2	0	0	100	0.1
3	89.9	10.1	0	90.0
4	60.1	39.9	0	60.3
5	28.5	71.5	0	28.4
6	83.0	2.8	0	83.0
7	88.5	5.0	0	88.6
8	95.8	0.6	0	93.9
9	98.0[b]			97.3
10	98.3[b]			97.8
11	. . .			85.5
12	12.0			11.5

[a] Samples 1 through 5 are synthetic mixtures of octadecylamine, dioctyldecylamine, and trioctyldecylamine. Samples 6 through 8 are technical-grade undistilled and distilled tallow amines previously analyzed by titration. Samples 9 and 10 are technical-grade distilled tallow primary amine acetates made from amines essentially free from secondary amine. Sample 11 is a technical-grade undistilled tallow primary amine acetate. Sample 12 is a crude N-octadecylacetamide.
[b] Per cent primary amine acetate determined by titration.

of the cinnamic anhydride solution and 1 drop of tri-n-butylamine (0.1 M, in acetonitrile). Mix the solution well and allow to stand at room temperature until the acylation is complete. Add 2 ml of 0.1 N aqueous sodium hydroxide, mix well, and then add 0.1 N sodium hydroxide approximately to volume. Allow to stand for 10 minutes to hydrolyze the excess reagent. Quantitatively transfer the solution with the aid of 15 ml of chloroform to a 125-ml separatory funnel (Teflon stopcock). Extract with three portions of chloroform (15, 15, and 10 ml), the first two portions having been used in the transfer. Combine the chloroform extracts in a separatory funnel and wash with 20 ml of water. Filter the extract through filter paper into a 50-ml flask. Rinse the filter paper with chloroform, adding the rinsings to the flask, and dilute to volume with chloroform. Measure the absorbance of this solution against chloroform in a 1-cm cell at the wavelength of maximum absorption for the amide. Prepare the blank solution by omitting only the amine; carry it through the entire procedure and obtain the absorbance.

Calculate the micromoles of amine contained in the 1-ml initial sample as follows:

$$\text{Micromoles of amine} = \frac{5 \times 10^4 (A_s - A_b)}{b \epsilon_{max}}$$

where A_s and A_b are the absorbances of the sample and blank final solutions, respectively, b is the path length, and ϵ_{max} is the molar absorptivity of the amide. Beer's law is followed by all the cinnamides in chloroform solution. Table 11-2 gives the necessary spectral data.

Table 11-2. Spectral Data for Amine Determination (4)

Cinnamide	λ_{max}, mμ	$\epsilon_{max} \times 10^{-4}$
n-Propyl	274	2.27
n-Butyl	274	2.22
tert-Butyl	273	2.18
n-Hexyl	274	2.27
n-Octyl	274	2.21
n-Dodecyl	274	2.27
Allyl	274	2.27
Benzyl	275	2.44
β-Phenethyl	274	2.31
Piperidyl	277.5	2.18
Diethyl	277	2.14
Dibenzyl	283	2.36
Phenyl	295	2.58
Morpholyl	280	2.17

The standard deviation is approximately 0.5 to 1.0% relative.

A reaction time of about 2 minutes should suffice for complete acylation unless the amine is sterically hindered.

The effect of water and methanol on the analysis was investigated. Even when the amine sample solution contains as much as 60% water, no interference is observed. Evidently the amine is so much more powerful as a nucleophile than is water that its rate of acylation permits the amide to be quantitatively formed before the reagent is hydrolyzed. Alcohols might be expected to interfere in two ways: they consume reagent, and they produce cinnamic esters which are extractable into chloroform. No interference is observed, however, if the hydrolysis time in the alkali is increased to permit preferential hydrolysis of the ester. This is possible because of the much greater susceptibility of esters, as compared with amides, to alkaline hydrolysis.

3. 2,4-Dinitrofluorobenzene Reaction—Primary and Secondary Amines.

This method consists of the reaction of the amine and 2,4-dinitrofluorobenzene in buffered solution to form the substituted 2,4-dinitrophenylamine:

$$RNHR' + \quad F \underset{}{\overset{NO_2}{\bigcirc}} NO_2 + NaOH \longrightarrow R-N \overset{R \quad NO_2}{\underset{}{\bigcirc}} NO_2$$

$$+ \quad NaF \quad + \quad H_2O \qquad (3)$$

After the reaction McIntire, Clements, and Sproull (5) converted the excess reagent to 2,4-dinitrophenol, separated the 2,4-dinitrophenylamine from the sodium salt of the 2,4-dinitrophenol, and measured the 2,4-dinitrophenylamine on the basis of its absorption. Kolbezan, Eckert, and Bretschneider (6) claimed that an inherent error in the method of McIntire et al. arises from the background absorption of 2,4-dinitrophenetole formed by the reaction of the dinitrofluorobenzene and ethyl alcohol, the reagent solvent specified. To overcome these difficulties, Kolbezan and his co-workers recommended that no dinitrofluorobenzene-reactive solvents such as alcohols be used in the analysis.

METHOD FROM MCINTIRE, CLEMENTS, AND SPROULL (5), AS MODIFIED BY KOLBEZAN, ECKERT, AND BRETSCHNEIDER (6)

Reagents. 2,4-Dinitrofluorobenzene solution. Dissolve 0.12 ml in 10 ml of absolute alchol:

Sodium bicarbonate solution, 0.1 M.

Sodium hydroxide solution, 0.2 N in 60% dioxane.

Procedure. Pipet carefully to the bottom of a 25-ml glass-stoppered cylinder 0.1 ml of aqueous amine solution containing 10 to 100 γ of amine, 0.05 ml of 2,4-dinitrofluorobenzene solution, and 0.1 ml of bicarbonate solution. Mix thoroughly and place the cylinder in a water bath for 20 minutes. Add 0.4 ml of 0.2 N sodium hydroxide in dioxane and continue heating for 60 minutes. Dilute to 10 ml with distilled water and extract with 10 ml of cyclohexane. (In the case of ethanolamine and other highly water-soluble amines, extract with tetrachloroethane.) Separate the liquid layers and read the absorbance of the organic solvent at the wavelength of maximum absorbance.

Kolbezan, Eckert, and Bretschneider (6) claimed that the absorbance of the reagent blanks (cyclohexane reference) was zero in the region 270 to 460 mμ. Spectral curves of the cyclohexane extracts of the reaction mixtures were identical to those of the pure crystalline derivatives dissolved in cyclohexane, with maxima near 330 mμ for primary amines and near 355 mμ for secondary amines. Recoveries of *n*-butylamine, isobutylamine, and 2-aminobutane were quantitative. Recoveries for isopropylamine and *tert*-butylamine were 85% and 60%, respectively, even when the reaction was carried out in sealed tubes with excess 2,4-dinitrofluorobenzene reagent. However, straight-line plots of absorbance versus concentration were obtained using the standard procedure.

4. Diazotization and Coupling—Primary Aromatic Amines.

Bandelin and Kemp (7) determined primary aromatic amines on the basis of diazotization and coupling with *N*-(1-naphthyl)ethylenediamine to yield a colored azo dye. The color produced in this reaction develops rapidly and reaches its full intensity in from 1 to 2 minutes.

METHOD FROM BANDELIN AND KEMP (7), AS ADAPTED BY SIGGIA (8)

Reagents. *N*-(1-Naphthyl)ethylenediamine, 0.1%.
Sodium nitrite, 0.1%.

Procedure. Dissolve or extract the sample containing approximately 10 μg of amine with 5 ml of 4 *N* sulfuric acid. Add 1 ml of 0.1% sodium nitrite and allow the solution to stand for 3 minutes to permit the diazonium salt development. Add 5 ml of 95% ethyl alcohol and allow the mixture to stand for 2 minutes. Add 1 ml of 0.1% *N*-(1-naphthyl)ethylenediamine. In some cases it is necessary to make the solution alkaline with sodium bicarbonate or sodium acetate or yet more strongly alkaline with sodium carbonate or even sodium hydroxide to cause coupling to take place. Care must be taken, however, to limit the alkalinity of the solution as much as possible because high alkalinity favors decomposition of diazonium compounds. Measure the absorbance of the solution and compare with standard solutions to obtain the original concentration of the amine.

5. 9-Chloroacridine Reaction—Primary Aromatic Amines.

Stewart, Shaw, and Ray (9) described a spectrophotometric method for the determination of small quantities of some primary aromatic amines on the basis of their reaction with 9-chloroacridine to give highly colored 9-aminoacridine hydrochlorides.

METHOD FROM STEWART, SHAW, AND RAY (9)

Apparatus. Perkin-Elmer Model 202 spectrophotometer or Beckman Model DU spectrophotometer with matched cells of 1-cm optical paths.

Procedure. Place 1 ml of an ethanolic solution of 9-chloroacridine (10^{-6} mole/ml) in a 10-ml volumetric flask. Add to this 1 ml of an ethanolic solution of amine (10^{-6} mole/ml). If the amine is present as a hydrochloride salt, add 1 ml of an aqueous solution of potassium bicarbonate (10^{-5} mole/ml) to free the amine. Adjust the pH to approximately 4 with 10% (v/v) aqueous hydrochloric acid. Shake the solution for 5 minutes at room tempeature. Add ethanol to volume and measure the absorbance at 435 mμ. Correct the absorbance measurements for reagent blanks in the procedure.

The product is highly colored orange. Reagent blank readings are very low. The absorption curves for equimolar concentrations of the various aromatic amines were almost identical. Standard curves can be prepared by plotting observed absorbance readings versus the volumes taken of equimolar concentrations of various amines. In all cases Beer's law holds for this system.

Quantitative data obtained by Stewart, Shaw, and Ray and shown in Table 11-3 reveal that the use of the procedure permits the determination of primary aromatic amines in the presence of primary, secondary, and/or tertiary aliphatic amines, secondary and tertiary aromatic amines, heterocycles, and carbonyl-containing compounds. Generally, secondary aromatic amines will not give colored solutions, but a few were found to yield highly colored products. However, these solutions exhibit maximum absorption at wavelengths other than 435 mμ and consequently do not interfere in the determination of primary aromatic amines.

The color stability was found to be good up to 1 week. Solutions of 9-chloroacridine must be prepared immediately before use because the acridine undergoes rapid ethanolysis in ethanol. Decomposition can be detected by the appearance of a brownish precipitate.

6. Copper-Salicylaldehyde Method—Primary Aliphatic Amines.
Critchfield and Johnson (10) presented a scheme involving the reaction of primary amines with an aqueous reagent containing cupric chloride, salicylaldehyde, and triethanolamine. The copper-salicylaldehydimine formed in the reaction is extracted into 1-hexanol. The amount of copper in the hexanol layer, which is determined colorimetrically by reaction with bis(2-hydroxyethyl)dithiocarbamic acid, is a measure of the primary amine in the sample.

METHOD FROM CRITCHFIELD AND JOHNSON (10)

Reagents. Triethanolamine. Distill 98% material under 1- to 2-mm pressure, using a column 6 inches long and 30 mm in diameter packed with 2-mm glass beads and heated by means of resistance wire. Use a 3-liter round-bottom distillation flask fitted with a thermometer well. Stir the contents of

Table 11-3.   Analysis of Known Primary Aromatic Amine Mistures for Primary Aromatic Amine Contents (9)

Components, concentration of mixture, 2.500×10^{-8} mole/ml	Primary aromatic amine	
	Found, mole/ml $\times 10^{-8}$	Per cent of theoretical theory

1. Aniline
   n-Butylamine
   N-Methylaniline
   Morpholine
	2.495	99.8

2. p-Methoxyaniline
   Di-n-butylamine
   Salicylic acid
   p-Dimethylaminobenzaldehyde
	2.500	100.0

3. p-Nitroaniline
   Quinoline
   Triethylamine
   N,N-Dimethylaniline
	2.500	100.0

the flask by means of a magnetic stirrer, and do not allow the kettle temperature to exceed 185°C during the distillation. An absorbance of 0.65 ± 0.02 for 0.372 mg of ethanolamine should be obtained by the procedure described below, when this material is used to prepare the copper-salicylaldehyde reagent.

Copper-salicylaldehyde reagent. Into a 100-ml glass-stoppered graduated cylinder measure of 15.0 ml of redistilled triethanolamine, 0.5 ml of salicylaldehyde, and 0.25 gram of cupric chloride dihydrate. Dilute to 100 ml with distilled water and mix the contents. This reagent is stable for at least a month; however, the reagent blank increases with age.

Bis(2-hydroxyethyl)dithiocarbamic acid reagent. Prepare a 2% by volume solution of carbon disulfide in methanol and a 5% by volume solution of diethanolamine in methanol. Prepare the reagent fresh daily by mixing equal volumes of the two components.

Calibration Curve.   Prepare a dilution of the pure compound in distilled water so that a 5-ml aliquot contains not more than the maximum sample size given in Table 11-4. To each of five 25-ml glass-stoppered graduated cylinders add 2 ml of the copper-salicylaldehyde reagent from a pipet. Transfer a

Table 11-4. Reaction Conditions for Determination of Primary Amines by Copper-Salicylaldehyde Method (10)

Compound	Primary amine, mg (maximum)	Time, minutes[a]
Aminoethylethanolamine	1.10	30– 60
N-Aminoethylmorpholine	0.85	15– 60
Amylamine	1.20	15– 45
Butylamine	0.70	15– 60
Ethanolamine	0.50	15– 60
Ethylamine	0.53	15– 60
2-Ethylhexylamine	1.40[b]	15– 60
Hexylamine	1.10	15– 60
Isoamylamine	1.10	15– 45
Isobutanolamine	0.60	60–120
Isobutylamine	0.90	15– 60
Isopropanolamine	0.60	15– 60
Methylamine	0.30	15– 60
Propylamine	0.73	15– 60
Propylenediamine	0.42	10– 20[c]

[a] Reaction time at 20 to 30°C unless otherwise specified.
[b] Make dilutions using a 10% solution of methanol.
[c] Perform reaction at 98° ± 2°C. Use 50-ml glass-stoppered graduated cylinders; do not stopper during reaction.

1.0-, a 2.0-, a 3.0-, and a 5.0-ml aliquot of the sample dilution to each of four of the 25-ml graduated cylinders, reserving one as a blank. Measure the absorbance of each standard at 430 mμ, using 1-cm cells and a suitable spectrophotometer.

Procedure. Add 2.0 ml of copper-salicylaldehyde reagent from a transfer pipet to each of two 25-ml glass-stoppered graduated cyclinders. Reserve one of the cylinders as a blank, and into the other measure an amount of sample calculated to contain not more than the maximum amount of primary amine listed in Table 11-4. The sample must not contain more than 0.01 mg of ammonia or 0.5 gram of secondary and tertiary amine. For samples of less than 0.01 mg use an aliquot of a suitable aqueous dilution. Dilute the contents of each graduate to the 10-ml mark with distilled water, stopper, and mix thoroughly. Allow the samples to react under the conditions specified in Table 11-4.

After the reaction is complete, add sufficient 1-hexanol to bring the total volume of liquid to 25 ml. Stopper the cylinders, shake vigorously 15 to 20 times, and allow the layers to separate. Add 5 ml of bis(2-hydroxyethyl)-dithiocarbamic acid reagent to each of two additional 25-ml glass-stoppered graduated cylinders. In this step it is important that the graduated cylinders and stoppers be clean and void of any metallic ions that react with this reagent. Pipet 5.0 ml of the hexanol layer from the graduated cyclinders in which the reaction was performed to the graduates containing the dithio-carbamic acid reagent. Add the hexanol dropwise to prevent the material from clinging to the walls of the pipet. Dilute the contents of each cylinder to the 25-ml mark with methanol, stopper, and mix the contents. Measure the absorbance of the sample versus the blank at 430 mμ, using 1-cm cells. Read the concentration of primary amine from the calibration curve.

A few primary amines do not react quantitatively with this reagent. In general, compounds that react incompletely can be divided into three classes: (a) aromatic amines such as aniline; (b) compounds that contain more than one primary amine group, for example, ethylenediamine and diethylenetri-amine; and (c) primary amines that are branched in the 2 position, such as tertiary and secondary butylamine and isopropylamine. Propylenediamine and 2-ethylhexylamine are exceptions to this generalization.

Ammonia interferes if more than 0.01 mg is present in the sample aliquot. This amount of ammonia corresponds to an absorbance of 0.03, which is within the experimental error of the method. The combined secondary and tertiary amine content of the sample must not exceed 0.5 gram; greater amounts of these amines tend to solubilize the copper complex in the aqueous layer. Strong oxidizing or reducing agents interfere by depleting the reagent. Compounds that form copper complexes soluble in 1-hexanol give high results; those forming water-soluble complexes, low results.

In Tables 11-5, 11-6, and 11-7 are listed data obtained by Critchfield and Johnson for the analysis of several mixtures of primary, secondary and tertiary amines. An accuracy within 5% can be anticipated by this method for the determination of primary amines in the presence of secondary and tertiary amines. Primary amine contents of less than 0.01% in the presence of secondary and tertiary amines cannot be determined by this method be-cause of the limitations on the sample size.

7. Dithiocarbamate Reaction—Secondary Amines. Several in-vestigators (11–13) have used colorimetric measurement of the copper dithiocarbamate complex as a basis for the determination of dimethylamine in various systems. Umbreit (14) defined conditions for more general appli-cability of these colorimetric methods. Secondary amines and carbon disulfide

Table 11-5.   Determination of Primary Amines in Presence of Secondary and Tertiary Amines (10)

Sample	Primary amine, % by weight		
	Added	Found[a]	Difference
Isopropanolamine in 2,5-dimethylpyrazine	0.12	0.12 (2)	0.00
	0.57	0.57 (2)	0.00
Ethanolamine in diethanolamine	2.35	2.16 (1)	−0.19
	0.36	0.32 (2)	−0.05
	0.12	0.13 (2)	+0.01
	0.09	0.13 (2)	+0.07
	2.4	2.5 (1)	+0.1
Ethanolamine in triethanolamine	0.31	0.31 (1)	0.00
	0.10	0.10 (1)	0.00
	0.98	0.90 (1)	−0.08
Ethanolamine in di- and triethanolamine	47.6	47.6 (2)	0.0
	21.3	21.2 (2)	+0.1
	29.2	28.7 (4)	−0.5
Butylamine in dibutylamine	1.07	1.14 (1)	+0.07
	0.56	0.53 (1)	−0.03
	0.22	0.15 (1)	−0.07
	0.49	0.54 (1)	+0.05

[a] Figures in parentheses represent numbers of determinations.

Table 11-6.   Determination of Ethanolamine in Presence of Di- and Triethanolamines (10).

Sample no.	Ethanolamine, % by weight		
	Van Slyke method	Copper salicylaldehyde method	Difference
1	11.5	12.1	+0.6
2	12.9	13.0	+0.1
3	12.5	12.5	0.0
4	11.7	12.2	+0.5
5	13.1	12.5	−0.6
6	10.8	11.1	+0.3
7	13.1	12.1	−1.0
		Average difference:	±0.5

Table 11-7. Control Laboratory Determination of Ethanolamine in Presence of Di- and Triethanolamine (10)

Sample	Operator	Ethanolamine, % by weight		Difference
		Van Slyke method	Copper salicylaldehyde method	
1	1	42.2	41.8	−0.4
2	1	39.1	41.2	+2.1
2	2	39.5	40.1	+0.6
3	1	38.3	36.6	−1.7
3	2	42.2	42.4	+0.2
4	1	34.3	35.4	+1.1
4	2	36.4	37.2	+0.8
5	2	14.4	14.5	+0.1
			Average difference:	±0.9

react to form dithiocarbamic acids. The dithiocarbamic acids and copper(II) react to form a yellow-colored salt which can be measured spectrophotometrically. Tertiary amines do not react with carbon disulfide. Primary amines undergo these reactions but give much lower color intensities. Therefore, primary amines in equivalent molar quantities will result in an error of 1% or less in the determination of secondary amines.

METHOD FROM UMBREIT (14)

Reagents. Carbon disulfide-pyridine-isopropyl alcohol. Accurately measure and mix 35 ml of carbon disulfide, 25 ml of pyridine, and 65 ml of isopropyl alcohol. When stored in a glass-stoppered reagent bottle, this is usable for at least 2 months.

Cupric chloride solution, 0.0013 M. Dissolve 0.1 to 0.12 gram of $CuCl_2 \cdot 2H_2O$ in 250 ml of water and dilute to 500 ml with pyridine.

None of the volume or weight measurements given is excessively critical as long as standards are used for comparison when any reagent solution is replaced.

Procedure. Transfer 1 ml of the sample solution to a 15 × 150-mm glass-stoppered test tube or other suitable reaction vessel. Add 4 ml of the carbon disulfide-pyridine-isopropyl alcohol reagent and 2 ml of the cupric chloride reagent. Agitate and allow the mixture to stand for 5 to 20 minutes at room temperature. (Some of the carbon disulfide settles out during this period.) Then add 3.0 ml of acetic acid (10% by volume of glacial acetic acid in water) and 3.0 ml of benzene. Agitate by inversion several times and allow the phases to separate. Remove 4.0 ml from the upper (organic) phase and dilute with 5 ml of isopropyl alcohol. After this solution has stood for 1 to $1\frac{1}{2}$ hours (diethylamine) or 20 minutes (N-methylaniline), measure the maximum absorbance at approximately 440 mμ.

Data for diethylamine and N-methylaniline, which were chosen as representative of aliphatic- and aromatic-substituted secondary amines, are summarized in Table 11-8. The response in both cases is linear over the concentration ranges indicated.

Table 11-8.   Analysis of Secondary Amines (14)

Taken, μg	Found, μg	Recovery, %
	Diethylamine	
11.3	11.0	97.3
22.7	22.4	98.7
45.4	45.4	100.0
68.0	67.3	99.0
90.7	90.5	99.8
	N-Methylaniline	
36.2	35.3	97.6
72.4	72.2	99.8
108.6	108.6	100.0
	Average:	99.1 $\pm$ 1.0

Results of the application of this method to primary-secondary and tertiary-secondary amine mixtures are summarized in Table 11-9.

Compounds tested specifically and found not to react under the conditions given are pyridine, triethylamine, tributylamine, N,N-dimethylaniline, 1,3-diphenylguanidine, diphenylamine, 3-carbonyl indoles, and ammonia. The secondary amines which do not react are all highly conjugated.

Table 11-9.   Determination of Secondary Amines in Mixtures (14)

$C_6H_5NHCH_3$ taken, μg	$C_6H_5NHCH_3$ found, μg	Error, μg
N-Methylaniline in *N-N*,dimethylaniline[a]		
18.2	19.0	+0.8
36.4	35.4	−1.0
N-Methylaniline + aniline[b]		
18.2	19.7	+1.5
36.4	37.0	+0.6
54.6	55.3	+0.7

Et_2NH taken, μg	Et_2NH found, μg	Error, μg
Diethylamine in triethylamine[c]		
11.3	11.5	+0.2
22.7	22.9	+0.2
34.0	33.7	−0.3
Diethylamine + ethylamine[d]		
11.3	13.3	+2.0
22.7	23.3	+0.6
34.0	35.1	+1.1

[a] 4.779 mg of $C_6H_5N(CH_3)_2$ taken.
[b] 100.3 μg of $C_6H_5NH_2$ taken.
[c] 3.615 mg of Et_3N taken.
[d] 64.8 μg of $EtNH_2$ taken.

8. Aconitic Anhydride Reaction—Tertiary Amines and Quaternary Amines.   Palumbo (15) found that *cis*-aconitic anhydride could be used for detecting tertiary aliphatic amines in the presence of primary and secondary amines. This qualitative procedure was adapted by Cromwell (16) for the quantitative colorimetric determination of trimethylamine. Sass and his coworkers (17) showed that the reaction could be used more generally for either tertiary amines, amine salts, or quaternary amines with a sensitivity of the order of 3 γ/ml of solution.

The probable reaction of tertiary amine and aconitic anhydride in acetic anhydride is shown in equation 4:

$$\text{(4)}$$

Primary and secondary amines do not show this reaction, presumably because amide formation removes them.

METHOD FROM SASS ET AL. (17)

Reagents. Aconitic anhydride solution. Dissolve aconitic anhydride (0.25 gram) that has been recrystallized from hot toluene solution in 40 ml of acetic anhydride and dilute to 100 ml with toluene. Age the reagent 24 hours before use.

Acetic anhydride, redistilled.

Toluene. Remove pyrrole and thiophene by washing with hydrochloric acid, sulfuric acid, and water. Distill the washed material out of calcium chloride.

Procedure. Dissolve samples containing tertiary amine or amine salts in toluene so that 1-ml aliquots will represent from 20 to 70 γ of amine. Add 1 ml of aconitic anhydride reagent, heat for 15 seconds on a boiling water bath, and allow to stand for 15 minutes. Add 5 ml of toluene, allow to stand for 15 minutes, and measure the developed color on a Klett-Summerson colorimeter, using filter No. 54, or in a spectrophotometer at 500 mμ.

Prepare a calibration curve by making dilutions of amine in toluene such that 2-ml aliquots represent from 5 to 70 γ of amine. Straight lines are obtained when absorbance is plotted against concentration of amine.

Results obtained by Sass and his coworkers are shown in Tables 11-10 and 11-11.

9. Chloranil Reaction—Tertiary Amines. Sass and his coworkers (17) also presented a colorimetric method for tertiary amines alone, based on the reaction of these amines and chloranil:

$$\text{(5)}$$

Table 11-10. Analytical Recovery of Amine from Prepared Mixtures of Amine and Amine Hydrochloride in Toluene by the _cis_-Aconitic Anhydride Method (17).

Mixture	Total amine, γ	
	Added	Found
Trimethylamine and hydrochloride	1050	1045
	745	740
	450	447
	230	234
Triethylamine and hydrochloride	980	970
	375	373
	561	564
	185	183
Tributylamine and hydrochloride	1127	1125
	877	868
	251	248
	750	748
	627[a]	629
Triamylamine and hydrochloride	1260	1249
	656	659
	875	869
	305	303
N-Methyldiisopropylamine and hydrochloride	650	655
	570	574
	478	475
	255	254
2-Diethylaminoethanol and hydrochloride	1042	1035
	755	751
	648	643
	285	287

[a] Tributylamine hydrofluoride.

Table 11-11. Analytical Results for Mixtures of Quaternary and Tertiary Amines by the *cis*-Aconitic Anhydride Method (17).

Mixture	Total amine, γ	
	Added	Found
Tetramethylammonium chloride and trimethylamine	500	501
	1000	997
	500	502
	250	252
Tetraethylammonium chloride and triethylamine	700	705
	920	915
	350	353
	150	151
Acetylcholine and trimethylamine	495	482
	999	993
	1003	1010
	499	505
Trimethylphenylammonium chloride and trimethylamine	496	494
	998	994
	502	496
	500	495
Tetrabutylammonium iodide and tributylamine	480	476
	984	882
	495	500
Tetra-*n*-propylammonium iodide and tri-*n*-propylamine	476	479
	940	960
	520	490

When this method is used in conjunction with the aconitic anhydride method, tertiary amines can be differentiated quantitatively from amine salts or quaternary amines.

METHOD FROM SASS ET AL. (17)

Reagents.   Chloranil, 1% in toluene.

Toluene. Remove pyrrole and thiophene by washing with hydrochloric acid, sulfuric acid, and water. Distill the washed material from calcium chloride.

Procedure. Weigh and dilute in toluene proportionate quantities of sample so that 3-ml aliquots contain tertiary amine in the range 100 to 800 γ. Pipet 3 ml of sample into a Klett tube, add 1 ml of chloranil solution, and heat on a boiling-water bath for 15 minutes. Cool for 5 minutes and measure the green color in a colorimeter, using a Klett filter No. 69, or in a spectrophotometer at 610 mμ.

Construct a calibration curve by treating 3-ml aliquots containing tertiary amine standard in the range 100 to 800 γ as just described.

Both this method and the aconitic anhydride method showed poor sensitivity for tertiary aromatic amines and no measurable color with amides. The straight-chain tertiary aliphatic amines from trimethylamine through triamylamine, including 2-dimethylaminoethanol, and N-methyldiisopropylamine, showed the same order of sensitivity. Triethanolamine showed poor sensitivity. Triisopropanolamine produced no color with chloranil. The chloranil method, although not as sensitive as the aconitic anhydride method, is useful for tertiary amines in the presence of amine salts.

Analytical results obtained by Sass and his coworkers using this method are shown in Tables 11-12 and 11-13.

10. *Complexation with Tetracyanoethylene—Tertiary Aromatic Amines.* Few colorimetric or spectrophotometric methods are available for the selective determination of a given tertiary aromatic amine in the presence of primary and secondary aliphatic amines. Schenk, Warner, and Bazzelle (18) based a method for these amines on complexation with tetracyanoethylene, illustrated in equation 6 for N,N-dimethylaniline:

$$C_6H_5NMe_2 + (NC)_2C{=}C(CN)_2 \rightarrow HCN + (NC)_2C{=}C(CN)C_6H_4NMe_2$$

$$(6)$$

Although primary and secondary amines also react, these reactions can be drastically reduced by acetylation.

METHOD FROM SCHENK, WARNER, AND BAZZELLE (18)

Apparatus. Beckman DU spectrophotometer with 1.0-cm cells.
Bausch and Lomb Spectronic 20 with 1.17-cm cells.

Reagent. Tetracyanoethylene solution, saturated in methylene chloride. The solutions can be used for 1 to 2 weeks.

Procedure: Tertiary Aromatic Amines in the Presence of Primary and Secondary Amines. Weigh the sample accurately into a 25-ml or larger volumetric flask and dissolve in methylene chloride. Take from this a 1- to 20-ml aliquot

Table 11-12.   Analytical Results for Mixtures of Amine and Amine Hydrochloride in Toluene by the Chloranil Method (17)

Mixture	Total amine added, γ	Free amine, γ	
		Added	Found
Trimethylamine and hydrochloride	1050	410	402
	745	0	3
	450	450	455
	230	100	96
Triethylamine and hydrochloride	980	650	643
	375	0	2
	561	560	556
	185	0	0
Tributylamine and hydrochloride	1127	500	503
	877	250	253
	251	0	2
	750	750	755
	627[a]	0	4
Triamylamine and hydrochloride	1260	575	569
	656	0	0
	875	875	870
	305	305	302
N-Methyldiisopropylamine and hydrochloride	650	650	645
	570	0	0
	478	275	272
	255	255	257
2-Diethylaminoethanol	1042	485	481
	755	0	2
	648	648	651
	285	285	288

[a] Tributylamine hydrofluoride.

containing about 5×10^{-3} mM of tertiary amine (Table 11-14). Transfer the aliquot to a 50-ml volumetric flask. Add 2 ml of acetic anyhdride to the flask for every estimated 5×10^{-3} mM of primary or secondary amine. Shake the flask and allow to stand for 5 minutes (15 minutes of acetylation time for 2,5-dimethylpiperazine). Add 5 ml of saturated tetracyanoethylene solution, giving a two- to threefold excess and a final concentration of about 0.009

Table 11-13. Analytical Results for Mixtures of Tertiary Amines in the Presence of Quaternary Amines by the Chloranil Method (17)

Mixture	Total amine added, γ	Free amine, γ	
		Added	Found
Trimethylamine and tetramethylammonium chloride	500	500	502
	1000	400	403
	500	0	2
	250	250	252
Triethylamine and tetraethylammonium chloride	700	700	695
	920	650	655
	350	0	3
	150	0	1
Trimethylamine and acetylcholine	495	495	490
	999	999	995
	1003	500	498
	499	0	2
Trimethylamine and trimethylphenylammonium chloride	496	496	500
	998	500	490
	502	250	248
	500	0	3
Tributylamine and tetrabutylammonium iodide	480	480	478
	984	465	424
	495	0	10
Tri-*n*-propylamine and tetra-*n*-propylammonium iodide	476	476	500
	940	454	411
	520	0	15

to 0.01 M tetracyanoethylene. More tetracyanoethylene may be added in the case of weakly colored π complexes. Dilute the solution to 50 ml with methylene chloride and allow to stand for 10 to 25 minutes or until the optimum absorbance is reached. (In the case of pyridine color development is slower and is dependent on temperature. Therefore the reaction temperature should be controlled to within $\pm 0.5°C$ for the calibration curve. Readings after 35 minutes are more stable than those obtained after 10 minutes.)

Determine the absorbance at the appropriate wavelength (Table 11-14) and read the concentration from a standard curve prepared from known

Table 11-14.   Lower Limit of Detection of Tertiary Aromatic Amines (18)

0.01 M tetracyanoethylene, 0.4 M acetic anhydride in methylene chloride, $A = 0.10$

Tertiary amine	Detected, M (wavelength, mμ)	Molar absorptivity[a] of final solution
Triphenylamine	1.6×10^{-4} (342)	7×10^2
Pyridine	2×10^{-3} (400[b])	5×10^1
2,5-Dimethylpyrazine	5×10^{-4} (460)	2×10^2
N,N-Dimethylaniline	8×10^{-4} (640)	1.5×10^2
1,10-Phenanthroline	2.9×10^{-5} (310[c])	3.5×10^3
Carbazole	1×10^{-3} (600[d])	1×10^2
Benzidine	1×10^{-5} (440)	1×10^4

[a] These molar absorptivities are the slopes of the Beer's law plots.
[b] Measured after 75-minute reaction at 26°C.
[c] The molar absorptivity of 1,10-phenanthroline alone at 310 mμ is 850.
[d] Triethyl phosphate solvent.

concentrations of the tertiary amine containing acetic anhydride at the same concentrations used for the unknowns.

Procedure: Tertiary Aromatic Amines in the Presence of Phenols by Extraction. Weigh the sample accurately in a volumetric flask and dissolve in pentane or, less preferably, in petane-methylene chloride or methylene chloride alone. Take from this 1 to 5 ml containing no more than 2 ml of methylene chloride to avoid water pickup in the extraction. The aliquot should contain about 1×10^{-3} mM of tertiary aromatic amine and no more than 0.1 mM of phenol. Transfer the aliquot to a 50- to 100-ml separatory funnel, add 10 ml of 10% aqueous sodium hydroxide, allow to come into contact for 30 seconds, add 10 ml of dry pentane, and allow to be in contact for 1 minute. Withdraw the aqueous layer and rinse the sides of the funnel twice with 2-ml portions of pentane. Withdraw any aqueous layer collected. Empty the contents of the funnel into a 50-ml volumetric flask and rinse the funnel twice with 2-ml portions of pentane. The total volume of the solution at this point should be no more than 23 ml to permit addition of sufficient methylene chloride to keep the tetracyanoethylene in solution.

If no primary or secondary amine is present, add 10 ml of saturated tetra-cyanoethylene in methylene chloride. Read the absorbance at the appropriate wavelength (Table 11-14). Prepare a standard curve, using the same proportions of pentane and methylene chloride as in the extraction and determination procedure for the unknown.

Results obtained by Schenk, Warner, and Bazzelle with this procedure are shown in Tables 11-15 and 11-16.

Table 11-15. Determination of Tertiary Aromatic Amines in the Presence of Primary and Secondary Amines (18)

0.02 M tetracyano-ethylene + amine, M	Primary and secondary amine, M	Absorbance		Molar absorptivity of final solution
		Primary + secondary amine N	No primary or secondary amine N	
N,N-Dimethylamine				
1.8×10^{-3}	9.2×10^{-3a}	0.33	0.33	1.8×10^2
Pyridine				
2.5×10^{-3}	9×10^{-3a}	0.17	0.17	6.8×10
2,5-Dimethylpyrazine				
4.6×10^{-3}	9.2×10^{-3a}	0.32	0.32	7.0×10
N,N-Dimethylaniline				
2×10^{-3}	7.8×10^{-3a}	0.25	0.25	1.2×10^2
Triphenylamine				
4×10^{-4}	1.6×10^{-4b}	0.26	0.26	6.5×10^2
1,10-Phenanthroline				
6×10^{-5}	1.6×10^{-4b}	0.20	0.20	3.3×10^3

[a] Mixture contains 2,6-diethylaniline, N-methylaniline, and diphenylamine. Final molarity of acetic anhydride is 0.4 M.
[b] Mixture contains 2,5-dimethylpyrazine, 2,5-dimethylpiperazine, aniline, and 3-picoline. Final molarity of acetic anhydride is 1.2 M.

11. Sulfonated Indicator Complexes with Quaternary Ammonium Compounds. Certain sulfonated indicators (e.g., bromophenol blue, bromothymol blue, bromocresol purple) produce with quaternary

Table 11-16.   Determination of Tertiary Amines after Extraction from Phenols (18).

75% methylene chloride-25% pentane solvent, 0.02 M tetracyanoethylene

		Absorbance	
Tertiary amine	Phenol interference	Ex-tracted	No phenols
N,N-Dimethylaniline, 0.001 M	Mixture,[a] 0.001 M	0.578	0.584
Carbazole,[b] 0.005 M	2,6-Dimethylphenol, 0.002 M	0.260[c]	0.253

[a] 2,6-Dimethoxyphenol, p-methoxyphenol, and $tert$-amylphenol.
[b] Solvent was 75% $(EtO)_3PO$—25% pentane.
[c] The absorbance of the carbazole-2,6-dimethoxyphenol before extraction was 0.97.

ammonium colors not normally associated with the indicators at any pH. The probable reaction involving bromophenol blue as typical is as follows:

(7)

where R is the quaternary cation.

Auerbach (19) based on the bromophenol blue reaction a method for field assays and laboratory control of antiseptics and germicidal solutions containing quaternary ammonium salts. The method depends on the reaction of quaternary nitrogen, in weakly alkaline solution, with bromophenol blue to form a product soluble in ethylene chloride, which will dissolve neither component before the combination occurs.

Quaternary cations that form salts extractable into organic solvents from alkaline aqueous solutions are of the type $[R_3R'N]^+$. Here R is CH_3 or a longer-chain alkyl group, and R' is C_4H_9 or a group with a longer chain than this.

Reagents. Bromophenol blue solution. Dissolve 40 mg of bromophenol blue powder in 100 ml of water containing 1 ml of 0.01 N sodium hydroxide. Prepare fresh daily.

Apparatus. Klett-Summerson colorimeter with a No. 60 filter.

Procedure. Place a sample containing 50 to 75 γ of the quaternary ammonium compound in 50 ml of water in a 125-ml separatory funnel. Avoid the use of ordinary stopcock grease; glycerol starch lubricant is satisfactory. Add 5 ml of the sodium carbonate solution, 1 ml of the bromophenol blue solution, and exactly 10 ml of benzene. Shake steadily for 2.5 to 3 minutes, allow the layers to partially separate (20 to 30 seconds), and then swirl the contents of the funnel. Allow to stand for several minutes or until well separated. Rinse a 15-ml centrifuge tube with a portion of the lower aqueous layer, discard this layer entirely, and then add the colored benzene layer to the tube. Stopper the tube with a clean rubber diaphragm stopper and centrifuge for a few minutes at about 1000 rpm to clarify. Transfer a portion of the colored layer to a dry Klett tube and read, using a No. 60 filter.

It may be preferable in some cases to measure the absorbance of the solution in a suitable spectrophotometer at 603 mμ versus a blank treated in the same manner exept that the sample is omitted.

Metcalfe (21) isolated and concentrated long-chain quaternary ammonium compound cations from solutions and complex mixtures by the use of a cellulose ion exchanger. The quaternary ammonium cations were removed from the exchanger with alcoholic acid and measured spectrophotometrically as the bromophenol blue-quaternary ammonium complex in chloroform at 603 mμ.

Fogh, Rasmussen, and Skadhauge (22) preferred bromocresol purple as the complexing agent and presented a method for analyzing cetylpyridinium chloride in concentrations ranging from 0 to 25 γ/ml. They also reported that the adsorption of quaternary compounds on glass (especially if very scratched) is significant and can cause large errors in the analysis. To overcome this difficulty, they recommended that tubes, cuvettes, and pipets be treated with Plexiglas.

Reagents. Bromocresol purple solution. Dissolve 0.2 gram in a few milliliters of 1 N sodium hydroxide solution and a little water. Dilute to about 200 ml and adjust the solution to pH 8.2. Make up to 300 ml with water. This solution is stable for 4 hours.

Buffer solution, 0.5 N disodium phosphate (pH 8.2).

Apparatus.   Beckman Model DU spectrophotometer with 1-cm cells.

Treat all glassware with a solution of Plexiglas in chloroform (1 to 2%). Fill each piece of glassware with the solution, empty, and dry inverted for 15 minutes. Since the film has a tendency to come off when the glassware is rinsed and dried after use, the glassware can be used only once before it is recoated.

Procedure.   Measure a sample containing up to 25 γ/ml of the quaternary compound and mix with 0.100 ml of indicator solution and 0.20 ml of 0.5 N disodium phosphate solution. Make up a similar mixture without the sample for use as a blank. Determine the absorbance of the sample mixture at 620 mμ versus the blank.

An experimental error of less than $\pm 2\%$ for concentrations of cetyl-pyridinium chloride between 10 and 25 γ/ml was claimed.

12. Quaternary Ammonium Compounds—Picrate Salts.   Picric acid was recommended by Sloneker and his coworkers (23) as a reagent for the determination of quaternary ammonium cations. In their method the quaternary ammonium picrate salt is separated from excess picric acid by chloroform extraction, and the absorbance of the extract is measured spectro-photometrically at 365 mμ.

METHOD FROM SLONEKER ET AL. (23)

Reagents.   Phosphoric acid, 3 M. Dilute 85% phosphoric acid with distilled water 1:5.

Sodium hydroxide solution, 3 M. Prepare and adjust so that a volume of it neutralized with an equal volume of 3 M phosphoric acid gives a salt solution of pH 3.4.

Picric acid solution, one-half saturated (solubility of picric acid, 1.2 grams/100 ml water). Store in a brown bottle.

Procedure.   Pipet 0.1 to 1 ml of a solution containing 10 to 100 μg of quaternary cation. Add distilled water to obtain a volume of 1.5 ml. Add 1.0 ml of 3 M phosphoric acid and 1.0 ml of 3 M sodium hydroxide solution. Add 0.1 ml of the picric acid solution and mix. Extract the precipitate from the aqueous solution with three 1.5-ml portions of chloroform. Use a test-tube mixer to obtain intimate dispersion of the immiscible liquids. After each extraction, allow the chloroform layer to clarify and transfer to a 10-ml volumetric flask with the aid of a capillary medicine dropper. Make the chloroform solution to volume and transfer to a colorimeter tube. Measure the absorbance of the picrate in the chloroform at 365 mμ. Keep to a minimum the amount of aqueous picric acid mechanically carried over to the 10-ml volumetric flask during the extraction.

Calibration. Prepared a calibration curve by substituting standards containing known amounts of quaternary ammonium cations for the sample.

The method was designed primarily for the determination of quaternary ammonium cations in the presence of polysaccharides. When sodium chloride was present, incomplete extraction of the picrate from the aqueous phase was noted. The standard curve passed through the abscissa rather than the origin. Sodium salts of phosphoric acid, amino acids, and amino sugars do not interfere.

Certain secondary amines can also be determined by this method and therefore can be considered interferences in the determination of quaternary ammonium compounds. Linear standard curves were obtained for cetyltrimethylammonium chloride, cetylpyridinium chloride, DL-coniine, dicyclohexylamine, dodecyltrimethylammonium chloride, and trimethylarachidyl-behenylammonium chloride. The secondary and tertiary amines of low-molecular-weight pyrrolidine and N-methylpyrrolidine could not be determined. The procedure should be applicable also to certain tertiary amines.

13. Imine—Hydrolysis to Carbonyl. Imines are easily hydrolyzed to the parent carbonyl compound:

$$RR'C{=}NR'' \xrightarrow{\text{H}_2\text{O}} RR'C{=}O + R''NH_2 \qquad (8)$$

Freeman (24) used reaction 8 to form the 2,4-dinitrophenylhydrazone derivatives of the carbonyl compounds; he weighed the derivatives produced and related their weight to the concentration of the imine. In place of the gravimetric determination, a colorimetric approach such as that described in Chapter 2, Section I.C.1, should be feasible and preferable for trace quantities of imines, with probably only slight modifications of the conditions. Hydrolysis of the imine and application of other spectrometric methods for carbonyl compounds given in Chapter 2 may also be useful in special situations.

II. GAS CHROMATOGRAPHIC METHODS*

A. Determination of Amines and Imines in Free Form by Gas Chromatography

The relative ease of analyzing for free amines directly by GC follows this sequence: primary < secondary < tertiary. Although the analysis of tertiary

* Written by Morton Beroza and May N. Inscoe.

amines presents no problems, the direct analysis of free primary and sceondary amines by GC is usually achieved with column packings having supports treated with alkali to repress possible ionization and with a silanizing agent to reduce adsorption. Recent examples of such analyses (compounds and packings) are as follows: aliphatic amines and imines on Chromosorb P coated with 20% Ucon LB-550X and treated with 20% alcoholic KOH (25); ethylamphetamine on 2% Carbowax 20M + 5% KOH or 10% Apiezon L + 10% KOH on acid-washed dimethylchlorosilane-treated Chromosorb G (26); amphetamine on a column of 15% Carbowax 6000 + 5% KOH on the above support presaturated with an ethereal solution of nicotine to minimize adsorption (27); pyridine bases on Celite or kieselguhr coated with 1% triethanolamine + 9% polyethylene glycol 1000, dioctyl sebacate, or silicone oil DC 550 (28); and diaminotoluenes on KOH-treated Chromosorb W coated with 5% Bentone 34 + 15% Hyprose SP-80 (29).

B. Determination of Primary Amino Groups by Gas Chromatography

METHOD OF HOFFMANN AND LYSYJ (30)

Primary amino groups react with nitrous acid to produce nitrogen stoichiometrically (Van Slyke method). The nitrogen is swept into a GC column by a helium carrier gas and is readily quantified with a thermal conductivity detector.

Apparatus.   A gas chromatograph, equipped with a 3-foot-$\frac{1}{4}$-inch o.d. copper column packed with 80/100 mesh 5A molecular sieve (held at 40°C) and a thermal conductivity detector, is used with helium as the carrier gas; the flow rate is 60 ml/minute.

Rubber-covered electric vibrator.

The reaction cell (Figure 11-1) consists of a small glass vessel attached to a glass T-joint. It is located between the helium source and the column, as shown in Figure 11-2, and is connected to the otherwise metal system with pressure rubber tubing. The arrangement allows the reaction cell to be included in the system or bypassed by manipulating the metal stopcocks.

Reagents.   Saturated aqueous sodium nitrite (deaired) and glacial acetic acid are used.

Procedure.   Introduce 1 to 5 mg of compound and 0.3 ml of the saturated solution of sodium nitrite into the reaction vessel and connect the vessel to the system with the pressure tubing. Close stopcock 3 (Figure 11-2) and open 1 and 2. Sweep out the reaction vessel with helium for 30 seconds. Close stop-

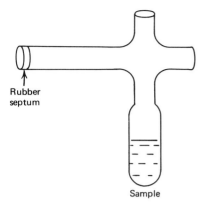

Rubber
septum

Sample

Figure 11-1. Reaction microcell (to scale).

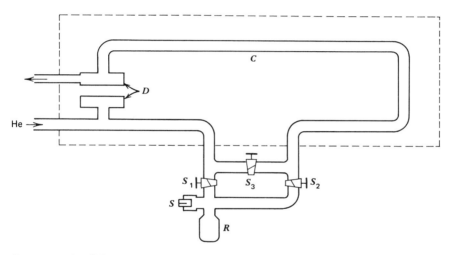

Figure 11-2. Schematic representation of reaction cell (R) in GC system. S = septum; S_1, S_2, S_3 = stopcocks; C = column; D = detector; He = helium.

cocks 1 and 2 and open 3. Inject 0.15 ml of the glacial acetic acid through the septum by means of a hypodermic syringe and shake the vessel with the rubber-covered vibrator for 1 minute. After waiting 10 minutes, close stopcock 3 and open 1 and 2 simultaneously for 30 seconds; then close 1 and 2 and open 3. The nitrogen appears as a symmetrical peak and is determined by the peak height method. Prepare a calibration curve using aliquots of a 5% glycine solution.

Table 11-17. Typical Procedures Used for Quantitative Determination of Amines as Amides by Gas Chromatography

Amide	Typical reagents	Types of compounds and references
Acetamide	CH_3COCl	Amphetamine[1]
	On-column $(CH_3CO)_2O$	Naphthylamine,[2] primary and secondary amines[3,4]
Propionamide	On-column $(CH_3CH_2CO)_2O$	Primary and secondary amines[4]
Trichloroacetamide	CCl_3COCl	Amphetamine[5]
Trifluoroacetamide	$(CF_3CO)_2O$	Primary and secondary amines[6,7]
		Catecholamines,[8,9] m- and p-xylylene diamines,[10] diamines from polyamide resins[11]
Trimethylsilamide	Bis(trimethylsilyl)acetamide	Diamines from polyamide resins,[12] long-chain (spingolipid) bases,[13] dopamine and tryptamine,[14] catecholamines[15]
Heptafluorobutyramide	$(C_3F_7CO)_2O$	Amphetamine and related amines[16,17]
Pentafluorobenzamide	Pentafluorobenzoyl chloride	Primary and secondary amines[18]
p-Bromobenzamide	p-Bromobenzoyl chloride	Amines from carbamate pesticides[19]
N-Carboethoxy	Ethyl chloroformate	Amino sugars[20]

[1] P. A. Toseland and P. H. Scott, *Clin. Chim. Acta*, **25**, 75 (1969).

[2] D. M. Marmion, R. G. White, L. H. Bille, and K. H. Ferber, *J. Gas Chromatog.*, **4**, 190 (1966).

[3] H. V. Street, *J. Chromatog.*, **37**, 162 (1968).

[4] M. W. Anders and G. J. Mannering, *Anal. Chem.*, **34**, 730 (1962).

[5] J. S. Noonan, P. W. Murdick, and R. S. Ray, *J. Pharmacol. Exptl. Therap.*, **168**, 205 (1969).

[6] M. Pailer and W. J. Huebsch, *Monatsh. Chem.*, **97**, 1541 (1966).

[7] W. J. Irvine and M. J. Saxby, *J. Chromatog.*, **43**, 129 (1969).

[8] D. D. Clarke, S. Wilk, S. E. Gitlow, and M. J. Franklin, *J. Gas Chromatog.*, **5**, 307 (1967).

[9] S. Kawai and Z. Tamura, *Chem. Pharm. Bull.* (*Tokyo*), **16**, 1091 (1968).

[10] S. Mori, M. Furusawa, and T. Takeuchi, *J. Chromatog. Sci.*, **8**, 477 (1970).

[11] S. Mori, M. Furusawa, and T. Takeuchi, *Anal. Chem.*, **42**, 138 (1970).

[12] S. Mori, M. Furusawa, and T. Takeuchi, *Anal. Chem.*, **42**, 959 (1970).

[13] H. E. Carter and R. C. Gaver, *J. Lipid Res.*, **8**, 391 (1967).

Results. The method was tested with glycine, glutamic acid, tyrosine, lysine, and valine. The percentage of primary amino nitrogen found was generally within 0.5% absolute of theory. Acidic amino acids were dissolved in 0.5 N sodium hydroxide to prevent immediate reaction upon contact with the sodium nitrite and consequent loss of nitrogen.

The apparatus of Norton, Turner, and Salmon (see Chapter 8, Section II) can be used in place of the original system described here, to simplify the manipulations involved.

Discussion. This general procedure has been applied to detect primary amine groups in stilbene derivatives (31).

C. Determination of Amines as Amides by Gas Chromatography

Amines and amine precursors are easily converted by means of a variety of agents to amides, which are readily determined by GC. Both polar and nonpolar liquid phases are used. Table 11-17 lists the amides formed from various compounds and the reagents used for the derivatization. (The derivatizing agents also generally react with free OH and other groups containing active hydrogens.) Acetamides and propionamides have been formed before GC and by on-column derivatization. In the latter procedure the anhydride derivatizing agent is injected after the sample or with it, and the derivative forms almost instantaneously on contact of the chemicals at the elevated temperature of the GC column. Trihaloacetamides form readily from the amine and the anhydride or acid chloride. Trifluoroacetamides, unlike the trifluoroacetates, have poor electron-capturing properties (32). The highly sensitive response of the electron capture detector to the derivatized catecholamines appears, therefore, to be due to the O-trifluoroacetyl rather than the N-trifluoroacetyl groups. In the analysis of diamines and ω-amino acids derived from homo- and copolymers on polyamide resins, trifluoroacetyl and trimethylsilyl derivatization was employed. The heptafluorobutyramides are useful derivatives; they have good stability and good electron-capturing properties, and they chromatograph well.

[14] W. J. A. VandenHeuvel, *J. Chromatog.*, **36**, 354 (1968).

[15] M. G. Horning, A. M. Moss, and E. C. Horning, *Biochim. Biophys. Acta*, **148**, 597 (1967).

[16] R. B. Bruce and W. R. Maynard, Jr., *Anal. Chem.*, **41**, 977 (1969).

[17] M. G. Horning, A. M. Moss, E. A. Boucher, and E. C. Horning, *Anal. Letters*, **1**, 311 (1968).

[18] A. Zlatkis and B. C. Pettit, *Chromatographia*, **2**, 484 (1969).

[19] R. L. Tilden and C. H. Van Middelem, *J. Agr. Food Chem.*, **18**, 154 (1970).

[20] M. D. G. Oates and J. Schrager, *J. Chromatog.*, **28**, 232 (1967).

D.   Determination of Amines as Nitrophenyl Derivatives by Gas Chromatography

Amines and amine precursors (e.g., carbamates) are converted in alkaline solution to nitrophenyl derivatives, which are determined with great sensitivity by electron capture GC. In this manner, 1-fluoro-2,4-dinitrobenzene has been used to determine amines derived from drugs in urine (33) and from carbamate pesticides (34), as well as *sec*-butylamine residues (35, 36). Picryl chloride has served to determine cyclohexylamine in cyclamates by forming the *N*-trinitrophenyl derivative (37), and 4-chloro-α,α,α-trifluoro-3,5-dinitrotoluene and α,α,α,4-tetrafluoro-3-nitrotoluene have been used to form 2,6-dinitro-4-trifluoromethyl- and 2-nitro-4-trifluoromethylanilines from carbamate pesticides, respectively (38).

E.   GC Determination of Amines by Condensation Reactions

Primary amines (catecholamines) have been converted to enamines or Schiff bases by aliphatic ketones or cyclobutanone, in which form they are readily analyzed by GC (39, 40). Primary aliphatic and aromatic amines can be determined quantitatively in submicrogram amounts after condensation with 2,5-hexanedione (41). Several biochemically important aminothiols and disulfides have been determined by GC after they were reacted with pivaldehyde to form thiazolidine or Schiff-base derivatives (42).

F.   Other Procedures for Determining Amines by Gas Chromatography

Long-chain (sphingomyelin) bases have been converted by periodic acid to aldehydes, which are determined by GC. The reaction is general for α-amino alcohols (43). Sphingomyelin bases were similarly cleaved to aldehydes, which were determined by GC as 1,3-dioxolane derivatives (44). Bromination of diphenylamine (45) and of aromatic herbicides to yield brominated anilines (46, 47) has served to determine residues of these compounds by electron capture GC. Urea herbicides may be determined as iodo derivatives by diazotizing the aniline derivatives of the herbicides and then treating them with KI and iodine as catalyst (48).

Cyanoethylation has been used to determine primary and secondary amino groups by GC; the cyanoethylation reaction products remain on the column, and the unreacted acrylonitrile is determined to give a measure of the amount reacted (49). The alkimino functional group has been determined by quaternization of the group with hydriodic acid, followed by thermal decomposition of the quaternary ammonium iodide(s) to yield the alkyl iodide(s), which are volatilized from the reaction mixture for determination by GC (50). An active hydrogen analysis by GC (see Chapter 8, Section II) is also responsive to amines and amides. Exhaustive methylation of tertiary amines (Hofmann reaction) produces the corresponding olefins; the reaction gave improved separations by GC and was suitable for quantitative analysis (51).

G. Determination of Amines as Salts or Complexes by Gas Chromatography

Alkyl (C_1 to C_5) amine hydrochlorides in aqueous solution may be analyzed directly by converting the salts to their free amines with a packing of 20% KOH on Chromosorb 101 in the injection port of the gas chromatograph, followed by analysis on a column containing 10% Amine 220 + 10% KOH on Chromosorb W (52). Before GC the amines are concentrated without loss as the salts; the amines in free form, although highly volatile, are not lost because they are not freed until within the gas chromatograph.

In another procedure five basic drugs were determined as ion-pair complexes (53). Salts of the drugs were adjusted to pH 5 with 0.1 N H_2SO_4 and then treated with bromothymol blue. The samples were extracted with chloroform for determination by GC. The complex decomposes to the amine in the gas chromatograph.

H. Determination of Quaternary Ammonium Compounds by Gas Chromatography

Quaternary ammonium compounds normally decompose at the temperatures needed for GC analysis. In an analysis of long-chain quaternary (alkyl- and/or benzyl-) ammonium halides, it was found that the the products of pyrolysis in the gas chromatograph had the t_R's of the tertiary amines formed by removal of alkyl or benzyl halides; furthermore, the homolog composition could be calculated from the t_R's and the areas of the product peaks (54).

Acetylcholine [$(CH_3)_3N(OH)CH_2CH_2OCOCH_3$] and related compounds are important in nerve transmission studies, and a number of methods have been devised to determine these quaternary amines. The compounds have been transformed by flash pyrolysis on a flash-heating ribbon to their volatile dimethylamine derivatives for quantitative analysis of several of the choline derivatives by GC (55). The dealkylation reaction is said to be general for alkylammonium chlorides up to tetraheptylammonium halide (56). Acetylcholine and related compounds have also been determined in the submicrogram range by N-demethylation with sodium benzenethiolate to give volatile tertiary amines that are readily determined by GC (57, 58).

I.   Determination of Organonitrogen Compounds by Gas Chromatography

Elemental analysis is usually a prerequisite to solving the structure of an organic molecule. If nitrogen is found, it is frequently desirable to determine the amount and/or types (functional groups). Commercial instrumentation, which includes mass spectrometers, has become available for elemental analysis, and many methods and types of apparatus for these analyses have been described. (See Section II of the Appendix.) In addition, GC detectors that respond to nitrogen compounds have also become available, and nanogram amounts of nitrogen compounds may be determined with them. (See Section II.D of the Appendix.) The response of the coulometric and electrolytic conductivity detectors for GC is highly specific for nitrogen; the response of the thermionic detector modified to determine nitrogen is moderately specific.

In addition to these procedures the amount of organonitrogen has been determined by subjecting samples to a Kjeldahl digestion with $CuSO_4$ as catalyst, decomposing the ammonium sulfate formed with platinum black in boiling sulfuric acid, collecting the gases in a 20-ml syringe, and determining the nitrogen by GC, using a hydrogen carrier gas and a thermal conductivity detector (59). A systematic analysis devised for distinguishing fourteen nitrogen functional groups in organic molecules utilizes different combinations of cleavage reactions and a GC study of the rate of formation of gaseous reaction products (60). The cleavage reactions were carried out in a microreactor attached to the gas chromatograph, reaction gases being collected in a syringe for introduction into the gas chromatograph. Although the system does not always work, the authors consider it an acceptable supplement to other methods of identification.

III.   ELECTROANALYTICAL METHODS*

A.   Coulometric Bromination of Aromatic Amines

Aromatic amines readily brominate at the unsubstituted ortho and para positions on the ring. This reaction has been known for some time, and several methods based on a bromate-bromide or bromine titration have been proposed (61). However, as stated in ref. 61, the method "does not enjoy wide popularity because there are so many interferences owing to the oxidizing and substitution properties of bromine." The major cause of these diffi-

---

* Written by Alan F. Krivis.

culties is the almost universal use of an excess of bromine (or bromine-forming reagents). In the presence of a large excess of bromine, for an extended period of time, it is likely that many undesirable reactions will take place.

Coulometric titration with bromine, however, minimizes many of these problems. The quantities of reagent are small and can be delivered with a very wide range of speeds. The conditions employed are adequate for the bromination of many aromatic amines. For example, the writer has been able to utilize a coulometric bromination to determine the residual unsubstituted anilino compound left in the product of the large-scale synthetic bromination of the aniline compound (62).

The procedure outlined in Chapter 1, Section III.C, for the analysis of phenols may be employed unchanged for aromatic amines. One important step to be taken is the determination of the stoichiometry of the reaction for the particular unknown involved; a suitable change may then be made in the calculations as outlined on p. 51.

In the event that this simplified procedure is not effective, another approach is possible. An excess of bromine may be generated, allowed to react for 1 minute with the unknown, and back-titrated coulometrically (63). The back-titration is based on the generation of Cu^{1+}, which reacts with the bromine. In this case, the starting solution contains Br^- and Cu^{2+}. During the first portion of the titration, the anode is in the titration cell and generates a slight excess of Br_2. After the 1-minute reaction time, the poles of the power supply are reversed to that the *cathode* is in the titration cell. Turning the power supply on now reduces Cu^{2+} to Cu^{1+}, and the latter reacts with the excess bromine.

B. Amperometric Titration of Aromatic Amines

Titration of aromatic amines with nitrite solution (equation 1) has not been widely used:

$$ArNH_2 + HCl + HNO_2 \rightarrow RN_2Cl + 2H_2O \tag{1}$$

This lack of popularity can be attributed to the fact that several of the recommended procedures require an external starch-iodide indicator and that the reaction may be slow. Substitution of an internal type of indicating system, and the addition of a catalyst to speed up the reaction, should lead to increased application. Such changes have been proposed (64).

The procedure to be described makes use of a biamperometric end-point detection system to determine the first excess of sodium nitrite. Essentially no steady current flows in the system until the first drop of excess reagent is added.

In addition to the use of a catalyst, the titration is carried out at room temperature. The reaction thus is much faster than when carried out at ice-bath temperatures without a catalyst.

EXPERIMENTAL METHOD [BASED ON THE PROCEDURE OF SCHOLTEN AND STONE (64)]*

Apparatus. Any polarograph, recording or manual, may be used for this analysis. (A manual instrument incorporating a galvanometer is most convenient.)

The electrodes are two identical platinum wires. Each may be purchased from a laboratory supply house or fabricated by sealing a short piece of platinum wire through the bottom of a glass tube. Electrical contact may be made through a mercury pool inside the glass tube. One electrode is connected to the positive lead of the polarograph; the other, to the negative lead.

Reagents. The reaction mixture is prepared in the cell from concentrated hydrochloric acid and potassium bromide.

The tirtrant is 0.1 M sodium nitrite solution.

Procedure. Weigh samples of about 4 mM of amine into 400-ml beakers. Add 250 ml of water, 20 ml of concentrated hydrochloric acid, and 1 gram of potassium bromide. Insert the glass paddle of a power stirrer and the platinum wire electrodes which have been cleaned in dichromate cleaning solution and well washed. Start the stirrer. Adjust the galvanometer to a sensitivity of 0.01 $\mu A/mm$ and apply 0.4 V across the electrodes. Although a current may flow at this point, it usually decreases to almost zero immediately. Add 0.1 M sodium nitrite solution (standardized against pure sulfanilic acid by this procedure), with the tip of the buret dipping into the solution, at such a rate that very little current flows. Near the end point the current increases, and at this point it is advisable to stop the titration to permit the galvanometer to become steady in order to locate the reference point on the current scale. Add the nitrite solution in small increments and observe the movement of the galvanometer needle.

The end point is indicated when the galvanometer shows a permanent deflection of about 0.1 μA away from the reference point previously found. Since the end point is not very sensitive, care is necessary in its vicinity. Subtract 0.05 ml from the volume added to correct for the excess sodium nitrite used.

* Reprinted in part with the permission of the copyright holder, the American Chemical Society.

C. Polarographic Determination of Secondary Amines

In Section B analytical use was made of the reaction of aromatic amines with nitrous acid to produce a diazonium salt. The particular procedure specified utilizes an amperometric measurement to detect the first excess of nitrite solution. Nitrous acid reacts with other amino compounds in addition to arylamines. For example, secondary amines can be converted to N-nitroso compounds by the action of nitrous acid:

$$R_2NH + HNO_2 \rightarrow R_2N—NO + H_2O \qquad (2)$$

Reaction 2 can be useful for the analysis of secondary amines in several possible ways. One might measure the excess of nitrite after complete reaction, detect the first excess of nitrite in a titration, or measure the quantity of N-nitroso compound produced by the reaction. The last-mentioned option has much to recommend it. In particular, only the secondary amine can produce the N-nitroso compound, and therefore this approach affords a specific measurement of the starting compound.

Quantitative measurement of the N-nitroso group can be accomplished by a polarographic method (65), and two procedures have been proposed to convert the amine to the nitroso derivative quantitatively (66, 67). Since the procedures are different, and one or the the other may have advantages in a specific case, both will be described.

EXPERIMENTAL METHOD [BASED ON THE PROCEDURE OF ENGLISH (66)]*

Apparatus. A recording polarograph is recommended for this analysis.

The optimum cell is an H-type (65), which is thermostated in a water bath (or jacket) at $25° \pm 0.1°C$.

Reagents. Sulfamic acid (NH_2HSO_3). Dissolve 15 grams in water and dilute to 100 ml.

Sodium hydroxide, 30% by weight.

Sodium hyposulfite. Commonly known as hydrosulfite, $Na_2S_2O_4$. High-grade commercial material is satisfactory.

Iodine. Approximately 0.2 N, prepared by dissolving 50 grams of KI and 26 grams of I_2 in 50 ml of water and diluting to 1 liter.

Gelatin. 0.5 gram of gelatin dissolved in 50 ml of warm water. This should be prepared daily.

Procedure. The sample should be in aqueous solution containing 15 to 25% of total bases. Transfer an aliquot, not exceeding 5 ml and containing

0.05 to 0.13 gram of amine, to an 8-inch test tube, dilute to 5 ml, and set in an ice bath. Add 3.0 ± 0.05 grams of sodium nitrite and 5 ml of glacial acetic acid and place in a water bath at 25° ± 2°C for 10 minutes. Immerse the tube to within about 2 inches of the top in an ice bath and, while swirling vigorously, add dropwise from a pipet sufficient sulfamic acid solution (about 17 ml) to destroy the excess nitrous acid as indicated by test with starch-KI paper. An excess of sulfamic acid will do no harm. Add a drop of phenolphthalein indicator solution, titrate to neutrality with the sodium hydroxide solution, and add 0.1 ml excess. Establishment of the correct alkalinity at this point is vital. Immerse the tube in a water bath at 60°C; after it has come to temperature, add 1.0 ± 0.05 gram of sodium hydrosulfite. Stopper the tube, invert it several times to dissolve the hydrosulfite completely, return it to the 60°C bath for 5 minutes, cool to 25°C, and dilute to 100 ml in a volumetric flask. Pipet 10 ml into a beaker, add 50 ml of water, 1 ml of glacial acetic acid, and 1 ml of starch indicator solution, and titrate to an exact end point with the iodine solution to destroy the excess hydrosulfite. Add 2.5 ml of concentrated hydrochloric acid and 1 ml of gelatin solution, and dilute to 100 ml in a volumetric flask.

Transfer an aliquot to the polarographic cell, deoxygenate by bubbling nitrogen through the solution for 10 minutes, and record the polarogram, starting at −0.3 V, versus a saturated calomel electrode. The curve is well defined; and, although the plateau is inclined at a much greater angle to the horizontal than is the condenser current line, it is reproducible and accurately measurable. The final solution is stable and may be retained even overnight before polarographing. The current for the nitroso reduction wave should be compared to a calibration curve to obtain the amount of amine in the original sample.

EXPERIMENTAL METHOD [BASED ON THE PROCEDURE OF LUND (67)]

Apparatus.   A recording polarograph is recommended for this analysis.

The optimum cell is an H-type (65), thermostated in a water bath (or jacket) at 25° ± 0.1°C.

Reagents.   Acetate buffer. Cautiously add 10 ml of glacial acetic acid and 10 ml of 0.2 M sodium hydroxide to 80 ml of water.

Sodium nitrite. Dissolve 20 grams of reagent-grade sodium nitrite in 80 ml of water.

Hydrochloric acid (4 M). Dilute 33.3 ml of concentrated hydrochloric acid (12 M) to 100 ml with distilled water.

Ammonium sulfamate. Dissolve 5 grams of ammonium sulfamate in 95 ml of water.

Procedure.   Add 1 ml of acetate buffer solution to 1 ml of amine solution containing an accurately weighed sample of amine (ca. 1 mg). Add 1 ml of nitrite solution, warm to 80°C for 15 minutes, and cool. Add 1 ml of 4 M hydrochloric acid solution, 5 ml of ethanol, and 1 ml of 5% ammonium sulfamate; transfer the mixture to the cell and dilute to a volume of 25.0 ml. Deoxygenate the solution by bubbling nitrogen through it for 10 minutes. Add 1.00 ml of ammonium sulfamate solution, divert the nitrogen to blanket the top of the solution, and record the polarogram, beginning at -0.6 V, versus the saturated calomel electrode. Measure the diffusion current for the nitroso reduction wave and compare it to a previously prepared calibration curve to obtain the amount of secondary amine in the sample.

IV.   NUCLEAR MAGNETIC RESONANCE METHODS*

A.   Amines

In nmr experiments the properties of amines are such that identification or analysis is somewhat difficult because the proton on the nitrogen tends to undergo exchange. The rate at which this occurs will determine the exact shape of the NH peak. The work of Anderson and Silverstein (68) applies nmr techniques to the determination of amines. Unfortunately the techniques used in determining the hydroxyl or alcohol functions do not apply in this case, and the problems are slightly more difficult.

Although solvents such as deuteriochloroform and carbon tetrachloride may be used in the determination of amines, this author found that the addition of trifluoroacetic acid to the amine was a better approach. The protonation of the amine is thereby accomplished, fixing the NH resonance in a low field position and eliminating possible overlapping of lines. By means of the integral the amine can be determined as primary, secondary, or tertiary.

The protons on the alpha carbon can also be utilized to determine or verify the amine functional group that is present. The amines studied by this method are dimethylaniline, benzylamine, and the following types of amines:

$$RCH_2CHN(CH_3)_2 \qquad\qquad (RCH_2CH_2)_2NH$$
$$|$$
$$CH_3$$

$$RCH_2CHNHCH_3 \qquad\qquad R_2CHCH_2NH_2$$
$$|$$
$$CH_3$$

* Written by Harry Agahigian.

Essentially, the splitting of the alpha protons, or the protons in the alpha substituent, is convenient for determining the relative amounts of primary, secondary, and tertiary amines. In some respects this direct analysis is more convenient than the usual reaction such as acetylation and diazotization. Although there is no problem with following the rate of formation of the reaction of a primary amine with benzaldehyde with nmr spectroscopy, the acetylation procedures might also easily be followed. The techniques reported by Babiec, Barrante, and Vickers (69) can also be applied in this case in which the trifluoroacetic acid function is prepared. The advantages of the nmr experiment are that derivatization is not required and that the preparation of the acid salt *in situ* is sufficient.

The work of Leader (70) utilizing hexafluoroacetone derivatization for the determination of alcohols also has applicability to the amines. In this case, the amines studied were methylamine, normal propylamine, isopropylamine, benzylamine, α-methylbenzylamine, ethanolamine, 3-amino-1 propanol-isopropanolamine, β-ethyoxyethylamine, 3-methoxypropylamine, and ethylenediamine. The series of aromatic amines studied consisted of o-chloroaniline, p-chloroaniline, p-aminophenol paraminobenzylic acid, sulfanilamide, 2-aminopyridine, 3-aminopyridine, and 4-aminopyridine. The secondary amines studied were dimethylamine, diethylamine, di-n-propylamine, diethanolamine, N-methylaniline, 2-methylaminoethanol, piperidine, morpholine and piperazine. The various amines can be identified because of the different chemical shift positions, the primary amines appearing at higher fields than the secondary amines.

The analysis of a mixture of diaminotoluene isomers by nmr spectrometry has been reported (71) and is an excellent example of the quantitative application of this technique. The nitrotoluenes have been studied by noting the methyl resonances of the various isomers and obtaining isomer ratios of the dinitrotoluene (72).

The aminotoluenes studied were the 3,4-diamino-, 2,5-diamino-, 2,3-diamino-, and 2,6-diaminotoluenes. The chemical shifts of the methyl resonances of each isomer are measured, and then the mixtures determined. The presence of nitrotoluenes in the presence of the diaminotoluenes is readily detected. The errors in such a determination are 0.5% absolute below 10% isomer content, 1.5% absolute from 10 to 50%, and 2.5% absolute above 50%.

A comprehensive study of nitrogen-containing compounds has been reported by Ma and Warnhoff (73). These authors point out the advantages of nmr analysis in determining the presence of a N-methyl function over the Herzig-Meyer (74) infrared techniques or Conroy's method (labelled methyl iodide quaternization (75). The weakness of the nmr method is the possibility of the overlapping of methyl resonances attached to other functional groups. Over 120 different types of ogranic nitrogen compounds are reported. The following ones are typical.

	In CDCl$_3$ for
Tertiary amines:	N—CH$_3$ Function
N,N-Dimethyl-n-butylamine	2.18 ppm
Methyldiethylamine	2.19
Tetramethylmethylenediamine	2.22
N-Methylpyrrolidine	2.33
N-Methylpiperidine	2.21
Nicotine	2.15
Scopoline	2.56
Codeine	2.42
Secondary aliphatic amines:	
N-Methyl-n-butylamine	2.43
N-Methylcyclopentylamine	2.39
N-Methylbenzylamine	2.42
Ephedirine	2.34
Aromatics and N-heteroaromatic amines:	
N-Methyltoluidine	2.79
N,N-Dimethylaniline	2.85
N-Methylphenylhydrazine	3.06

The N-methyl protons of the aliphatic tertiary amines exhibit a diamagnetic shift of 0.6 to 0.8 ppm in trifluoroacetic acid. The secondary amines are deshielded in acid from 0.45 to 0.6 ppm. This difference is sufficient to distinguish secondary and tertiary amines. The aromatic compounds containing tertiary and secondary amines are essentially the same in behavior as the aliphatic amines. The amides and imides give much smaller shift increments in trifluoroacetic acid, and there is no coupling in the acid solution. These differences might be used as criteria for distinguishing these various functional groups, although other spectroscopic methods may prove superior.

Quaternary ammonium salts containing N-methyl functions have also been reported, as well as methoxyl groups, C-methyls, and acetyl. The use of trifluoroacetic acid on methoxyl groups does not yield dramatic changes in the O-methyl resonance. There may be some overlapping, but use of another spectroscopic method might eliminate any ambiguity. The spectra were obtained in solution concentrations varying from 10% to 20% (w/v). The authors also report data for N-methylene and N-methine compounds. A similar study is reported by Freifelder, Mattoon, and Kriese (76) on N-substituted methylamines.

Using the Varian coding system, Slomp and Lindberg (77) tabulated a variety of organic compounds that contain nitrogen. The chemical shift data for indoles, hydrazines, and aryl- and unsaturation-substituted nitrogen compounds are reported.

B.   Amino Acids and Peptides

The study of amino acids and peptides by the method reported by Bovey and Tiers (78) requires the use of trifluoroacetic acid as a solvent. The advantage of this approach over those using water or D_2O is that accurate chemical shifts can be obtained. Trifluoroacetic acid can serve as the standard. The solutions are prepared at a concentration of 20% (w/v). Although glycine, cysteine, and cystine are less soluble, spectra can be obtained. The nmr spectra were obtained at 40 MHz. Solutions are made up so that 100 mg is dissolved in 0.5 ml of trifluoroacetic acid containing tetramethylsilane. Care is taken that the trifluoroacetic acid does not absorb water.

The compounds studied were glycine, diglycine, triglycine, tetraglycine, DL-alanine, L-leucine, DL-isoleucine, L-arginine, DL-phenylalanine, L-histidine, DL-methionine, L-glutamic acid, N-acetyl-DL-alanine, L-proline, glycyl-L-proline, betaine, and others. Bovey and Tiers reported aromatic amino acids, serine and threonine, and their glycyl peptides, as well as sulfur-containing amino acids and their peptides. The use of 60 MHz enhances the sensitivity and spectral differences because of chemical shift changes related to higher field strengths.

C.   Imines

Ethylenimines and imino compounds have been studied by Bottini and Roberts (78a), Kotera, Okada, and Miyazaki (78b), and Reinecke and Kray (78c). Ethylenimine derivatives of phosphonitrilic trimers were reported by Kobayashi, Chasin, and Clapp (78d), who established the various isomers present by obtaining the ^{31}P and ^1H spectra. In a similar study reported by Ottmann et al. (78e), all of the isomers in the substitution on the P_3N_3 ring were identified. The isomers studied were $(PN)_3Cl_5(NC_2H_4)$, $(PN)_3Cl_4$-$(NC_2H_4)_2$, $(PN)_3Cl_3(NC_2H_4)_3$, $(PN)_3Cl_2(NC_2H_4)_4$, $(PN)_3Cl(NC_2H_4)_5$, and $(PN)_3(NC_2H_4)_6$. Also reported were various methoxyl, dimethylamine, and thioether derivatives.

Reinecke and Kray (78c) reported data for 1-azabicycloalkane enamines and iminium salts. The nmr data established the double-bond position of the enamine derivatives and the iminium perchlorates. These data are applicable to the study of related nitrogen-containing compounds.

V. RADIOCHEMICAL METHODS*

Radioreagent techniques for determining organic functional groups were first applied extensively to the quantitative estimation of micro and sub-micro amounts of amino acids by derivatization of their amino groups. Since the introduction of p-iodobenzenesulfonyl-^{131}I chloride for this purpose, a variety of other reagents and radiochemical methods have been used for derivatizing both primary and secondary amines. Tertiary amines, although much less amenable to derivative formation, have also been determined. The labeled reactants most widely used are the chlorides of sulfonic and carboxylic acids, acetic anhydride, and dinitrofluorobenzene. Several macro as well as various micro and submicro quantitative methods have been developed for determining specific compounds and for resolving mixtures of labeled derivatives.

A. Conversion to Substituted Aryl Sulfonamides

The advantages of p-iodobenzenesulfonyl-^{131}I chloride (pipsyl chloride) as a radioreagent were well recognized in the first reports of its use for determining amino acids (79–82). A five- to tenfold excess quantitatively converts primary and secondary aliphatic amines to the corresponding sulfonamides when an emulsified mixture of reagent, sample, and Na_2CO_3 or $NaHCO_3$ solution is heated for a short time at 90 to 100°C. Excess sulfonyl chloride is readily hydrolyzed to the sulfonic acid, which may be separated from the products by ion-exchange resins or appropriate extractions.

Determination of individual amines in a mixture is possible by chromatographic separation or crystallization methods after the addition of carrier. Since ^{131}I is a gamma emitter, one of the easily resolved pure beta emitters (e.g., ^{35}S) may be used to label the carrier derivative. The second isotope serves to identify a specific derivative and to monitor carrier recovery and purification. The labeled sulfonic acid, precursor of the radioreagent, is prepared conveniently from p-diazobenzenesulfonic acid and readily available $K^{131}I$.

The very reactive nature of the reagent is somewhat disadvantageous in that primary amines may be partly converted to disubstituted derivatives. Since significant disubstitution occurs only after monosubstitution is complete, use of a series of successive partial reactions, with periodic removal of the monopipsyl compound, has been proposed (81) to avoid low results for certain amino acids. The 8-day half-life of iodine-^{131}I requires concurrent assay of a source prepared from the same lot of reagent used with samples.

---

* Written by D. Campbell.

In the simplest form of the pipsyl chloride method for amino group analysis, no carrier derivative is added and the treated sample is handled quantitatively throughout the series of operations used to remove excess reagent (80). An aliquot of the solution freed of sulfonic acid is subjected to chromatographic analysis to separate the derivatives, and the activity due to each isolated substance is determined by a suitable radioassay technique. A known weight of a derivative prepared from the same lot of reagent is chromatographed and assayed in an identical manner. The number of millimoles, M_s, of amine in the specimen used for analysis is given by

$$M_s = \frac{A}{S_r NF} \tag{1}$$

where A = activity in the portion of derivative isolated (μCi),
$\quad S_r$ = specific activity of the radioreagent (μCi/mM),
$\quad N$ = number of amine groups in the compound,
$\quad F$ = fraction of final sample solution taken for chromatographic separation.

The need for quantitative handling of the specimen solution may be obviated by applying the principle of reverse isotope dilution to the determination of the labeled derivative. After conversion of the amine to the substituted sulfonamide, a known weight of nonradioactive derivative, which is very large in comparison with the amount of labeled derivative present, is added. The weight of inactive derivative used is the smallest practicable amount required for subsequent purification steps. Excess reagent is then removed by using ion-exchange resins (79) or extraction techniques (81) without regard for small to moderate losses of material. The derivative is purified to constant specific activity by crystallization (81). A more rigorous criterion for the purity attainable with a given solvent, however, is agreement of the specific activity of the derivative in the filtrate with that of the recovered product (83). Chromatographic separation methods are not required, and the specific activities of derivative and radioreagent are determined by direct radioassay. In this variation,

$$M_s = \frac{M_d}{S_r/S_f - 1} \tag{2}$$

where M_d is number of millimoles of derivative added, and S_f is specific activity of the purified derivative (μCi/mM). This technique is especially useful with amino acids, since the addition of a large amount of racemic carrier ensures determination of the total of both of the enantiomorphs which may be present after protein hydrolysis.

The advantages of chromatographic separations may be combined with the convenience of purification methods which do not require quantitative handling by labeling the carrier derivative with a second isotope, to serve as an indicator. Sulfur-35, a pure beta emitter, is useful for this purpose: its beta particle is easily resolvable from the gamma radiation of ^{131}I, and pipsyl-^{35}S chloride may be prepared from $H_2^{35}SO_4$ through acetylsulfanilic acid. A definite quantity of the pipsyl-^{35}S derivative of specific activity S_d μCi/mM is added immediately after derivatization, so that $M_d \gg M_s$. The ^{35}S activity introduced is then $(M_d)(S_d)$, and the fraction of derivative recovered at any stage of purification is the fraction of added activity remaining at that point.

The second isotope is also useful for identifying a specific pipsylamine derivative in a chromatographically separated mixture, as well as for monitoring the chromatographic purification of a known derivative. The criterion of purity is attainment of a constant value for the ratio of the activity of sulfur to that of iodine. Derivatives are commonly eluted from chromatograms before radioassay. When G-M counting is employed for assays on planchets, a 3-mil aluminum absorber is useful for differentiating the ^{35}S-^{131}I pair. Iodine-131 may be determined by liquid scintillation counting with an efficiency approaching 100% (84).

The excellent sensitivity of the radioreagent-reverse isotope dilution method and the high selectivity afforded by an indicator isotope permit analysis of mixtures of primary and secondary amines on the micro scale. The approach has been used extensively for determining various amino acids in biological samples (85–88), and a scheme has been presented for estimating eleven of these compounds in 1 mg of protein (86). After its application to histamine, the dipipsyl derivative of which required four crystallizations to attain constant specific activity, the method was proposed for determining any amine yielding a crystalline substituted p-iodobenzenesulfonamide (89). The technique has also been used for estimating microgram amounts of 2,4-dioxypyrimidine and its 5-methyl derivative (90). Paper chromatography for resolving pipsyl derivatives has been supplemented by liquid-column (91) and thin-layer (92) separation methods. A paper chromatographic procedure has also been developed for establishing the radiochemical purity of the reagent (93).

Recently, 1-(dimethylamino)naphthalene-5-sulfonyl chloride (dansyl chloride), labeled with ^{14}C in the methyl group, has been applied to amino acids (94, 95).

Treatment of an amine with nonradioactive p-bromobenzenesulfonyl chloride (brosyl chloride), followed by isolation of the derivative on a paper chromatogram for subsequent determination of bromine by neutron activation analysis, has been proposed as a quantitative method (96). This technique, derivative activation chromatography, does not involve the use of a radioreagent but does require a neutron source of high flux, such as a reactor, if the potential sensitivity of activation analysis is to be fully realized (Cf. Chapter 2).

B.   Conversion to Carboxylic Acid Amides

Separation of p-iodobenzamides-^{131}I by liquid-column chromatography, followed by automatic scanning of the column with a solid scintillation detector, has been suggested for the quantitative determination of amines (cf. Chapter 1) (97). Although the only separation of such substituted benzamides reported is that of the derivatives of o- and m-toluidine, illustrated in Figure 11-3, the method is applicable in principle to any amine quantitatively forming an amide with p-iodobenzoyl chloride which can be isolated on a column. The specific activity of the radioreagent is determined in the same manner after formation of a suitable derivative. The amount of amine in the sample used for analysis is then calculated from equation 1. Although the report does not include procedures for the quantitative derivatization of amines or for the removal of excess reagent, these operations would not be expected to present experimental difficulties.

The radioreagent-direct isotope dilution method using inactive 3-chloro-4-methoxybenzoyl chloride for quantitative derivatization, followed by addition of the ^{36}Cl-labeled derivative and purification, is applicable to primary

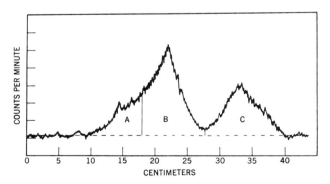

Figure 11–3.   Recorder trace of developed chromatographic column. Radioactivity in arbitrary units is plotted against distance from top of column (97). Zones A and B, N-p-iodobenzoyl-o-toluidine-^{131}. Zone C, N-p-iodobenzoyl-m-toluidine-^{131}. Reprinted from *Anal. Chem.*, 27, 1898 (December 1955). Copyright 1955, by the American Chemical Society. Reprinted by permission of the copyright owner.

and secondary amines as well as to hydroxy compounds (98) (cf. Chapter 1). Ethylenediamine and aniline were determined by treatment with a 50% excess of nonradioactive reagent in pyridine for 30 minutes at room temperature. The reaction between the organic chloride and primary amines must take place without heating to avoid low results due to the formation of disubstituted derivatives. The amount of substituted amide produced is then determined by direct isotope dilution, using the pure derivative prepared from labeled reagent. The results of five analyses of aniline, using equation 5 of Section V, Chapter 1, end-window G-M counting, and the melting point depression constant for measurement of the purity of the derivative, are given in Table 11-18. The N-phenyl-3-chloro-4-methoxybenzamide-^{36}Cl was crystallized twice from absolute ethanol and once from ca. 3:2 ethanol-water to give a product melting within 0.2°C of the pure amide. Purification of derivatives by procedures more rigorous than several crystallizations may be necessary when substantial amounts of homologs are present. An advantage of the technique is its applicability in the presence of moderate amounts of water. Another desirable aspect of the method is that the only radioactivity added to the sample is in the form of the derivative to be determined. There are therefore no labeled contaminants to be removed and radiochemical purity corresponds to chemical purity.

Acetic anhydride reacts quantitatively with many primary and secondary amines, frequently under mild conditions, to yield substituted acetamides. Quantitative derivatization of most secondary amines may be achieved at elevated temperatures. Acetic anhydride is often preferable to acid chlorides as a radioreagent for determining primary amines because of its lesser tendency to form diacyl compounds. It is also amenable to use with the

Table 11-18. Determination of Aniline Using 3-Chloro-4-methoxybenzoyl-^{36}Cl Chloride

Weight taken, mg	Weight recovered, mg	Percent recovered
94.8	95.1	100.3
98.5[a]	98.0	99.5
98.0[a]	97.8	99.8
99.1[a]	98.0	98.9
98.4[a]	97.5	99.1

[a] To the sample were added 5 mg of *o*- and 5 mg of *p*-toluidine.

reference-indicator radioisotope technique in analyses for micro and sub-micro amounts of amines. For this purpose, the isotopic modification employed for converting the amine to the substituted acetamide may be labeled with ^{14}C and that used to prepare the carrier (indicator) derivative may be labeled with tritium.

Application of the ^{14}C-labeled reagent is illustrated by the determination of neomycins A and B and neamine by acetylation of their primary amine groups (99). A specimen of 10 to 20 mg is dissolved in 10 ml of 0.01 N aqueous NaOH. To a 1-ml aliquot is added 0.1 ml of 3 M aqueous K_2HPO_4 and 0.1 ml (ca. 2 meq) of acetic-1-^{14}C anhydride of specific activity 50 $\mu Ci/ml$. The solution is then shaken for 30 minutes, although reaction is complete in a much shorter time. The derivatives are separated chromatographically on Whatman No. 4 paper, using a descending system consisting of n-butanol-water-piperidine in a proportion of 84:16:2 by volume. The spots are cut out, placed in vials, and heated for 2 hours at 60°C in the presence of 0.4 ml of water and 1.6 ml of ethanol. A mixed solvent scintillator solution is then added, and the radioactivity in the vials is counted. The amount of each substance present is determined by concurrent analysis of a sample containing known quantities of each amino compound.

Amino groups in lipids have been determined in a similar manner (100). A submicro method using ^3H-labeled carrier as indicator has also been described for determining various amino acids in 2 μg of protein (101). Thyroxine (3,5,3',5'-tetraiodothryonine) has been determined by means of the ^3H-labeled reagent with either ^{14}C (as the N-acetyl derivative) or ^{131}I serving as indicator (102). In addition, 3-iodotyrosine and 3,5-diiodotyrosine have been estimated, using the tritiated anhydride for derivative formation and acetyl-^{14}C carrier as indicator (103). The monoacetyl derivatives of thyroxine and triiodothryonine have been recommended over the diacetyl derivatives on the basis of stability, solubility in polar solvents, and ease of purification (104).

Tritiated acetic anhydride is readily separated from nonvolatile labeled impurities by distillation under vacuum (105).

The preparation of benzyloxycarbonyl-α-^{14}C chloride and its application to the optical analysis of amino acids have been described (106). The sensitivity of the method was shown to be at least 0.001% in determining the L-enantiomorph of asparagine in the optically pure D-form. The absence of racemization in the benzyloxycarbonylation of amino acids was also confirmed.

C. Conversion to Dinitrophenyl Amines

Primary amines, as well as secondary amines which are not sterically hindered, react quantitatively with excess 2,4-dinitrofluorobenzene (DNFB) in the presence of NaHCO₃ at moderate temperatures to yield secondary and tertiary dinitrophenyl amines, respectively. The reaction products are readily separable from unconsumed reagent, which may be hydrolyzed and extracted as the sodium salt of 2,4-dinitrophenol. The conditions favoring quantitative derivatization of primary and secondary amino groups were well established after introduction of the compound for the spectrophotometric determination of amino acids (107) and before the availability of the labeled reagent.

Derivatives of reactive amines are formed quantitatively within 20 minutes at 60°C by treating 10 to 100 μg of the compound contained in 0.1 ml of water with 0.05 ml of a solution of 0.12 ml of reagent in 10 ml of absolute ethanol, using 0.1 ml of 0.1 M aqueous NaHCO₃ solution as buffer (108). Excess reagent is hydrolyzed by adding 0.4 ml of 0.2 N NaOH solution and continuing heating for 60 minutes. After the addition of water, the derivative is extracted from the aqueous phase with an organic solvent. Among primary and secondary amines reacting under these conditions are ethanolamine, aniline, dodecylamine, di-n-butylamine, benzyl n-butylamine, and d-desoxyephedrine. Diisopropylamine and dicyclohexylamine, however, do not yield derivatives when treated in this manner. Dry dioxane has been proposed in place of ethanol-aqueous NaHCO₃ as a reaction solvent to avoid the formation of ethyl 2,4-dinitrophenyl ether (109).

The reagent may be prepared labeled generally with tritium at the three possible positions or uniformly with ¹⁴C. The availability of both isotopic forms permits convenient use of the reference-indicator isotope principle, which has been employed in determining submicrogram quantities of amino acids (110, 111). For this application, the radioreagent used to derivatize the amino acids was labeled with tritium as the reference isotope and that used to prepare indicator derivatives was labeled with ¹⁴C. Derivatives of the several acids were separated by paper chromatography, and radioassays were made by gas counting after combustion of the paper segments containing the isolated compounds. The methods developed were useful for determining as little as 10 ppm of an amino acid. Recently, two-dimensional thinlayer chromatography using silica gel has been employed for separating ¹⁴C-labeled dinitrophenyl amino acids (112).

The ¹⁴C-labeled compound has been used for estimating terminal amine groups in nylon and thus for calibration of a spectrophotometric method (113). A polymer sample of up to 125 mg, containing not more than 10 μeq of amine, was dissolved by refluxing in 15 ml of a solvent composed of LiBr (1 gram) in absolute ethanol-water, 5:0.3 by volume, until solution was complete. After addition of 1.5 ml of 0.05 N aqueous HCl, the solution was refluxed for an additional minute. A 1-ml aliquot of ethanolic radioreagent solution containing 0.1 gram of DNFB-¹⁴C of specific activity 1 μCi/mM was added immediately, followed by 0.15 to 0.20 gram of solid $NaHCO_3$. After the solution had been heated at 80°C for 40 minutes, the polyamide was precipitated in a tenfold volume of water to remove excess DNFB; precipitation was twice repeated, using formic acid as solvent and a twentyfold volume of water. Radioassays of reagent and of dried acetylated product were made by liquid scintillation counting after oxygen flask combustion.

Although labeled DNFB is a radioreagent of demonstrated value, its wider application to amines has perhaps been limited by its cost in comparison with that of radioactive acetic anhydride. Another disadvantage to the use of DNFB-^{3}H is the possible loss of ^{3}H activity, which may occur by exchange if the tritium-labeled reagent or its derivatives are subjected to strongly acid conditions. A further restriction on the use of DNFB as a radioreagent is the severe quenching effect which nitro groups have in liquid scintillation counting. This interference becomes particularly troublesome in experiments using two isotopes, such as the method for amino acids, which employs tritium as reference isotope and ^{14}C as indicator isotope. The effect may be overcome, however, by burning the sample before radioassay, which is then made either by liquid scintillation counting or by gas-phase techniques.

The oxygen flask (Schöniger) method offers a very convenient means for converting a variety of combustible organic materials labeled with ^{14}C and tritium to a standard form for radioassay. It is well adapted to routine analysis and is especially advantageous when used in conjunction with liquid scintillation counting. Ethanolamine and 2-phenylethylamine are effective absorbents for $^{14}CO_2$. They are often dissolved in an hydroxylated solvent such as methanol or 2-methoxyethanol for addition to the flask. When absorption of $^{14}CO_2$ is complete, the toluene solution of scintillators is introduced. An aliquot of the solution is then removed for radioassay.

An alternative, and perhaps more common, procedure is addition of both absorbent and scintillators dissolved in a mixture of toluene and methanol. Absorbent-scintillator systems used with ^{14}C can usually be employed as well with samples containing both ^{14}C and tritium; for materials containing only tritium, the base is omitted. A disadvantage in adding scintillators to the flasks, however, is the difficulty encountered in cleaning the flasks without the use of organic solvents. Such solvents must be avoided, since a serious explosion may result in the subsequent combustion from the presence of residual organic vapors. Consequently, a preferred technique is addition to the flask of only the solution of the absorbent in the hydroxylated solvent. An aliquot comprising the major portion of the absorbent solution is then removed and added to a liquid scintillation counting vial containing a concentrated toluene solution of the scintillators.

D. Other Methods

A number of other procedures employing radioisotopes have been developed for determining specific substances, principally in biochemical applications. Representative methods are summarized in Table 11-19. In the procedure developed for neomycin on cotton (114), the primary amino groups of the compound are first converted to substituted dithiocarbamic acid groups, which are then treated with ammoniacal ^{110m}AgNO$_3$ to form ^{110m}Ag$_2$S:

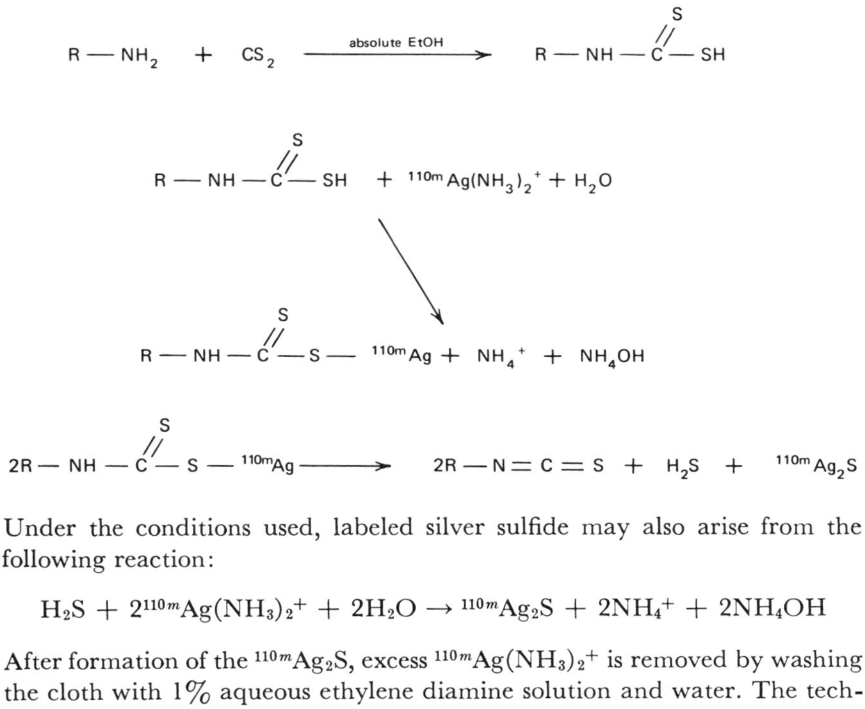

$$R - NH_2 \;+\; CS_2 \quad\xrightarrow{\text{absolute EtOH}}\quad R - NH - \overset{\overset{\displaystyle S}{\|}}{C} - SH$$

$$R - NH - \overset{\overset{\displaystyle S}{\|}}{C} - SH \;+\; {}^{110m}Ag(NH_3)_2{}^+ + H_2O$$

$$R - NH - \overset{\overset{\displaystyle S}{\|}}{C} - S - {}^{110m}Ag + NH_4{}^+ + NH_4OH$$

$$2R - NH - \overset{\overset{\displaystyle S}{\|}}{C} - S - {}^{110m}Ag \longrightarrow 2R - N = C = S + H_2S + {}^{110m}Ag_2S$$

Under the conditions used, labeled silver sulfide may also arise from the following reaction:

$$H_2S + 2^{110m}Ag(NH_3)_2{}^+ + 2H_2O \rightarrow {}^{110m}Ag_2S + 2NH_4{}^+ + 2NH_4OH$$

After formation of the ^{110m}Ag$_2$S, excess ^{110m}Ag(NH$_3$)$_2{}^+$ is removed by washing the cloth with 1% aqueous ethylene diamine solution and water. The technique is standardized for a specific lot of ^{110m}AgNO$_3$ by treating fabric samples containing known amounts of neomycin with the solution of labeled reagent and preparing a graph relating net count rate to amount of amine deposited. The reaction with carbon disulfide to yield a substituted dithiocarbamic acid and the decomposition of the silver salt of the N-monoalkyl acid to form the isothiocyanate are characteristic of primary aliphatic amines.

On the basis of this series of reactions, the method has been suggested for the determination of other primary aliphatic amines under suitable conditions.

Table 11-19. Additional Compounds Determinable by Radioisotope Methods

Compound or moiety determined	Radioreagent used	Reaction or method	Substance assayed	References
Amino acids	^{64}Cu(OAc)$_2$	Formation of copper complex	Copper complex	115
Amino acids	^{64}Cu$_3$(PO$_4$)$_2$	Formation of copper complex	Copper complex	116, 117
Amino acids	None	Neutron activation of copper complex	Copper complex	118
Amino acids	None	Simple isotope dilution	Amino acids	119–121
N-Terminal amino acids	C$_6$H$_5$NC^{35}S	Reverse isotope dilution	Phenylthiohydantoin	122
Amino acids	BrCH$_2$COOH-^{14}C	Formation of —CH$_2$—COOH derivative	—CH$_2$—COOH derivative	123
L-Amino acids	^{14}C-Amino acid	Direct isotope dilution	Corresponding tRNA[a] derivative	124
Thiamine	Na$_2$^{35}SO$_3$	Formation of sulfonic acid	Sulfonic acid	125
Neomycin	^{110m}AgNO$_3$	Via dithiocarbamate	Silver sulfide	114
Galactosamine and glucosamine	Respective ^{14}C-amine	Formation of the α-naphthylthiourea	Substituted α-naphthylthiourea	126
Chlortetracycline	Methyl-^{14}C iodide	Reverse isotope dilution	Quaternary ammonium derivative	127
Strychnine	Methyl-^{14}C iodide Methyl-^{3}H iodide	^{14}C as recovery indicator	Quaternary ammonium derivative	128
Urea	Xanthydrol-9-^{14}C	Formation of dixanthylurea	Dixanthylurea	129

[a] Transfer ribonucleic acid.

Its applicability to secondary aliphatic amines has also been indicated (114). High sensitivity, however, may be difficult to attain where the presence of reducing agents leads to the formation of metallic silver, or where silver ion is strongly adsorbed on the substrate. Good sensitivity would normally be expected, since up to two equivalents of silver are insolubilized for each equivalent of amine.

REFERENCES

Absorption Spectrophotometric

1. K. Whetsel, W. E. Roberson, and M. W. Krell, *Anal. Chem.*, **30**, 1594 (1958).
2. F. H. Lohman and W. E. Norteman, Jr., *Anal. Chem.*, **35**, 707 (1963).
3. A. J. Milun, *Anal. Chem.*, **29**, 1502 (1957).
4. W. H. Hong, and K. A. Connors, *Anal. Chem.*, **40**, 1273 (1968).
5. F. C. McIntire, L. M. Clements, and M. Sproull, *Anal. Chem.*, **25**, 1757 (1953).
6. M. J. Kolbezan, J. W. Eckert, and B. F. Bretschneider, *Anal. Chem.*, **34**, 583 (1962).
7. F. J. Bandelin and C. R. Kemp, *Ind. Eng. Chem., Anal. Ed.*, **18**, 470 (1946).
8. S. Siggia, *Quantitative Organic Analysis via Functional Groups*, 3rd Ed., John Wiley, New York, 1963, p. 511.
9. J. T. Stewart, T. D. Shaw, and A. B. Ray, *Anal. Chem.*, **41**, 360 (1969).
10. F. E. Critchfield and J. B. Johnson, *Anal. Chem.*, **28**, 436 (1958).
11. H. C. Dowden, *Biochem. J.*, **32**, 455 (1938).
12. L. Nebbia and F. Guerrieri, *Chim. Ind. (Milan)*, **35**, 896 (1953).
13. E. L. Stanley, H. Baum, and J. L. Gove, *Anal. Chem.*, **23**, 1779 (1951).
14. G. R. Umbreit, *Anal. Chem.*, **33**, 1572 (1961).
15. M. Palumbo, *Farm. Sci. Tec. (Pavia)*, **3**, 675 (1948).
16. B. T. Cromwell, *Biochem. J.*, **46**, 578 (1950).
17. S. Sass, J. J. Kaufman, A. A. Cardenas, and J. J. Martin, *Anal. Chem.*, **30**, 529 (1958).
18. G. H. Schenk, P. Warner, and W. Bazzelle, *Anal. Chem.*, **38**, 907 (1966).
19. M. E. Auerbach, *Ind. Eng. Chem., Anal. Ed.*, **15**, 492 (1943); **16**, 739 (1944).
20. S. Siggia, *Quantitative Organic Analysis via Functional Groups*, 3rd Ed., John Wiley, New York, 1963, p. 555.
21. L. D. Metcalfe, *Anal. Chem.*, **32**, 70 (1960).
22. J. Fogh, P. O. H. Rasmussen, and K. Skadhauge, *Anal. Chem.*, **26**, 392 (1954).
23. J. H. Sloneker, J. B. Mooberry, P. R. Schmidt, J. E. Pittsley, P. R. Watson, and A. Jeanes, *Anal. Chem.*, **37**, 243 (1965).
24. S. Freeman, *Anal. Chem.*, **25**, 1750 (1953).

Gas Chromatographic

25. A. Di Lorenzo and G. Russo, *J. Gas Chromatog.*, **6**, 509 (1968).
26. A. H. Beckett, L. G. Brookes, and E. V. B. Shenoy, *J. Pharm. Pharmacol.*, **21**, Suppl. 151S (1969).

27. D. B. Campbell, *J. Pharm. Pharmacol.*, **21**, 129 (1969).

28. A. A. F. van der Meeren and A. L. Th. Verhaar, *Anal. Chim. Acta*, **40**, 343 (1968).

29. C. E. Boufford, *J. Gas Chromatog.*, **6**, 438 (1968).

30. E. R. Hoffmann and I. Lysyj, *Microchem. J.*, **6**, 45 (1962).

31. J. Franc, V. Kovar, and F. Mikes, *Mikrochim. Acta*, **1966**, 133.

32. D. D. Clarke, S. Wilk, and S. E. Gitlow, *J. Gas Chromatog.*, **4**, 310 (1966).

33. T. Walle, *Acta Pharm. Suecica*, **5**, 367 (1968).

34. E. R. Holden, W. M. Jones, and M. Beroza, *J. Agr. Food Chem.*, **17**, 56 (1969).

35. E. W. Day, Jr., F. J. Holzer, J. B. Tepe, J. W. Eckert, and M. J. Kolbezen, *J. Assoc. Offic. Anal. Chemists*, **51**, 39 (1968).

36. I. C. Cohen and B. B. Wheals, *J. Chromatog.*, **43**, 233 (1969).

37. S. W. Gunner and R. C. O'Brien, *J. Assoc. Offic. Anal. Chemists*, **52**, 1200 (1969).

38. D. G. Crosby and J. B. Bowers, *J. Agr. Food Chem.*, **16**, 839 (1968).

39. P. Capella and E. C. Horning, *Anal. Chem.*, **38**, 316 (1966).

40. S. Kawai and Z. Tamura, *Chem. Pharm. Bull. (Tokyo)*, **15**, 1493 (1967).

41. T. Walle, *Acta. Pharm. Suecica*, **5**, 353 (1968).

42. E. Jellum, V. A. Bacon, W. Patton, W. Pereira, Jr., and B. Halpern, *Anal. Biochem.*, **31**, 339 (1969).

43. W. J. Baumann, H. H. O. Schmid, and H. K. Mangold *J. Lipid Res.*, **10**, 132 (1969).

44. R. V. Panganamala, J. C. Geer, and D. G. Cornwell, *J. Lipid Res.*, **10**, 445 (1969).

45. W. H. Gutenmann and D. J. Lisk, *J. Agr. Food Chem.*, **11**, 468 (1963).

46. W. H. Gutenmann and D. J. Lisk, *J. Agr. Food Chem.*, **12**, 46 (1964).

47. W. H. Gutenmann and D. J. Lisk, *J. Gas Chromatog.*, **4**, 424 (1966).

48. I. Baunok and H. Geissbuehler, *Bull. Environ. Contam. Toxicol.*, **3**, 7 (1968).

49. S. I. Obtemperanskaya and N. D. Khoe, *Zh. Analit. Khim.*, **24**, 1588 (1969).

50. N. D. Cheronis and T. S. Ma, *Organic Functional Group Analysis by Micro and Semi-micro Methods*, Interscience, New York, 1964, pp. 622–626.

51. H. B. Hucker and J. K. Miller, *J. Chromatog.*, **32**, 408 (1968).

52. G. R. Umbreit, R. E. Nygren, and A. J. Testa, *J. Chromatog.*, **43**, 25 (1969).

53. P. F. G. Boon and A. W. Mace, *J. Chromatog.*, **41**, 105 (1969).

54. L. D. Metcalfe, *J. Am. Oil Chemists' Soc.*, **40**, 25 (1963).

55. P. I. A. Szilagyi, D. E. Schmidt, and J. P. Green, *Anal. Chem.*, **40**, 2009 (1968).

56. D. E. Schmidt, P. I. A. Szilagyi, and J. P. Green, *J. Chromatog. Sci.*, **7**, 248 (1969).

57. D. J. Jenden, I. Hanin, and S. I. Lamb, *Anal. Chem.*, **40**, 125 (1968).

58. I. Hanin and D. J. Jenden, *Biochem. Pharmacol.*, **18**, 837 (1969).

59. J. Franc, B. Trtik, and K. Placek, *J. Chromatog.*, **36**, 1 (1968).

60. J. Franc and F. Mikes, *J. Chromatog.*, **26**, 378 (1967).

Electroanalytical

61. S. Siggia, *Quantitative Organic Analysis via Functional Groups*, 3rd Ed., John Wiley, New York, 1963, p. 450.

62. A. F. Krivis, *in* C. A. Streuli and P. R. Averell, eds., *Analytical Chemistry of Nitrogen and Its Compounds*, John Wiley, New York, 1971.

63. R. P. Buck and E. H. Swift, *Anal. Chem.*, **24**, 449 (1952).

64. H. G. Scholten and K. G. Stone, *Anal. Chem.*, **24**, 749 (1952).

65. I. M. Kolthoff and J. J. Lingane, *Polarography*, 2nd Ed., Interscience, New York, 1952.

66. F. L. English, *Anal. Chem.*, **23**, 344 (1951).

67. H. Lund, *Acta Chem. Scand.*, **11**, 990 (1957).

Nuclear Magnetic Resonance

68. W. R. Anderson, Jr., and R. M. Silverstein, *Anal. Chem.*, **37**, 14 (1965).

69. J. S. Babiec, Jr., J. R. Barrante, and G. D. Vickers, *Anal Chem.*, **40**, 610 (1968).

70. G. W. Leader, *Anal. Chem.*, **42**, 17 (1970).

71. A. Mathias, *Anal. Chem.*, **38**, 1931 (1966).

72. A. Mathias, *Anal. Chim. Acta*, **35**, 376 (1966).

73. J. C. N. Ma and E. W. Warnhoff, *Can. J. Chem.*, **43**, 1849 (1965).

74. F. Pregl and J. Grant, *Quantitative Organic Microanalysis*, Blakiston, Philadelphia, 1946, p. 156.

75. H. Conroy, P. R. Brook, M. K. Rout, and N. Silverman, *J. Am. Chem. Soc.*, **80**, 5178 (1958).

76. M. Freifelder, R. W. Mattoon, and R. W. Kriese, *J. Org. Chem.*, **31**, 1196 (1966).

77. G. Slomp and J. G. Lindberg, *Anal. Chem.*, **39**, 60 (1967).

78. F. A. Bovey and G. V. D. Tiers, *J. Am. Chem. Soc.*, **81**, 2870 (1959).

78a. A. T. Bottini and J. D. Roberts, *J. Am. Chem. Soc.*, **80**, 5203 (1958).

78b. K. Kotera, T. Okada, and S. Miyazaki, *Tetrahedron Letters*, **941** (1967).

78c. M. G. Reinecke and L. R. Kray, *J. Org. Chem.*, **31**, 4215 (1966).

78d. Y. Kobayashi, L. A. Chasin, and L. B. Clapp, *Inorg. Chem.*, **2**, 212 (1963).

78e. G. Ottmann, H. Agahigian, H. Hooks, G. Vickers, E. Kober, and R. Ratz, *Inorg. Chem.*, **3**, 753 (1964).

Radiochemical

79. A. S. Keston, S. Udenfriend, and R. K. Cannan, *J. Am. Chem. Soc.*, **68**, 1390 (1946).

80. A. S. Keston, S. Udenfriend, and M. Levy, *J. Am. Chem. Soc.*, **69**, 3151 (1947).

81. A. S. Keston, S. Udenfriend, and R. K. Cannan, *J. Am. Chem. Soc.*, **71**, 249 (1949).

82. A. S. Keston, S. Udenfriend, and M. Levy, *J. Am. Chem. Soc.*, **72**, 748 (1950).

83. M. M. Rapport and B. Lerner, *J. Biol Chem.*, **232**, 63 (1958).

84. B. A. Rhodes, *Anal. Chem.*, **37**, 995 (1965).

85. S. Udenfriend, *J. Biol. Chem.*, **187**, 65 (1950).

86. S. F. Velick and S. Udenfriend, *J. Biol. Chem.*, **190**, 721 (1951).

87. S. Udenfriend and S. F. Velick, *J. Biol. Chem.*, **190**, 733 (1951).

88. S. F. Velick and L. F. Wicks, *J. Biol. Chem.*, **190**, 741 (1951).

89. R. W. Schayer, Y. Kobayashi, and R. L. Smiley, *J. Biol. Chem.*, **212**, 593 (1955).

90. J. R. Fresco and R. C. Warner, *J. Biol. Chem.*, **215**, 751 (1955).

91. M. C. Corfield, J. C. Fletcher, and A. Robson, *Chem. Ind. (London)*, **1956**, 661.

92. M. Cole and J. C. Fletcher, *Biochem. J.*, **102**, 825 (1967).

93. C. N. Turcanu and I. Zamfir, *Studii Cercetari Chim.*, **16**, 215 (1968); *Chem Abstr.*, **69**:83197t.

94. G. Rapoport, M. F. Glatron, and M. M. Lecadet, *Compt. Rend., Ser. D*, **265**, 639 (1967); *Chem. Abstr.*, **68**:10140p.

95. R. F. Chen, *Anal. Biochem.*, **25**, 412 (1968).

96. J. M. Steim and A. A. Benson, *Anal. Biochem.*, **9**, 21 (1964).

97. W. M. Stokes, W. A. Fish, and F. C. Hickey, *Anal. Chem.*, **27**, 1895 (1955).

98. P. Sorensen, *Anal. Chem.*, **27**, 388 (1955).

99. D. G. Kaiser, *Anal. Chem.*, **35**, 552 (1963).

100. H. K. Mangold, *Fette, Seifen, Anstrichmittel*, **61**, 877 (1959).

101. J. K. Whitehead, *Biochem. J.*, **68**, 662 (1958).

102. J. K. Whitehead and D. Beale, *Clin. Chim. Acta*, **4**, 710 (1959).

103. D. Beale and J. K. Whitehead, *Clin. Chim. Acta*, **5**, 150 (1960).

104. G. A. Hagen, L. I. Diuguid, B. Kliman, and J. B. Stanbury, *Anal. Biochem.*, **33**, 67 (1970).

105. H. H. Henderson, F. Crowley, and L. E. Gaudette, *in* S. Rothchild, ed., *Advances in Tracer Methodology*, Vol. 2, Plenum, New York, 1965, pp. 83–86.

106. W. R. Waterfield, *J. Chem. Soc.*, **1964**, 541.

107. F. Sanger, *Biochem. J.*, **39**, 507 (1945).

108. F. C. McIntire, L. M. Clements, and M. Sproull, *Anal. Chem.*, **25**, 1757 (1953).

109. M. J. Kolbezen, J. W. Eckert, and B. F. Bretschneider, *Anal. Chem.*, **34**, 583 (1962).

110. J. K. Whitehead, *Biochem. J.*, **80**, 35P (1961).

111. D. Beale and J. K. Whitehead, in *Tritium in the Physical and Biological Sciences*, Vol. I, IAEA, Vienna, 1962, pp. 179–190.

112. F. Drawert, O. Bachmann, and K.-H. Reuther, *J. Chromatog.*, **9**, 376 (1962).

113. R. G. Garmon and M. E. Gibson, *Anal. Chem.*, **37**, 1309 (1965).

114. W. Seaman and D. Stewart, Jr., *Am. Dyestuff Reptr.*, **54**, 104 (1965).

115. T. Wieland, K. Schmeiser, E. Fischer, and H. Maier-Leibnitz, *Naturwissenschaften*, **36**, 280 (1949).

116. S. Blackburn and A. Robson, *Chem. Ind. (London)*, **1950**, 614.

117. S. Blackburn and A. Robson, *Biochem. J.*, **54**, 295 (1953).

118. G. A. Brodskaya, T. I. Chanysheva, and A. I. Chanyshev, *Radiats. Narusheniya Tverd. Telakh Zhidk.*, **1967**, 141; *Chem. Abstr.*, **69**:64467q.

119. T. Wada, *Nippon Nogeikagaku Kaishi*, **30**, 229 (1956); *Chem. Abstr.*, **51**:1361c.

120. T. Wada, *Nippon Nogeikagaku Kaishi*, **31**, 743 (1957); *Chem. Abstr.*, **52**:12969i.

121. T. Wada, *Dai-2-Kai Nippon Isotope Kaigi Hobunshu*, **2**, 513 (1958); *Chem. Abstr.*, **55**:7531f.

122. G. L. Callewaert and C. A. Vernon, *Biochem. J.*, **107**, 728 (1968).

123. H. J. Goren, D. M. Glyck, and E. A. Barnard, *Arch. Biochem. Biophys.*, **126**, 607 (1968).

124. I. B. Rubin and G. Goldstein, *Anal. Biochem.*, **33**, 244 (1970).

125. A. A. Noujaim, W. V. Kessler, J. E. Christian, and A. M. Knevel, *J. Pharm. Sci.*, **53**, 455 (1964).

126. E. R. Graham and A. Neuberger, *Biochem. J.*, **109**, 645 (1968).

127. C. E. Breckinridge, Jr., and J. E. Christian, *J. Pharm. Sci.*, **50**, 777 (1961).

128. R. A. Wiley and J. L. Metzger, *J. Pharm. Sci.*, **56**, 144 (1967).

129. M. Herbain and D. Bertin, *Bull. Soc. Chim. Biol.*, **41**, 621 (1959).

12 Hydrazines and hydrazides

I. ABSORPTION SPECTROPHOTOMETRIC METHODS*

Measurement after Chemical Reaction

1. Hydrazines—Reaction with Carbonyl Compounds. The reaction of a carbonyl compound and a hydrazine (see equation 1, Section I, of Chapter 2) can be used as the basis for the determination of either reactant. The application of the reaction to the determination of carbonyl compounds was described in Chapter 2.

Pesez and Petit (1) observed that a characteristic color results upon the addition of a solution of *p*-dimethylaminobenzaldehyde in ethyl alcohol and hydrochloric acid to hydrazine in dilute hydrochloric acid solution. They postulated that the product is an azine which isomerizes reversibly in an acid to a *p*-quinone:

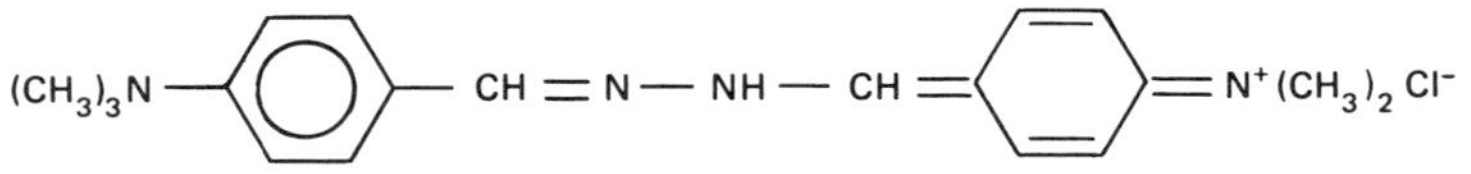

These acid salts are water soluble and obey Beer's law.

Watt and Chrisp (2) based a spectrophotometric procedure for hydrazine on this observation.

<small>METHOD FROM WATT AND CHRISP (2)</small>

Apparatus. Beckman Model DU spectrophotometer with Corex 1-cm cells.

Reagent. Color reagent, 0.4 gram of *p*-dimethylaminobenzaldehyde, 20 ml of ethanol, and 2.0 ml of concentrated hydrochloric acid.

---

* Written by J. Gordon Hanna.

298

Procedure.   Add 10 ml of the color reagent to aliquots of the hydrazine solution selected so that the final hydrazine concentration is within the range 0.02 to 0.8 ppm and dilute the resulting mixture to a total volume of 25.0 ml with 1 N hydrochloric acid solution. Prepare a blank consisting of 10 ml of the color reagent in 25.0 ml of solution 1 M with respect to hydrochloric acid. Read the absorbance of the sample versus the blank at 458 mμ.

Prepare a standard curve of absorbance versus concentration based on application of the procedure to known concentrations of hydrazine.

At room temperature color develops immediately and is stable after a period of 10 minutes. It was claimed that over the hydrazine concentration range 0.06 to 0.47 ppm the relative error does not exceed 1%.

McKennis and Yard (3) reported that methylhydrazine gives a color similar to the *p*-dimethylaminobenzaldehyde-hydrazine reaction product. By analogy, they proposed that the methylhydrazine product responsible for the absorption of visible light is

They found that microgram quantities of methylhydrazine adhered to Beer's law and were readily determined. Because of the greater convenience in preparation and the fact that similar results were obtained with both, aqueous acidic solutions (4) were preferred to the alcoholic acidic solutions of *p*-dimethylaminobenzaldehyde prepared according to Watt and Chrisp (2).

METHOD FROM MCKENNIS AND YARD (3) FOR METHYLHYDRAZINE

Reagent.   *p*-Dimethylaminobenzaldehyde. Dissolve 0.200 gram in 5 ml of 2 N sulfuric acid.

Apparatus.   Beckman Model DU spectrophotometer with 1-cm cells.

Procedure.   Dilute aliquots of solutions of methylhydrazine representing 25 to 80 γ to 5.5 ml with water and develop the color by adding 0.5 ml of the aqueous *p*-dimethylaminobenzaldehyde solution. Allow to stand for 15 minutes for the color to reach a maximum. Measure the absorbance versus a blank at 458 mμ. Prepare a calibration curve, using standard solutions of methylhydrazine sulfate.

The fact that the maximum absorbance of the methylhydrazine-*p*-dimethylaminobenzaldehyde condensation product is so closely similar to the maximum absorbance of the hydrazine-*p*-dimethylaminobenzaldehyde condensation product prohibits the use of the method for the determination of one in the presence of the other.

2. Maleic Hydrazide. Wood (4) reduced and hydrolzyed maleic hydrazide in water, alkali, and zinc to split off hydrazine, which he then distilled and determined colorimetrically, using p-dimethylaminobenzaldehyde. The method is not generally applicable, although Wood noted that bicyclodisuccinhydrazide also yields hydrazine under the conditions used. Phthalhydrazide does not give hydrazine under these conditions.

II.   GAS CHROMATOGRAPHIC METHODS*

Hydrazine and hydrazine derivatives have been determined directly by GC on alkali-treated packings. Treatment of the packing with methanolic KOH after coating with liquid phase gave more symmetrical peaks and better separations than alkali treatment of the support, followed by coating with liquid phase (5).

III.   ELECTROANALYTICAL METHODS†

A.   Direct Polarographic Determination of Hydrazines

Hydrazine and its organic derivatives are of importance in many applications, primarily because of their facile oxidation. It is this ease of oxidation which permits the analysis of hydrazines by many different routes, including direct polarographic oxidation.

In a strongly alkaline medium (0.1 M sodium hydroxide) hydrazine (6) and a variety of organic hydrazines and hydrazides (7) produce well-defined, reversible, anodic polarographic waves. The wave for hydrazine itself is sufficiently separated (Table 12-1) from the waves for organic hydrazines to permit measurement of inorganic hydrazine contamination of the latter compounds.

One of the interesting features observed was that the half-wave potentials of alkyl-substituted hydrazines varied linearly with molecular weight (Figure 12-1).

The use of 0.1 M sodium hydroxide as a background electrolyte has some disadvantages. The hydrazines are kept in their free base form and thus may

---

* Written by Morton Beroza and May N. Inscoe.
† Written by Alan F. Krivis.

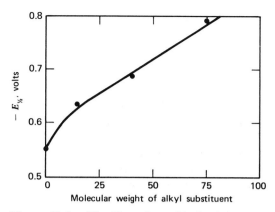

Figure 12–1. The $E_{1/2}$ values of hydrazines versus substitutent molecular weight.

Table 12-1. Polarographic Values for Substituted Hydrazines in 0.1 M NaOH

Compound	$-E_{1/2}$, V[a]
Hydrazine	0.548
Monomethylhydrazine	0.634
n-Propylhydrazine	0.689
n-Hexylhydrazine	0.791
1,1-Dimethylhydrazine	0.694
1,2-Dimethylhydrazine	0.698
1,2-Diisobutylhydrazine	0.757
Phenylhydrazine	0.752
Acetic hydrazide	0.619

[a] Versus a saturated mercurous sulfate electrode.

be volatilized more readily or may decompose more rapidly. However, these problems are outweighed by the well-defined waves which are produced in this solution. With care, excellent precision may be obtained.

EXPERIMENTAL METHOD [ADAPTED FROM THE STUDY OF KRIVIS AND SUPP (7)]*

Apparatus. Any polarograph, recording or manual, may be used in conjunction with a thermostated (25°C) H-cell containing a mercurous sulfate reference electrode (8, 9).

* Reprinted in part with the permission of the copyright owner the Academic Press.

Solutions.   The background electrolyte solution is a 0.1 M sodium hydroxide solution containing 0.005% gelatin. The gelatin is best added to the caustic solution immediately before use. For convenience, the gelatin may be prepared as a fresh stock solution (0.5 to 1.0%) and the proper volume added to the hydroxide solution in the cell.

Procedure.   Stock solutions of the hydrazine samples should be prepared immediately before use and deoxygenated with nitrogen presaturated with a portion of the stock solution.

The correct amount of background solution should be transferred to the H-cell and bubbled with nitrogen for 10 minutes. A suitable-sized aliquot of the degassed hydrazine stock solution should be added to the cell, the contents mixed, and the polarogram run immediately. The final concentration of hydrazine should be about 1 mM. The diffusion current for the oxidation wave of the hydrazine unknown should be calculated and compared to a calibration curve to determine the percentage of hydrazine present in the unknown.

B.   Coulometric Titration of Hydrazines and Hydrazides

Bromine, in acid solution, oxidizes hydrazines and hydrazides as shown in equations 1 and 2:

$$N_2H_4 + 2Br_2 \rightarrow N_2 + 4HBr \tag{1}$$

$$\underset{\displaystyle \substack{O \\ \parallel}}{R C N_2 H_3} + 2Br_2 + H_2O \rightarrow \underset{\displaystyle \substack{O \\ \parallel}}{R C OH} + N_2 + 4HBr \tag{2}$$

If the conditions are maintained properly, the reaction is quantitative and very rapid. Since the solution is acidic, the unknowns are present as salts and thus are less prone to decompose or become lost because of volatility.

The coulometric generation of bromine, as detailed in Section III, Chapter 1, for the titration of phenols, works very well in the analysis of almost all types of hydrazino compounds (10, 11, 12). Although the stoichiometry indicated in equations 1 and 2 is applicable to the majority of unknowns, there are a number of exceptions and this aspect should be clarified with known materials.

Hydrazines may be titrated in any of the solvents listed below (11) under "Experimental Method." Hydrazides must *not* be run in the acetic acid solvent (12), however, because acetic acid can react with hydrazides to form diacylhydrazines, which are not oxidizable under the conditions of the analysis. For this reason the aqueous or methanolic HCl-KBr titration media are recommended as the more generally applicable solvents.

The lack of reactivity of diacylhydrazines can help in another way, however. A major contaminant of most hydrazides is the symmetrically substituted diacylhydrazine derived from the parent hydrazide. The analysis described is specific for the hydrazide; the diacylhydrazine does not react and therefore is not an interference.

EXPERIMENTAL METHOD [ADAPTED FROM KRIVIS ET AL. (12)]*

Apparatus. A coulometric constant-current power supply and timer should be used for this analysis. The unit may be either laboratory constructed or purchased from a laboratory supply house. The biamperometric detection circuit may be wired from a 10-0-10-μA meter-relay, 1.5-V battery, and a variable resistor as shown in Figure 1-5 of Chapter 1, Section III.

The generating electrodes may be a coiled platinum anode and a platinum foil cathode inside an isolation chamber. The latter is a glass tube with a very fine frit sealed in its bottom.

The detecting system consists of two small platinum wire electrodes with 0.2 V impressed across them from the battery-variable resistor network. The meter should be wired in series with one of the electrodes.

Titration Solvents.

(*a*) Acetic acid, 60%; methanol, 26%; aqueous KBr, 14% (1 *M*); mercuric acetate, 0.006 *M*.

(*b*) 0.3 *M* HCl-0.1 *M* KBr, aqueous.

(*c*) 0.3 *M* HCl-0.1 *M* KBr, 85% methanol.

Procedure. Place a magnetic stirring bar in a 100-ml beaker, transfer an accurately weighed sample containing approximately 1 mg of unknown to the beaker, and add 50 ml of suitable bromination solution. Dip the electrodes into the solution, start the stirrer, and titrate with a generation current sufficient to give about a 200-second titration. The end point is reached when a stable 5-μA current is shown on the detection circuit meter. Run a blank titration and subtract the blank time from the sample titration. Calculate the amount of unknown present by means of Faraday's law.

Calculation.

$$\text{Per cent of unknown} = \frac{(i)\ (t)\ (\text{mol. wt. of unknown})\ (100)}{(4)\ (\text{sample weight})\ (96487)}$$

C.   Potentiometric Titration of Hydrazines

Hydrazines and their derivatives may be oxidized by a wide variety of agents. Perhaps the best one available for titrimetric use is potassium iodate. Standard solutions of iodate are reasonably stable, and several different end-point detection systems can be utilized (13–15).

A remarkably simple method involves potentiometric titration using a platinum-calomel electrode pair (15). The sample is tritrated in acid solution at ice-bath temperatures until a significant change in potential is observed; the change at the end point is usually so great that a plot of the curve is not necessary.

The reaction followed by hydrazine is as follows:

$$N_2H_4 + KIO_3 + 2HCl \rightarrow KCl + ICl + N_2 + 3H_2O$$

and most monosubstituted alkylhydrazines also have the same stoichiometry. Some aromatic hydrazines vary from this stoichiometry, however, as do certain disubstituted hydrazines (15). For example, some 1,1-disubstituted materials involve only two electrons instead of four. Therefore, it is wise to establish the stoichiometry of the reaction of the particular unknown before samples are run.

EXPERIMENTAL METHOD [ADAPTED FROM THE STUDY OF MCBRIDE, HENRY, AND SKOLNIK (15)]*

Apparatus.   A direct-reading pH meter should be used to monitor the emf changes.

A platinum-calomel electrode pair should be employed as the detecting system. The platinum electrode may be in any of the forms obtainable commercially (e.g., button, disk, foil, or helix), but a small foil electrode is both inexpensive and quite satisfactory. A commercial calomel reference electrode also is suitable if the saturated potassium chloride solution in the salt bridge is replaced with 1 M potassium chloride.

Procedure.   Position a beaker containing 15 ml of distilled water and a magnetic stirring bar in an ice-brine bath, which, in turn, is rested on a magnetic stirrer. Allow the beaker and contents to cool. Transfer an aliquot of a sample solution and 50 ml of cold, concentrated hydrochloric acid to the beaker. Make sure that sufficient time is allowed for cooling the resulting solution to $-5°C$. Immerse a platinum-calomel electrode pair in the solution and titrate with 0.1 N potassium iodate solution until the largest emf change

* Reprinted in part with the permission of the copyright owner, the American Chemical Society.

per smallest increment of titrant occurs. The change at the end point is very abrupt and very large.

Two important points should be emphasized. The temperature must not be allowed to rise above 0°C during the titration, and the hydrochloric acid concentration must be maintained at 5 to 6 M, as a minimum.

REFERENCES

Absorption Spectrophotometric

1. M. Pesez and A. Petit, *Bull. Soc. Chim. France*, **1947**, 122.
2. G. W. Watt and J. D. Chrisp, *Anal. Chem.*, **24**, 2006 (1952).
3. H. McKennis, Jr., and A. S. Yard, *Anal. Chem.*, **26**, 1960 (1954).
4. P. R. Wood, *Anal. Chem.*, **25**, 1879 (1953).

Gas Chromatographic

5. C. Bighi and G. Saglietto, *J. Gas Chromatog.*, **4**, 303 (1966).

Electroanalytical

6. S. Karp and L. Meites, *J. Am. Chem. Soc.*, **84**, 906 (1962).
7. A. F. Krivis and G. R. Supp, *Microchem. J.*, **14**, 603 (1969).
8. I. M. Kolthoff and J. J. Lingane, *Polarography*, 2nd Ed., Interscience, New York, 1952.
9. L. Meites, *Polarographic Techniques*, 2nd Ed., Interscience, New York, 1965.
10. L. Szebelledy and Z. Somogyi, *Z. Anal. Chem.*, **112** (1938).
11. E. C. Olson, *Anal. Chem.*, **32**, 1545 (1960).
12. A. F. Krivis, E. S. Gazda, G. R. Supp, and P. Kippur, *Anal. Chem.*, **35**, 1955 (1963).
13. R. A. Penneman and L. F. Audrieth, *Anal. Chem.*, **20**, 1058 (1948).
14. I. M. Kolthoff, *J. Am. Chem. Soc.*, **46**, 2009 (1924).
15. W. R. McBride, R. A. Henry, and S. Skolnik, *Anal. Chem.*, **25**, 1042 (1953).

13 Diazonium salts

I. ABSORPTION SPECTROPHOTOMETRIC METHODS*

A. Direct Spectrophotometric Determination of Stabilized Diazonium Compounds

Rosenberger and Shoemaker (1) found that stabilized diazonium compounds which give 4-(dialkylamino)benzenediazonium cations can be determined by direct measurement of the absorbance at 380 mμ in aqueous solution. Beer's law is followed in the concentration range 1 to 10 ppm. The method should be applicable to other benzenediazonium cations with a calibration curve established at the proper absorption maximum. The absorption band in the 350- to 400-mμ region is attributed to the —N=N— chromophore group. None of the decomposition products or other expected impurities absorb at this wavelength.

B. Measurement after Chemical Reaction

Coupling Reaction. Traces of diazoniium compounds can be determined spectrophotometrically on the basis of the coupling reaction in which a suitable dye is produced and measured. Siggia (2) recommended phloroglucinol, naphthol, or R-salt as the coupling reagent. Although phloroglucinol and naphthol are more sensitive as a result of the intensities of the dyes formed, R-salt is more convenient because of its solubility and its reactions are usually run in aqueous solution.

* Written by J. Gordon Hanna.

METHOD FROM SIGGIA (2)

Reagents.
   Coupling reagents:
      R-salt solution, 1% aqueous.
      Phloroglucinol solution, 0.5% aqueous.
      Naphthol solution, 1% in acetone.
   Alkaline agents:
      Sodium acetate solution, 5% aqueous or alcoholic.
      Sodium carbonate solution, 5% aqueous.
      Sodium hydroxide solution, 5% aqueous or alcoholic.

Procedure: Aqueous Systems.   Using R-salt as the coupler, add an amount sufficient to react with all the diazonium compound in the sample. Add sodium bicarbonate solution or sodium acetate solution to promote full development of color. Allow to stand at room temperature until maximum color has developed (usually 5 to 10 minutes). Measure the absorbance of the sample solution versus a blank and determine the concentration from a calibration curve prepared using known concentrations of standard diazonium compound.

Some diazonium compounds require stronger alkaline conditions than are supplied by the bicarbonate or acetate solutions (pH 7 to 8.5). Sodium carbonate (pH 7 to 10) or sodium hydroxide (pH above 10) can be used in these cases. However, sodium carbonate and especially sodium hydroxide must be used with care because the high alkalinity produced by these bases can cause decomposition of diazonium compounds.

Procedure: Alcoholic Systems.   Dissolve the sample in alcohol and use either phloroglucinol or naphthol as the coupler. Use only alcoholic sodium acetate or alcoholic sodium hydroxide as the alkaline agent.

II.   ELECTROANALYTICAL METHODS*

Polarographic Determination of Diazonium Salts

Diazonium salts are readily reduced by chemical means, for example, titanous chloride solution (2). From this behavior, one might predict that diazonium salts could be reduced polarographically, and such has been found to be the case with a wide range of these salts (3).

* Written by Alan F. Krivis.

The polarographic behavior of diazonium salts is complex. From two to four waves are found in acidic to slightly alkaline media (pH 4 to 9). Above pH 9 to 10, diazotates and isodiazotates are formed and these are not reducible. The total diffusion current for all the waves corresponds to a four-electron reduction, so that the overall reaction is as follows:

$$ArN_2^+ + 4e + 3H^+ \rightarrow ArNHNH_2$$

Several of the waves mentioned above shift in half-wave potential as a function of pH. The potential of the first wave, however, is independent of pH, and this wave can serve as a means for the quantitative analysis of diazonium salts. If the compound is run in a buffer at ca. pH 7, the first wave is well separated from the other waves and the diffusion current can easily be measured at -0.3 V versus the saturated calomel electrode. This measurement also forms the basis for an amperometric titration of phenolic compounds with a standard solution of diazonium salt (4).

EXPERIMENTAL METHOD

Apparatus. Any polarograph, recording or manual, may be used for this analysis. The method involves only two measurements at -0.3 V so that a simplified apparatus is more than adequate.

The optimum cell is an H-cell with a saturated calomel reference electrode and a dropping mercury indicating electrode (DME). However, a much simpler cell with a mercury pool reference may be used if desired (5). The glassware for either of these cells can be purchased from several laboratory supply houses, including Sargent-Welch Scientific Co., 7300 N. Linder Ave., Skokie, Illinois 60076.

Reagents and Solutions. The test solution should contain a combined buffer-background electrolyte. The actual pH can be varied quite widely to optimize conditions for a particular unknown; buffers from pH 4 (acetate) to 8 (borate) are useful. A solution containing 0.05 M KH_2PO_4-0.03 M NaOH-0.02 M KCl has a pH approximating 7 and can serve as the supporting electrolyte for many compounds.

Procedure. Transfer an accurately weighed sample containing about 0.1 mM of diazonium salt to a 100-ml volumetric flask. Add sufficient buffer solution to dissolve the sample and dilute to the mark with additional buffer solution. Mix thoroughly and transfer an aliquot to the polarographic cell. Deoxygenate with pure nitrogen by bubbling for 10 minutes, insert the DME in the cell, and divert the nitrogen to blanket the surface of the solution. Measure the current at -0.3 V. In a separate experiment, measure the residual current also at -0.3 V. (This is accomplished by transferring only the buffer solution to the cell, deoxygenating, and then measuring the current.) Subtract the residual current from the sample current to obtain the diffusion current for the unknown. Compare this diffusion current to a previously prepared calibration curve to determine the amount of diazonium salt in the sample.

REFERENCES

Absorption Spectrophotometric

1. H. M. Rosenberger and C. J. Shoemaker, *Anal. Chem.*, **31**, 204 (1959).
2. S. Siggia, *Quantitative Organic Analysis via Functional Groups*, 3rd Ed., John Wiley, New York, 1963, p. 549.

Electroanalytical

3. R. M. Elofson, R. L. Edsberg, and P. A. Mecherly, *J. Electrochem. Soc.*, **97**, 166 (1950).
4. R. M. Elofson and P. A. Mecherly, *Anal. Chem.*, **21**, 565 (1949).
5. I. M. Kolthoff and J. J. Lingane, *Polarography*, 2nd Ed., Interscience, New York, 1952.

14 Isocyanates and isothiocyanates

I. ABSORPTION SPECTROPHOTOMETRIC METHODS*

A. Direct Infrared Measurement

David (1) used the first overtone NCO stretching vibration at 2.642 μ in the near-infrared region for the quantitative determination of mixtures that contain isocyanates. Bailey, Kirss, and Spaunburgh (2) obtained kinetic data on the reactivity of the isocyanate group, using the intense absorption of the group at 4.5 μ.

Lord (3) quantitatively differentiated toluene-2,4-diisocyanate and toluene-2,6-diisocyanate in mixtures by infrared measurement, employing two approaches, one for mixtures containing 5 to 95% of the 2,4-isomer and the other for mixtures containing more than 95% of this isomer. The first method consists of the measurement of the 2,4-isomer at 12.35 μ and the 2,6-isomer at 12.80 μ of cyclohexane-diluted samples. Correction factors were applied for the slight overlap of the two isomer bands. The other method (used when there is more than 95% of the 2,4-isomer) consists of the measurement of undiluted samples at 12.70 μ for the 2,6-isomer.

Whiffen, Tarkington, and Thompson (4) used the absorption exhibited by isothiocyanates at 1710 cm^{-1} to determine dodecyl isothiocyanate in mixtures also containing dodecyl thiocyanate.

B. Measurement after Chemical Reaction

1. Hydrolysis of Isocyanates to Amines, Followed by Diazotization and Coupling. Marcali (5) developed a method for the determination of toluene-2,4-diisocyanate in air, based on hydrolysis of the diisocyanate

---

* Written by J. Gordon Hanna.

310

to the corresponding diamine, followed by diazotization of the amine and coupling with N-1-naphthylethylenediamine. The final color was measured spectrophotometrically at 550 mμ. The absorbing medium for collecting the diisocyanate from the air was an aqueous mixture of acetic and hydrochloric acids. Meddle, Radford, and Wood (6) used a similar procedure for aromatic isocyanates in air but employed an acidic dimethylformamide solution as the absorber. In both of these methods aromatic primary amines present in the atmosphere interfere by yielding high results.

Meddle and Wood (7) discovered that isocyanate from the atmosphere collected in dimethylformamide containing only 1,6-diaminohexane is not converted into amine even after subsequent addition of hydrochloric acid but yields a product that could not be diazotized and coupled to give a color. On the basis of this observation they then developed a method whereby the samples of test atmosphere are drawn simultaneously through two different absorbing solutions. One sample is collected in a dimethylformamide solution of 1,6-diaminohexane, and the other in a dimethylformamide solution of 1,6-diaminohexane and hydrochloric acid. Any aromatic amine present or produced in each absorbing solution is diazotized and coupled with N-1-naphthylethylenediamine to form the colored complex, which is measured. The difference between the resulting absorbances gives a measure of the isocyanate content of the air.

METHOD FROM MEDDLE AND WOOD (7)

Reagents. Dilute hydrochloric acid. Dilute 15 ml of concentrated acid to 100 ml with water.

1,6-Diaminohexane in dimethylformamide, 70 μg/ml.

Diazotization solution. Dilute 3 grams of sodium nitrite and 5 grams of sodium bromide to 100 ml with water.

Sulfamic acid solution, 10% (w/v) aqueous.

N-1-Naphthylethylenediamine solution. Dissolve 0.75 gram of the amine dihydrochloride in water, add 2 ml of concentrated hydrochloric acid, and dilute to 100 ml with water. Prepare a fresh solution after 2 days.

Procedure. Add 3 ml of the diaminohexane-dimethylformamide solution to each of two absorber tubes. Then add 2 ml of hydrochloric acid solution to one of them (A). Simultaneously draw 10 liters of the test atmosphere through each absorbing solution at the rate of 1 liter/minute. Allow the absorbing solutions to stand for 10 minutes for any reaction to reach completion. Add 2 ml of hydrochloric acid solution through the top of the inlet tube of absorber tube B, containing only diaminohexane-dimethylformamide solution. Lift the inlet tubes so that the sinters in each absorber tube are clear of the respective solutions and expel the liquid trapped in each sinter as completely as possible.

Ensure that the temperature of the absorbing solutions is not above 20°C; then add 0.5 ml of the diazotization solution to each. Shake the mixtures and allow them to stand for 2 minutes. Add 0.5 ml of sulfamic acid to each absorber tube and shake until the effervescence has ceased; 2 minutes later, add 0.5 ml of the naphthylethylenediamine solution to each tube and shake well. Allow the colors to develop for 10 minutes before measuring the absorbance of each solution in 20-mm cells versus water as reference.

Subtract the absorbance of the solution in tube B from that of the solution in tube A to obtain the absorbance of the isocyanate in tube A. Determine the amount of isocyanate by reference to the calibration curve.

Calibration Curve. To a series of 6 absorber tubes, each containing 3 ml of the diaminohexane-dimethylformamide solution and 2 ml of the hydrochloric acid solution, add 0, 0.01, 0.02, 0.03, 0.04, and 0.05 ml of standard isocyanate (30 mg in 100 ml of toluene) solution, covering the range 0 to 15 μg from a micrometer syringe. Develop the color in each tube as described in the procedure. Measure the color at an appropriate wavelength (550 mμ for toluene-2,4-diisocyanate). Plot micrograms of isocyanate versus absorbance.

2. Isocyanates—Reaction with Butylamine and Determination of Excess Butylamine with Malachite Green.

Kubitz (8) determined residual isocyanate in urethane-based polymers by allowing it to react with excess standard solution of *n*-butylamine in tetrahydrofuran. He then determined the excess unreacted butylamine colorimetrically with malachite green, with which it forms a colorless derivative:

$$RNCO + R'NH_2 \rightarrow RNHCONHR' \tag{1}$$

$$(C_{23}H_{25}N_2C_2HO_4)H_2C_2O_4 + 8n\text{-}C_4H_9NH_2 \rightarrow$$

$$2C_{23}H_{25}N_2NHC_4H_9 + 3(C_4H_9NH_2)_2H_2C_2O_4 \tag{2}$$

METHOD FROM KUBITZ (8)

Reagents. Tetrahydrofuran, sodium dried and distilled. The peroxide content at the time of use must be less than 5 ppm in terms of active oxygen.

Pyridine. It must contain 0.05 to 0.15% water to control the color intensity of the pyridine-malachite green solution.

n-Butylamine in tetrahydrofuran. Purify *n*-butylamine by distillation from solid potassium hydroxide and collect the fraction boiling between 76 and 78°C.

(*a*) Standard solution for use within 2 days: Add 30 ml of dried and distilled tetrahydrofuran to a 50-ml volumetric flask, weigh 5 drops of *n*-butylamine from a tared Lunge bottle into the flask, and record the loss in weight of the bottle (approximately 0.1 gram). Dilute to 50 ml with tetrahydrofuran. Take a 10-ml aliquot of this solution and dilute to 250 ml with tetrahydrofuran. The concentration of the final solution should be near 1.2 μM per milliliter of tetrahydrofuran.

(*b*) Standard solution stable for a month or more: Add 0.25 grams of *n*-butylamine to 1 gal of dried distilled tetrahydrofuran. Let the solution stand for at least 3 days before use. Pipet 25.00 ml of the solution into a 50-ml Erlenmeyer flask and titrate with 0.01 N hydrochloric acid to a bromophenol blue end point. The concentration should fall in the range of 1.0 to 1.2 μM *n*-butylamine per milliliter.

Malachite green in pyridine. Recrystallize malachite green [oxalate salt, $(C_{23}H_{25}N_2C_2HO_4)_2H_2C_2O_4$] from hot water and dry in air for several days. Weigh 0.45 gram of the recrystallized malachite green into 450 grams of pyridine containing 0.05 to 0.15% water. This solution should contain sufficient reagent to give a transmittance of 25 to 35% when 5 ml of the solution is mixed with 5 ml of tetrahydrofuran. The solution must stand for 24 hours after preparation or after readjustment before use.

Apparatus.   Photometer, Cenco-Sheard-Sanford or equivalent, with 610-mμ filter, 1-cm cells.

Calibration.   Pipet the following 5-ml mixtures of tetrahydrofuran and standard *n*-butylamine into a 1-cm cell and to each add 5 ml of the solution of malachite green in pyridine:

Tetrahydro-furan, ml	Standard butylamine, ml
5	0
4	1
3	2
2	3
1	4
0	5

Mix the solution in the cell with a thin glass rod until it appears homogeneous. Do not place the solution in the light path of the Photelometer until it is time to take the reading. At the end of 3 minutes $\pm$ 5 seconds from the time of the start of the addition of malachite green solution to the cell, read the per cent trasmittance compared with a blank (5 ml of tetrahydrofuran and 5 ml of pyridine). Make a plot of per cent transmittance versus concentration in micromoles of butylamine per 5 ml.

Procedure. Weigh a sample of polymer into the 25-ml glass-stoppered flask (use 0.05 gram for 0.4 to 1.0% isocyanate and 0.10 gram for 0 to 0.4% isocyanate). From a buret or pipet, add 15.0 ml of the standard *n*-butylamine solution to the flask and shake mechanically for the following periods of time, depending on the speed or accuracy required:

—NCO Recovery as a Function of Shaking Time

Shaking time, hours	Approx. fraction of —NCO recovery
0.5	0.62
1.5	0.82
3–4	1.00

Withdraw 5 ml of the unreacted butylamine solution from the 25-ml flask, using a 5-ml pipet with the tip wrapped like a swab with a thin layer of glass wool. Remove the glass wool and add the 5-ml portion to the 1-cm absorption cell. Pipet 5 ml of the malachite green solution into the cell and mix. At the end of 3 minutes ± 5 seconds from the time of the start of the malachite green addition, read the per cent transmittance versus a blank (5 ml of tetrahydrofuran and 5 ml of pyridine). From the calibration plot determine the concentration of unreacted butylamine. Determine the difference between the initial and final *n*-butylamine concentrations and multiply this difference by 3 to obtain the change in *n*-butylamine concentration of the 15-ml aliquot.

The 95% confidence limit for the average of a dulpicate determination of less than 0.60% free isocyanate in a cured elastomer was ±0.033%, calculated on the basis of quadruplicate determinations of 4 samples by 2 analysts on 2 different days.

Water, weak amines, ureas, and urethanes did not interfere. Strong and weak tertiary amines and amines that might be present in urethane-based polymers, such as *o*-, *p*-, and *m*-toluidines, which have ionization constants less than 1×10^{-6}, did not react with the malachite green.

II. GAS CHROMATOGRAPHIC METHODS*

Isothiocyanates, for example those in grape and mustard seeds, have been determined directly (9). The use of Telfon supports has been found advantageous in the analysis of isocyanates (10, 11).

* Written by Morton Beroza and May N. Inscoe.

III. RADIOCHEMICAL METHODS*

A. Conversion of Isocyanates to Substituted Ureas

Interest in determining the isocyanate group radiochemically derives principally from the presence of very low concentrations of this structure in various polyurethanes prepared by condensation polymerization. Residual concentrations are characteristically less than 200 μM/gram in intermediates and are usually less than 20 μM/gram in final products. Although cross-linked polyurethanes, such as foams and elastomers, commonly exhibit no appreciable solubility in organic solvents, essentially linear polymers are usually completely soluble in dimethylformamide (DMF). These linear polymers may be prepared with terminal isocyanate groups useful in subsequent cross-linking reactions yielding elastomers.

A method employing n-butylamine-1-^{14}C as radioreagent has been described for determining the —NCO group in a class of such DMF-soluble isocyanate-terminated polyurethanes (12). A 1-gram sample of polymer is dissolved at room temperature in 25 ml of a DMF solution containing 2.0 ml of a 0.5 M solution of the labeled amine in chlorobenzene. After the polymer has dissolved, 1.0 ml of nonradioactive n-butylamine and 4 ml of chlorobenzene are added. Excess amine is removed by a single precipitation in water. Radioassays of the derivatized polyurethane and of the radioreagent, as N-(n-1-butyl-^{14}C)-4-chlorobenzamide, are made by liquid scintillation counting in a 2:1 toluene-DMF medium. The sensitivity of the method is approximately 1 μM NCO/gram. The moles per gram, E, of combined butylamine-^{14}C in the separated polymer is given by

$$E = \frac{S_p}{S_a} \tag{1}$$

where S_p is the specific activity of the product (μCi/gram), and S_a is the specific activity of the amine (μCi/mole). The NCO content, C, of the original polyurethane (moles/gram) is obtained from

$$C = \frac{E}{1 - EW} \tag{2}$$

where W, the molecular weight of butylamine, represents the increase in average functional equivalent weight of the polymer resulting from derivatization. The principles of radioreagent methods for determining low concentrations of functional groups in elastomers have been discussed in detail (13).

* Written by D. Campbell.

B. Conversion of Isothiocyanates to the Labeled Form

A radioisotope approach potentially useful for determining the isothio-cyanate group in low-molecular-weight compounds is isotope exchange of elemental ^{35}S with the sulfur of the isothiocyanate:

$$R\text{---}N\text{=}C\text{=}S + {}^{35}S_8 \rightleftharpoons R\text{---}N\text{=}C\text{=}{}^{35}S + {}^{35}S_8$$

Phenyl isothiocyanate and ethyl isothiocyanate attain 100% exchange, that is, reach equilibrium specific activity, within 6 hours at 180°C in the melt and in decalin, respectively (14, 15). Under these conditions, however, slight decomposition occurs. The exchange reaction, which is useful only in the absence of other exchangeable forms of sulfur, may be followed readily by paper chromatography of aliquots taken periodically from the system (16).

REFERENCES

Absorption Spectrophotometric

1. D. J. David, *Anal. Chem.*, **35**, 37 (1963).
2. M. E. Bailey, V. Kirss, and R. G. Spaunburgh, *Ind. Eng. Chem., Anal. Ed.*, **48**, 794 (1956).
3. S. S. Lord, Jr., *Anal. Chem.*, **29**, 497 (1957).
4. D. H. Whiffen, P. Tarkington, and H. W. Thompson, *Trans. Faraday Soc.*, **41**, 200 (1945).
5. K. Marcali, *Anal. Chem.*, **29**, 552 (1957).
6. D. W. Meddle, D. W. Radford, and R. Wood, *Analyst*, **94**, 369 (1969).
7. D. W. Meddle and R. Wood, *Analyst*, **95**, 402 (1970).
8. K. A. Kubitz, *Anal. Chem.*, **29**, 814 (1957).

Gas Chromatographic

9. P. Rigolier, *Rev. Franc. Corps Gras*, **15**, 683 (1968).
10. D. L. Andersen, *J. Assoc. Offic. Anal. Chemists*, **53**, 1 (1970).
11. G. W. Ruth, *J. Gas Chromatog.*, **6**, 513 (1968).

Radiochemical

12. D. R. Campbell, *Microchem. J.*, **14**, 536 (1969).
13. D. R. Campbell and W. C. Warner, *Rubber Chem. Technol.*, **42**, 1 (1969).
14. E. N. Guryanova and L. S. Kuzina, *Zh. Fiz. Khim.*, **28**, 2116 (1954).
15. L. S. Kuzina and E. N. Guryanova, *Zh. Fiz. Khim.*, **33**, 2030 (1959).
16. J. Moravek, Z. Nejedly, and J. Filip, *Science*, **138**, 146 (1962).

15 Mercaptans

I. ABSORPTION SPECTROPHOTOMETRIC METHODS*

Measurement after Chemical Reaction

1. Conversion to Nitroso Derivative and Determination of the Combined Nitrous Acid. Saville (1) converted mercaptans to their nitroso derivatives:

$$RSH + HONO \rightarrow RSNO$$

which yield an equivalent of nitrous acid on mercuric ion-assisted hydrolysis. He used the liberated nitrous acid to diazotize sulfanilamide, which, in turn, coupled with *N*-naphthylethylenediamine to form an intensely colored azo dye. The intensity of the color of the dye was directly related to the concentration of mercaptan in the original sample.

METHOD FROM SAVILLE (1)

Reagents. Solution A: Mix 1 volume of 0.01 M aqueous sodium nitrite with 9 volumes of 0.2 to 1.0 N sulfuric acid.

Solution B: 0.5% ammonium sulfamate in water.

Solution C: Mix 1 volume of 1.0% aqueous mercuric chloride with 4 volumes of 3.4% sulfanilamide in 0.4 N hydrochloric acid.

Solution D: Prepare a 1% solution of *N*-1-naphthylethylenediamine in 0.4 N hydrochloric acid fresh daily.

Procedure. Add 1 ml of the mercaptan solution (0.02 to 5 mM) in water or in aqueous ethanol to 5 ml of solution A in a 25-ml volumetric flask. Allow to stand for 0.5 to 5 minutes, depending on the type of mercaptan present, and then add 1 ml of solution B. Stopper and shake well for a few seconds

---

* Written by J. Gordon Hanna.

317

to ensure removal of excess nitrous acid. After 1 to 2 minutes, rapidly add 10 ml of solution C. Add solution D to bring the volume of the flask to the mark. The color will develop in 3 to 5 minutes. After 10 minutes read the absorbance with a yellow-green filter or at about 550 mμ. Calculate the quantity of mercaptan from a previously prepared calibration chart.

2. Reaction with Silver Ions. The quantitative replacement of the thiol hydrogen atom with a heavy metal ion has been a favorite basis for the determination of mercaptans:

$$RSH + M^+ \rightarrow RSM + H^+$$

Kunkel, Buckley, and Gorin (2) based a spectrophotometric method on the decomposition of silver dithizonate by thiols to form free dithizone, which has an absorption maximum at 615 mμ.

METHOD FROM KUNKEL, BUCKLEY, AND GORIN (2)

Apparatus. Beckman Model DU spectrophotometer with 1-cm Corex cells.

Reagents. Dithizone solution. Dissolve 0.1 gram in 1 liter of carbon tetrachloride.

Silver dithizonate solution. Vigorously shake 200 ml of the carbon tetrachloride solution of dithizone with 1 gram of silver nitrate in 100 ml of distilled water and 5 ml of 6 M sulfuric acid. Wash the orange-red carbon tetrachloride layer with water, separate, filter, and dilute with sufficient carbon tetrachloride to give an absorbance of 5.0 at 460 mμ.

Procedure. Add exactly 3 ml of the silver dithizonate solution to a 1-cm absorption cell. Add 0.01 to 1 ml of the sample (the sample size is limited by the size of the cell) and stir to mix. After 5 minutes, read the absorbance at 615 mμ versus a blank consisting of the dithizonate solution, which has an almost negligible absorbance at 615 mμ. Correct for the dilution caused by the addition of the sample by adjusting the value of the absorbance to that which it would be if the total volume were 3.00 ml. Read the amount of mercaptan from a standard curve.

Data obtained by Kunkel, Buckley, and Gorin (2) are shown in Table 15-1. The absorbance values obtained with hexanethiol (sample I) were plotted against the amounts taken, 0.002 to 0.08 mM, and a good straight line was obtained. This was then used as a calibration curve for the analysis of the other samples.

3. Tertiary Mercaptans—Determination as Thionitrites. Ashworth and Keller (3) converted mercaptans to thionitrites and subsequently

Table 15-1. Summary of Mercaptan Analyses (2)

Thiol	Purity, %	Number of determinations	Purity found,%, ± average deviation
Dodecanethiol	91.4	8	91.9 ± 2
Hexanethiol			
Sample I	99.9[a]		
Sample II	96.3	8	96.5 ± 0.7
tert-Hexanethiol	97.4	7	97.5 ± 1.6
Butanethiol			
Sample I	99.6[a]		
Sample II	94.5	8	94.2 ± 1.1

[a] American Petroleum Institute Certified standards, others by the method of Kolthoff and Harris (4).

measured the resonating thionitrite species in the ultraviolet. Primary mercaptans produce no significant absorption under the experimental conditions used, and secondary mercaptans show about 30% of the molar absorptivity of the tertiary.

METHOD FROM ASHWORTH AND KELLER (3)

Apparatus.   Cary Model 11 spectrophotometer with 2-cm fused silica cells.

Procedure.   Weigh the sample containing 1.0 mM of tertiary mercaptan to the nearest 0.1 mg into a 100-ml volumetric flask. Dissolve in carbon tetrachloride and dilute to volume. Pipet a 5-ml aliquot into a 125-ml separatory funnel containing about 25 ml of water. Pipet 20 ml of carbon tetrachloride into the separatory funnel. Carry out the following manipulations in a dark room provided with incandescent illumination only.

Pipet 1 ml of 50% sodium nitrite solution and 2 ml of 6 N hydrochloric acid into the separatory funnel, shake vigorously for 30 seconds, vent the funnel, and continue shaking for an additional 30 seconds. Draw off the carbon tetrachloride layer into a second separatory funnel, add 25 ml of distilled water, and shake vigorously. Draw off the carbon tetrachloride layer into a third separatory funnel, add 15 ml of 5% ammonium carbonate solution, and shake vigorously. Filter the carbon tetrachloride through a small pledget of cotton packed in the stem of the funnel to remove suspended water. Record the absorbance of the carbon tetrachloride solution from 400–

270 mμ in a 2.0-cm cell, using carbon tetrachloride in the reference cell. Measure the absorbance of the thionitrite at the absorbance maximum (about 344 mμ). Reagent and matrix blanks are normally zero.

Calibration.   Prepare a standard solution of the tertiary mercaptan to be measured in carbon tetrachloride. Process an aliquot of this solution containing about 0.005 mM of mercaptan as directed above and calculate the absorptivity at the absorbance maximum (344 mμ). Calculate unknowns, using this absorptivity.

The absorbance peak of the *tert*-butyl thionitrite occurs at 344 mμ. The *sec*-butyl thionitrite has a peak at 340 mμ and an absorptivity about 30% of that for the tertiary thionitrite. No characteristic absorption is obtained for *n*-butyl thionitrite and isobutyl thionitrite. Normal mercaptans and iso-mercaptans have molar absorptivities no greater than about 2% of the values reported for tertiary mercaptans, which can be determined with little interference from these isomers.

Results for six replicates show that the procedure is reproducible to within about 2% of the amount measured for tertiary mercaptans. Trace acids remaining in the carbon tetrachloride will cause poor reproducibility. For this reason the steps involving water and bicarbonate washes have been incorporated into the procedure.

The molar absorptivities of a random selection of fourteen thionitrites for normal, iso-, secondary, and tertiary mercaptans are given in Table 15-2.

II.   GAS CHROMATOGRAPHIC METHODS*

In the analysis of sulfur compounds amenable to separation by GC, considerable specificity and enhanced sensitivity can be obtained by using detectors highly specific in their response to sulfur compounds (5, 6). (See also the discussion of flame photometric and coulometric detectors in Section II.D of the Appendix.) The electron capture detector will also respond to some sulfur compounds (e.g., thiols, sulfides, polysulfides, and $\diagdown$P=S).

Many sulfur compounds are readily determined by GC directly. As usual, the more polar the compound, the more uncertain the quantitative aspects of the analysis become. Thus the relative order of difficulty in chromatography is as follows: sulfides < sulfones < sulfoxides, which is in accord with the polarity of these compounds (other things being equal). Thiols, sulfides, disulfides, and trisulfides are readily chromatographed directly

---

* Written by Morton Beroza and May N. Inscoe.

Table 15-2. Molar Absorptivities (ϵ) of Thionitrites (3)

Mercaptan	Peak wavelength, mμ	ϵ	Solvent
n-Butyl	344	10	Carbon tetrachloride
	229 (slope)	130	*n*-Hexane
Isobutyl	344	15	Carbon tetrachloride
n-Amyl	344	10	Carbon tetrachloride
Isoamyl	344	11	Carbon tetrachloride
n-Decyl	344	9	Carbon tetrachloride
n-Octyl	344	9	Carbon tetrachloride
sec-Butyl	344	200	Carbon tetrachloride
	229 (slope)	1660	*n*-Hexane
	340	165	*n*-Hexane
Cyclohexanethiol	344	238	Carbon tetrachloride
	340	220	*n*-Hexane
	229 (slope)	2700	*n*-Hexane
tert-Butyl	229	9360	*n*-Hexane
	343	707	*n*-Hexane
	339	978	Chloroform
	343	687	Isooctane
	344	771	Carbon tetrachloride
tert-Amyl	344	780[a]	Carbon tetrachloride
tert-Octyl	344	897	Carbon tetrachloride
tert-Dodecyl	344	918[a]	Carbon tetrachloride
tert-Tetradecyl	344	936[a]	Carbon tetrachloride
tert-Hexadecyl	344	950[a]	Carbon tetrachloride

[a] Corrected for typical assay of mercaptan reported by manufacturer, Phillips Petroleum Co. All mercaptans reported in this table are commercial grade.

(7–9). The direct determination of H_2S and mercaptans in aqueous alkaline solution has been accomplished by injecting the sample on an acetic acid-coated Chromosorb packing located ahead of the analytical column. The liberated H_2S and mercaptans then chromatograph normally (10).

A.   Determination of Mercaptans by Derivatization and Gas Chromatography

Mercaptans (as well as phenols)in water are converted to their penta-fluorobenzyl ethers for determination by electron capture GC (11). Amino-thiols and aminodisulfides react rapidly with pivaldehyde to form volatile neopentylidene derivatives (thiazolidine or Schiff base) suitable for quantitative determination by GC (12). Thiophene, thionaphthene, and other sulfur compounds, including thiols, have been determined quantitatively by hydrogenolysis over Raney nickel in solution or in the vapor phase; the hydrocarbon products were determined by GC (13, 14).

B.   Subtraction of Sulfur Compounds for Analyses by Gas Chromatography

A gas sampling train was used to adsorb higher concentrations of sulfur compounds, which were determined by GC; H_2S was retained by $CdSO_4$, R—SH by $Hg(CN)_2$, and RSR and RSSR by $HgCl_2$ solution (15). Metallic precipitates that formed were treated with acid to regenerate the sulfur compounds (except for RSSR, which formed mercaptans) for determination by GC. In another study $HgCl_2$ was used to remove methyl sulfide from an aqueous solution (16); although a short column of $HgCl_2$ on firebrick removes methyl sulfide, as well as other organic sulfur compounds, in GC, it also quantitatively adsorbs olefins and acetylene (15).

III.   ELECTROANALYTICAL METHODS*

A.   Amperometric Titration of Mercaptans with Mercuric Ion

The amperometric titration of mercaptans with standard silver solution has been described (17). Both the experimental arrangement and the procedure are quite simple (17). There are certain shortcomings, however, to the use of a standard silver titrant which are a function of (a) the stability of the silver mercaptides formed, and (b) the precipitation of silver halides; in

---

* Written by Alan F. Krivis.

particular, biological samples may cause difficulties from either or both of these factors. The end result is a significant deviation from the true value for the mercaptan. A discussion of some of these problems has been published (18).

The use of mercuric ion may alleviate the difficulties mentioned above (19). In general, mercury mercaptides are much more strongly associated than silver mercaptides, and many of the mercury species also are soluble. The fact that the halide complexes of mercury are soluble permits the use of mercuric chloride as a titrant.

The procedures described below under "Experimental Method" permit the ready determination of a range of types of mercaptans, at low concentrations of materials. The important (and yet difficult to analyze) biological mercapto compounds are especially emphasized.

The details of the analysis should be adjusted to the particular type of compound being titrated. For this reason, two different procedures are listed. One is oriented toward high-molecular-weight polymers; the other, toward small molecules. The former are best handled in a phosphate buffer, and the latter in a borate buffer.

EXPERIMENTAL METHOD [ADAPTED FROM KOLTHOFF, STRICKS, AND MORREN (19)]*

Apparatus. Any polarograph, recording or manual, may be used for this analysis. The titration vessel is a beaker (120 ml) fitted with a rubber stopper with holes for the buret tip, nitrogen inlet tube, agar salt bridge, and indicating electrode. A saturated calomel reference electrode is used for this titration (20).

A rotating platinum electrode (including a synchronous rotator) may be purchased from a variety of laboratory supply houses, including Sargent-Welch Scientific Co., 7300 N. Linder Ave., Skokie, Illinois 60076.

Reagents. An aqueous phosphate buffer containing 0.01 M NaH_2PO_4-0.08 M Na_2HPO_4 and 0.5 M potassium chloride is used as the titration medium for proteins. An aqueous solution of 0.05 M borax-0.1 M potassium chloride is preferred for lower-molecular-weight compounds such as cysteine and reduced glutathione.

The titrant is 0.001 M mercuric chloride, 1 ml of which corresponds to 0.033 mg sulfhydryl, 0.121 mg cysteine, or 0.307 mg glutathione.

Procedure: Cysteine. Introduce 30 ml of borate buffer into the titration vessel; fit the stopper, the salt bridge, the nitrogen inlet tube, and the rotating

---

* Reprinted in part with the permission of the copyright owner, the American Chemical Society.

electrode; and bubble nitrogen through the solution for 10 minutes. Add sufficient sample to make the solution ca. 10^{-5} M in cysteine. Continue to bubble nitrogen through the solution.

Begin rotation of the electrode and titrate with the mercuric chloride titrant at a an applied potential of -0.2 V. Plot the current versus the volume of titrant. (A reverse L-shaped curve should result, with the intersection of the two lines being the end point.)

The electrode should be cleaned after each titration by dipping in warm nitric acid (diluted 1:1 with water) for about 5 minutes. Rinsing with distilled water prepares the electrode for the next titration.

Procedure: Proteins. Transfer 30 ml of phosphate buffer to the titration vessel; fit the stopper, the salt bridge, and the nitrogen inlet tube; and bubble nitrogen through the solution for 10 minutes. Add sufficient sample to make the solution ca. 10^{-5} M in sulfhydryl. Continue to bubble nitrogen through the solution. Insert the platinum electrode and begin rotation. Titrate with the mercuric chloride titrant at an applied potential of -0.2 V. Plot the current versus the volume of titrant. (A reverse L-shaped curve should result, with the intersection of the two lines being the end point.)

The electrode operates best for these titrations with a thin coating of mercury on its surface. This coating is produced in the normal course of events during the first few titrations. Therefore, the acid bath, as specified for the cysteine titration above, should *not* be used for protein titrations. A simple rinse with distilled water is all that is needed after each titration.

B. Coulometric Titration of Mercaptans

In recent years, a major improvement in silver titrations, which entails the electrolytic generation of silver ion *in situ*, has been developed. Electrolysis of a suitable solution with a silver wire anode and platinum cathode can produce the silver ion needed. Electrolytic generation of a reagent obviates the need for preparation, standardization, storage, controlled introduction, and volumetric measurement of a titrant.

If the electrolysis is carried out with a known, controlled constant current, the quantity of silver ion generated may be calculated from Faraday's laws and the length of time that the current has been turned on (cf. Section III, Chapters 1 and 12). In a coulometric titration such as the following, the real experimental measurement is time, which can be measured very precisely, and the real reagent is the electron, which can be dispensed very easily from an electronic power supply.

Although a coulometric titration eliminates the classical buret, a means for detecting the end point of the titration must still be found. For mercaptan analyses, an amperometric detection system based on a gold-platinum electrode pair is recommended (21).

The procedure to be described was developed originally for the determination of mercaptans in petroleum stocks (21); therefore, the solvent system is essentially nonaqueous. However, modification of the solvent system may be carried out to suit the particular sample being measured.

EXPERIMENTAL METHOD [ADAPTED FROM LEISEY (21)]*

Apparatus. A coulometric constant-current power supply and timer should be used for this analysis. The unit may be either laboratory constructed or purchased from a laboratory supply house. The amperometric detection circuit may be wired from a galvanometer (sensitivity of 0.02 μA per division), a 1.5-V battery, and a variable resistor as shown in Chapter 1, Section III.

The generating electrodes are heavy wires of silver (anode) and platinum (cathode). A constant length of wire (about 25 mm) should be immersed in the titration medium. This can be arranged by sealing one end of the wire through the bottom of a glass tube so that the desired length of wire extends beyond the bottom of the tube. Electrical contact may be made with the portion of the wire which is on the inside of the glass tube. The wire-tube assembly is mounted so that the entire length of exposed wire is immersed under the surface of the titration medium. It is not necessary to isolate the cathode in a separate chamber as discussed in Chapters 1 and 12; both electrodes can be in direct contact with the titration solvent.

The detecting system consists of wires of platinum (anode) and gold (cathode) with 0.25 V impressed across them from the battery—variable resistor network. A fixed length of wire (about 25 mm) should be exposed to the solution as described above for the generating electrodes. The galvanometer should be wired in series with the cathode.

The electrodes are most conveniently supported from above by means of a stopper of some sort. A more complete discussion of this apparatus was given in Chapter 1, Section III.C.

Titration Solution. The titration medium can be prepared by mixing 100 ml of 95% ethanol, 50 ml of benzene, 0.5 gram of ammonium nitrate, and 2 ml of concentrated ammonium hydroxide. Larger volumes may be prepared by mixing these components in the proportions specified.

Procedure. Place a magnetic stirring bar into a 250-ml beaker, transfer an accurately weighed sample containing about 1 mg of unknown to the beaker,

* Reprinted in part with the permission of the copyright owner, the American Chemical Society.

and add 150 ml of the titration solution. Dip the electrodes into the solution, start the stirrer, and titrate with a generation current sufficient to give about a 200-second titration. The end point is reached when a stable 0.2-μA current is shown on the detector circuit galvanometer. Run a blank titration and subtract the blank time from the sample titration. Calculate the amount of unknown present from Faraday's law.

Calculation.

$$\text{Per cent of unknown} = \frac{(i)\ (t)\ (\text{mol. wt. of unknown})\ (100)}{(\text{sample weight})\ (96487)}$$

IV. NUCLEAR MAGNETIC RESONANCE METHODS*

Leader (22) reports the reaction of mercaptans to yield an adduct with hexafluoroacetone; the ^{19}F nmr data were obtained at 56.4 MHz. The sample preparation and the analysis are essentially identical to those discussed in Chapter 1, Section IV.A. The ^{19}F chemical shifts of the mercaptan adduct are 0.04 to 1.46 ppm from hexfluoroacetone $\cdot$ H_2O. The mercaptans reported were ethanethiol, propanethiol, 1-butanethiol, 2-methylpropanethiol, 2-mercaptoethanol, 3-mercaptopropanol, ethyl thioglycolate, benzylmercaptan, 2-propanethiol, 2-butanethiol, *tert*-butylmercaptan and benzenethiol.

The method of Butler and Mueller (23) utilizes the addition of trichloroacetyl isocyanate to a solution containing the thiol and forming the monothiocarbamate:

$$\underset{\displaystyle CCL_3C—NCO}{\overset{\displaystyle O}{\overset{\displaystyle \|}{}}} + RSH \rightarrow \underset{\displaystyle CCl_3C—N—C—SR}{\overset{\displaystyle O\quad H\quad O}{\overset{\displaystyle \|\quad\ |\quad\ \|}{}}}$$

The advantages of this method are that derivatization occurs rapidly and the progress of the reaction is easily followed. The thiol solution is prepared in deuteriochloroform at approximately 10% concentration. The spectra are obtained at 60 MHz. The thiol proton appears in the range from 0.8 to 3.00 ppm from tetramethylsilane. The formation of the thiocarbamate causes the disappearance of the thiol resonance. The formation of the electronegative thiocarbamate group produces the paramagnetic shift of the alpha protons and a simplification of the nmr spectrum. The rate of exchange of the thiol proton is such that coupling with adjacent protons is observed; this is essentially eliminated, however, upon derivatization. The paramagnetic shift of

---

* Written by Harry Agahigian.

Table 15-3.   α-Hydrogen Chemical Shifts for Thiols and Isocyanate Derivatives

Compound	α-Hydrogen chemical shift	
	Thiol	Derivative
CH_2=$CHCH_2SH$	3.12	3.58
CH_2=CCH_2SH ⎜ CH_3	3.08	3.62
$CH_3CH_2CH_2CH_2SH$	2.47	2.94
$(CH_3)_2CHCH_2SH$	2.37	2.85
$CH_3CHCH_2CH_3$ ⎜ SH	2.84	3.55
$CH_3(CH_2)_8CH_2SH$	2.48	2.93
CH_3CHCH_2SH —CH_2— ⎜ ⎜ SH —CH —	2.73 3.03	3.22 3.75
O ‖ $CH_3CCH_2CH_2SH$	2.63	3.18
CH_2=$CHCH_2SCH_2CH_2CH_2SH$	2.60	3.05
CH_2=$CCH_2SCH_2CHCH_2SH$ ⎜ ⎜ CH_3 CH_3	2.50	3.02

the alpha protons of the methylene is 0.45 to 0.55 ppm, while the shift of the alpha methene proton is 0.71 to 0.72 ppm. The thiols studied are shown in Table 15-3.

V.   RADIOCHEMICAL METHODS*

The extreme sensitivity of radioreagent methods has made these techniques especially valuable for determining very low concentrations of mercapto groups in specimens of biological origin. Of particular interest is the

* Written by D. Campbell.

estimation of the thiol structure in various proteins, in which it is present as the combined amino acid, L-cysteine (L-2-amino-3-mercaptopropanoic acid). Certain of the methods widely used for determining mercapto groups in proteins are adaptable as well to the analysis of low-molecular-weight thiols. Radiometric titration and direct isotope dilution are also useful approaches with macro amounts of mercaptans. Isotope exchange with elemental sulfur is potentially applicable to certain aromatic thiols.

A.   Conversion to S-Succinimido or S-Succinyl Derivatives

After the initial report (24) of rapid and quantitative adduct formation between N-ethylmaleimide (NEM) and aliphatic thiols in neutral or slightly acid solution at room temperature, extensive analytical studies demonstrated the usefulness of this reaction in conventional methods (25, 26). Incorporation of ^{14}C into the maleic moiety (commonly at the 2, 3 positions) yields a radioreagent widely applicable to micro and submicro amounts of thiols. Thus, L-cysteine is converted to S-(N-ethylsuccinimido)-L-cysteine (27):

$$
\begin{array}{c}
\overset{\displaystyle NH_3^+}{\underset{\displaystyle |}{}} \\
{}^-OOC - CH - CH_2 - SH \quad + \quad \substack{{}^{14}CH - C \diagup{}^{O} \\ || \qquad \diagdown N - C_2H_5 \\ {}^{14}CH - C \diagdown_{O}}
\end{array}
$$

$$
\begin{array}{c}
\overset{\displaystyle NH_3^+}{\underset{\displaystyle |}{}} \qquad\qquad\qquad {}^{14}CH_2 - C\diagup{}^{O} \\
{}^-OOC - CH - CH_2 - S - {}^{14}CH - C\diagdown N - C_2H_5 \\
\diagdown_O
\end{array}
$$

This reaction has been employed extensively for determining mercapto groups in flour protein (28–31) and does not appear to require precise pH control in this application. Although (**I**) was initially reported to be stable under conditions used to subsequently hydrolyze the polypeptide (32), later work indicated that a significant proportion (ca. 20%) of the adduct was further hydrolyzed under the same conditions to yield S-succinyl-L-cysteine (**II**) and ethylamine (33). To obviate empirical corrections required by partial conversion of (**I**) to (**II**) (33, 34), hydrolysis under conditions sufficiently rigorous to convert (**I**) to (**II**) quantitatively has been proposed (28). This complete conversion may be effected by hydrolyzing the protein containing the L-cysteine-NEM adduct (**I**) in 6 N hydrochloric acid at 120°C for 22 hours, or at 110°C for 72 hours (35), in a sealed tube after de-gassing (29, 30). Addition of carrier (**I**) before hydrolysis is desirable when only extremely small quantities of L-cysteine are present. The hydrolysis products are readily separable by paper chromatography for subsequent radioassay.

The method has been standardized (30) for a given lot of NEM-^{14}C by treating known amounts of L-glutamyl-L-cysteylglycine (glutathione) with the radioreagent and, after hydrolysis, preparing a graph relating the amount of thiol taken to the observed activity of the chromatographically isolated *S*-succinyl derivative. The most reliable background correction was obtained by chromatographing a mixture of NEM-^{14}C and nonradioactive (**II**) and measuring the activity of the spot of the latter compound.

In addition to its proven usefulness in protein analysis, labeled NEM appears promising as well for determining very small amounts of free low-molecular-weight mercaptans. With excess NEM the reaction is rapid in neutral or slightly acidic solution. Thus, L-cysteine reacts quantitatively within 2 minutes at pH 5.4 to 6.6 (25, 36). Thioglycolic acid, mercapto-ethanol, and 2-amino-4-mercaptobutanoic acid (homocysteine) also form adducts rapidly (26). Quantitative hydrolysis of the adducts to *S*-succinyl derivatives is not required in principle with low-molecular-weight compounds, but may assist in separating adducts from excess reagent by chromatographic methods. The thiol-NEM reaction yields a derivative possessing a (new) center of asymmetry, and this factor must be considered in separation methods. The dependency of reaction rates on pH and the fact that reactions are faster in water than in ethanol (36) suggest a polar rather than a free radical mechanism of adduct formation. The sulfide, sulfite, and thiosulfate anions also react with NEM (37).

Labeled NEM may be purified by extraction from 50% aqueous ethanol into ethyl chloride at -14°C, drying with $CaCl_2$, evaporating to dryness, and subliming the residue at about 45°C (38).

B. Conversion to Organomercury Derivatives

The mercapto group reacts readily with various organomercury compounds to form the —Hg—S— structure, and these substances are perhaps the most thiol-specific of available reagents for this group in proteins. Since both mercury and an organic moiety become bound to the sulfur, either may be labeled. Mercury-203, a gamma emitter having a half-life of 47 days, is useful when self-absorption effects or strongly colored samples make direct liquid scintillation counting methods disadvantageous. This radioisotope also offers higher sensitivity than is attainable with ^{14}C-labeled reagents, since solid scintillation detectors permit the use of much larger samples; analytical specimens, in addition, may be recovered unchanged. Radioassay of a portion of the [203]Hg-labeled reagent at the time of sample analysis is the most reliable method of half-life correction. Organomercury compounds which have been used for quantitative determination of thiol groups in proteins include methylmercuric-[203]Hg bromide (39), phenylmercuric-[203]Hg acetate (40, 41), methylmercuric-^{14}C iodide (41), and salts of *p*-chloromercuribenzoic acid-[14]C (42).

The applicability of phenylmercuric-[203]Hg acetate and of methylmercuric-^{14}C iodide to the thiol group in an insoluble protein has been investigated in detail (41). A simplified modification based on measurement of the decrease in radioactivity of a solution of either labeled reagent after treatment of the specimen proved useful when submicro analyses were not required. Methods were also developed for determining [203]Hg and ^{14}C incorporated into specimens containing submicro amounts of mercapto group. In these more sensitive variations, treated samples freed of excess radioreagent by repeated washings were assayed by both direct solid and liquid scintillation counting and after conversion of the radioisotopes to soluble form by oxygen-flask combustion. The three procedures gave comparable results for the insoluble substrate investigated; however, direct solid scintillation counting was preferred for submicro amounts of thiol. The techniques involving assay of specimens after derivatization are not limited in principle to insoluble materials, however, and may be extended to soluble samples provided that a means is available for quantitative separation of excess reagent and recovery of a representative portion of the substrate.

Thiol groups in several electrophoretically separable proteins have also been detected by autoradiography after treatment of the sample with 1,3-(chloromercuri)-2-methoxypropylurea-[203]Hg (chlormerodrin-[203]Hg) (43).

C. Other Methods

The reaction of labeled tetraethylthiuram disulfide with the mercapto group at pH 8 to 9 to liberate the radioactive anion of diethyldithiocarbamic acid has been proposed as a sensitive and specific method for determining the thiol structure (44). The radioreagent used is uniformly labeled with [35]S. In the presence of an excess of reagent, low-molecular-weight mercaptans form predominantly the corresponding disulfides:

$$2R\text{---}SH + \left[(C_2H_5)_2N\text{---}\overset{\overset{\text{35}S}{\|}}{C}\text{---}{}^{35}S \right]_2 \rightarrow$$

$$R\text{---}S\text{---}S\text{---}R + 2\left[(C_2H_5)_2N\text{---}\overset{\overset{\text{35}S}{\|}}{C}\text{---}{}^{35}S \right]^- + 2H^+$$

Mercapto groups present in proteins react to yield the diethylthiocarbamyl disulfide derivative (45):

$$R\text{—SH} + \left[(C_2H_5)_2N\text{—}\overset{\overset{\displaystyle ^{35}S}{\|}}{C}\text{—}^{35}S \right]_2 \rightarrow$$

$$R\text{—S}\text{—}^{35}S\text{—}\overset{\overset{\displaystyle ^{35}S}{\|}}{C}\text{—}N(C_2H_5)_2 + [(C_2H_5)_2N\text{—}\overset{\overset{\displaystyle ^{35}S}{\|}}{C}\text{—}^{35}S]^- + H^+$$

The labeled anion produced in each reaction is decomposed in a solution buffered at pH 3.8 to yield diethylamine and carbon disulfide-^{35}S. The latter is allowed to diffuse into alkaline piperidine to reform a labeled dithio- carbamate, which is then isolated for radioassay. The method has been standardized for a specific lot of radioreagent by treating known amounts of a thiol-containing compound (e.g., L-glutamyl-L-cysteylglycine) and pre- paring a graph relating the count rate of the isolated pentamethylene dithio- carbamate to the amount of thiol taken.

The reagent uniformly labeled with ^{35}S is conveniently prepared and offers excellent sensitivity, although the 87-day half-life of this radioisotope requires concurrent assay of a reference source prepared from the same lot of reagent that is employed in the analysis. If ^{14}C is used to obviate half-life corrections, the label must be incorporated exclusively into the —C(S)—S— moiety to retain the advantage inherent in the ease of separation of the carbon disulfide from the treated sample. The possible isotope effect involved in the use of the ^{14}C-labeled compound, however, must be assessed initially. The method has been used to determine mercaptan sulfur in as little as 10 μg of certain proteins. The technique is subject to interference from sub- stances capable of oxidizing diethyldithiocarbamate anion to tetraethyl- thiuram disulfide in alkaline solution.

Both isotope dilution and radiometric titration have been used for deter- mining benzenethiol in coal tar acids. After addition of benzenethiol-^{35}S in the isotope dilution method (46), mercuric chloride was added to precipitate the labeled derivative, which was isolated, purified, and assayed. The anal- ysis was performed radiometrically (47) by adding excess ^{110m}AgNO$_3$ solution and titrating the unconsumed radioreagent with ammonium thiocyanate. Silver-110m has been proposed as a radiometric titrant generally applicable to mercaptans (48).

Mercapto groups have been determined in L-cysteine and in L-glutamyl- L-cysteylglycine by treating the substances on glass-fiber paper with 1:1 1% KMnO$_4$-1% H$_2$SO$_4$, followed by gaseous H^{36}Cl (49). The Mn^{36}Cl$_2$ formed from the MnO$_2$ is then assayed by G-M counting. The technique, which can detect a few milligrams of sulfur-containing amino acids, is not specific for the thiol group, since a positive response is also given by methionine.

A rapid method for determining mercapto groups in serum is based on their reaction with $^{110m}Ag^+$, which has been incorporated into an ion-exchange resin at a concentration of 0.5 $\mu M/ml$ (50). To 1 ml of serum is added 0.15 ml of 1.5 N NH$_4$OH; after 2 minutes, the mixture is passed through a column consisting of 1 ml of the $^{110m}Ag^+$-containing resin. The column is then eluted three times with 1 ml of 0.15 N NH$_4$OH, and the radioactivity in the fractions is counted by a liquid scintillation technique.

Isotope exchange of the mercaptan sulfur of 2-mercaptobenzothiazole (MBT) with elemental ^{35}S at elevated temperatures is potentially useful for determining this thiazole derivative in the absence of other exchangeable forms of sulfur:

Sulfur in the thiazole ring does not participate in the exchange reaction. The process has been studied extensively in both polar and nonpolar solvents (51, 52). The apparent rate constant for exchange is dependent on the concentration of MBT but is independent of the concentration of elemental sulfur. No significant difference in the constant was observed for reactions conducted in either polar or nonpolar media. The reaction may be followed by removing aliquots from the homogeneous system and separating the components by paper chromatography (53). The substances are readily separable by chemical means for measurement of the equilibrium specific activity of the MBT.

REFERENCES

Absorption Spectrophotometric

1. B. Saville, *Analyst*, **83,** 670 (1958).
2. R. K. Kunkel, J. E. Buckley, and G. Gorin, *Anal. Chem.*, **31,** 1098 (1959).
3. G. W. Ashworth and R. E. Keller, *Anal. Chem.*, **39,** 373 (1967).
4. I. M. Kolthoff and W. E. Harris, *Ind. Eng. Chem., Anal. Ed.*, **18,** 161 (1946).

Gas Chromatographic

5. H. V. Drushel, *Anal. Chem.*, **41,** 569 (1969).
6. H. V. Drushel and A. L. Sommers, *Anal. Chem.*, **39,** 1819 (1967).

7. M. H. Brodnitz and C. L. Pollock, *Food Technol.*, **24**, 78 (1970).

8. R. W. Freedman, *J. Gas Chromatog.*, **6**, 495 (1968).

9. J. P. Zikakis and R. L. Salsbury, *J. Dairy Sci.*, **52**, 2014 (1969).

10. H. D. LeRosen, *Abstracts of Papers*, American Chemical Society, 148th National Meeting, Chicago, September 1964, p. 26B.

11. F. K. Kawahara, *Anal. Chem.*, **40**, 1009 (1968).

12. E. Jellum, V. A. Bacon, W. Patton, W. Pereira, Jr., and B. Halpern, *Anal. Biochem.*, **31**, 339 (1969).

13. R. Staszewski, J. Janak, and T. Wojdala, *J. Chromatog.*, **36**, 429 (1968).

14. J. Uhdeova, M. Hrivnac, M. Dodova, R. Staszewski, and J. Janak, *J. Chromatog.*, **40**, 359 (1969).

15. M. Feldstein, S. Balestrieri, and D. A. Levaggi, *J. Air Control Assoc.*, **15**(5), 215 (1965).

16. R. Bassette and C. H. Whitnah, *Anal. Chem.*, **32**, 1098 (1960).

Electroanalytical

17. I. M. Kolthoff and W. E. Harris, *Ind. Eng. Chem.*, *Anal Ed.*, **18**, 162 (1946).

18. I. M. Kolthoff and J. Eisenstädter, *Anal. Chim. Acta*, **24**, 83 (1961).

19. I. M. Kolthoff, W. Stricks, and L. Morren, *Anal. Chem.*, **26**, 366 (1954).

20. I. M. Kolthoff and J. J. Lingane, *Polarography*, 2nd Ed., Interscience, New York, 1952.

21. F. A. Leisey, *Anal. Chem.*, **26**, 1607 (1954).

Nuclear Magnetic Resonance

22. G. W. Leader, *Anal. Chem.*, **42**, 17 (1970).

23. P. E. Butler and W. H. Mueller, *Anal. Chem.*, **38**, 1409 (1966).

Radiochemical

24. E. Friedmann, D. H. Marrian, and I. Simon-Reuss, *Brit. J. Pharmacol.*, **4**, 105 (1949).

25. E. Roberts and G. Rouser, *Anal. Chem.*, **30**, 1291 (1958).

26. N. M. Alexander, *Anal. Chem.*, **30**, 1292 (1958).

27. D. G. Smyth, A. Nagamatsu, and J. S. Fruton, *J. Am. Chem. Soc.*, **82**, 4600 (1960).

28. R. Trachuk and I. Hylnka, *Cereal Chem.*, **40**, 704 (1963).

29. C. C. Lee and T.-S. Lai, *Cereal Chem.*, **44**, 620 (1967).

30. C. C. Lee and T.-S. Lai, *Can. J. Chem.*, **45**, 1015 (1967).

31. C. C. Lee and T.-S. Lai, *Cereal Chem.*, **46**, 598 (1969).

32. C. C. Lee and E. R. Samuels, *Can. J. Chem.*, **39**, 1152 (1961).

33. C. C. Lee and E. R. Samuels, *Can. J. Chem.*, **40**, 1040 (1962).

34. C. C. Lee and E. R. Samuels, *Can. J. Chem.*, **42**, 164 (1964).

35. D. G. Smyth, O. O. Blumenfeld, and W. Konigsberg, *Biochem. J.*, **91**, 589 (1964).

36. C. C. Lee and E. R. Samuels, *Can. J. Chem.*, **42**, 168 (1964).

37. R. J. Ellis, *Biochem. J.*, **110**, 43P (1968).

38. M. Flavin, *Anal. Biochem.*, **5**, 60 (1963).

39. P. J. Geiger, O. K. Reiss, J. Wein, and L. Hellerman, *Federation Proc.*, **19**, No. 1 (Part A), A-5, P1656 (1960).

40. P. H. Springell and S. J. Leach, *Nature*, **208**, 1326 (1965).

41. S. J. Leach, A. Meschers, and P. H. Springell, *Anal. Biochem.*, **15**, 18 (1966).

42. V. G. Erwin and P. L. Pedersen, *Anal. Biochem.*, **25**, 477 (1968).

43. L. P. Stratton and E. Frieden, *Nature*, **216**, 932 (1967).

44. A. H. Neims, D. S. Coffey, and L. Hellerman, *J. Biol. Chem.*, **241**, 3036 (1966).

45. A. H. Neims, D. S. Coffey, and L. Hellerman, *J. Biol. Chem.*, **241**, 5941 (1968).

46. I. Matsushima, *Coal Tar*, *Japan*, **14**, 646 (1962); *Chem. Abstr.*, **60**:2334f.

47. I. Matsushima, *Coal Tar*, *Japan*, **14**, 649 (1962); *Chem. Abstr.*, **60**:2334g.

48. P. Bebesel and I. Sirbu, *Rev. Chim.* (*Bucharest*), **11**, 288 (1960); *Chem. Abstr.*, **58**:5034a.

49. J. Z. Beer, A. Z. Budzynski, and K. Malwinska, *Chem. Analit.*, **12**, 1055 (1967); *Chem. Abstr.*, **68**:36642d.

50. R. Muenze, *Wiss. Z. Karl-Marx-Univ. Leipzig, Math.-Naturw. Reihe*, **18**, 621 (1969).

51. G. A. Blokh, E. A. Golubkova, and G. P. Miklukhin, *Dokl. Akad. Nauk SSSR*, **86**, 569 (1952).

52. E. N. Guryanova and V. I. Vasilyeva, *Zh. Fiz. Khim.*, **28**, 60 (1954).

53. J. Moravek, Z. Nejedly, and J. Filip, *Science*, **138**, 146 (1962).

16 Dialkyl disulfides

I. ABSORPTION SPECTROPHOTOMETRIC METHODS*

Measurement after Chemical Reaction

Reduction to Mercaptans. Disulfides can be quantitatively reduced to mercaptans:

$$RSSR + 2[H] \rightarrow 2RSH$$

Kolthoff and his coworkers (1) passed a solution containing the mercaptan through amalgamated zinc in a Jones-type reduction column. They then determined the resulting thiol by an amperometric titration; it appears, however, that the end determination could be performed spectrometrically instead by one of the methods outlined in Chapter 15.

Siggia and Stahl (2) used sodium borohydride as the reducing agent, destroyed the excess borohydride with sodium hydroxide and nitric acid, and then titrated the mercaptan produced potentiometrically with standard silver nitrate solution. Again, one of the spectrometric end determinations described in Chapter 15 appears feasible. If the resulting mercaptan can be distilled, it can be concentrated in this manner and the technique used for the determination of trace quantities of disulfides.

II. ELECTROANALYTICAL METHODS†

A. Coulometric Bromination of Disulfides

The disulfide functional grouping is one of that unusual class of species which may be analyzed by either oxidation or reduction. The reaction with

---

* Written by J. Gordon Hanna.
† Written by Alan F. Krivis.

bromine (3) may be cited as a typical example of oxidation:

$$RSSR + 5Br_2 + 4H_2O \rightarrow 2RSO_2Br + 8HBr$$

or

$$RSSR + 5Br_2 + 6H_2O \rightarrow 2RSO_3H + 10HBr$$

Reference 3 describes the titration of disulfides with a bromate-bromide mixture.

Direct coulometric bromination has many advantages, however, over bromination by means of a buret titration, as discussed in Section III of Chapters 1 and 12. For these reasons, a coulometric bromination of disulfides was proposed some time ago (4).

EXPERIMENTAL METHOD [BASED ON THE STUDY OF SEASE, NIEMAN, AND SWIFT (4)]*

Apparatus. A coulometric constant-current power supply and timer should be used for this analysis. The unit may be either laboratory constructed or purchased from a laboratory supply house. The biamperometric detection circuit may be wired from a 1.5-V battery, a variable resistor, and a micro-ammeter (10-0-10 scale is useful); the circuit (and its use) is identical to that described in Chapter 1, Section III.C.

Both of the generating electrodes are fabricated of platinum. The anode is a coil of wire (16 gauge) wrapped around the isolation chamber for the cathode. The isolation chamber is a glass tube with a very fine frit sealed in its bottom. The cathode is a coil of platinum inserted into the isolation chamber. A portion of the titration solvent can be used in the isolation chamber.

The biamperometric electrodes are small platinum wires. These may be sealed through the bottoms of glass tubes so that a fixed length of wire is immersed in the titration solution. The variable resistor-battery network (see Figure 1-5) should be adjusted to apply 0.2 V across the two electrodes.

Titration Solvent. There is a great deal of latitude in the composition of a solvent suitable for coulometric bromination. Bromine is produced with 100% efficiency in many different solvents.

For the present purposes, a good general-purpose solvent consists of 60% glacial acetic acid, 26% methanol, and 14% 1 *M* potassium bromide

---

* Reprinted in part with the permission of the copyright owner, the American Chemical Society.

(aqueous). Modification of this solvent for a particular analysis may be worth while.

Procedure. Place a magnetic stirring bar into a 100-ml beaker and transfer an accurately weighed sample containing about 0.005 to 0.01 mM of disulfide compound to the beaker. Add 50 ml of the titration solvent, dip the electrodes and isolation chamber into the solution, and start the stirrer. Make sure that enough solvent is in the isolation chamber so that its level is above that of the test solution.

Titrate at a generating current of about 20 mA. (Adjust the sample size and/or the generating current to give a titration time of about 200 seconds.) The end point is reached when a stable 5-μA current is shown on the detecting circuit meter. Run a blank titration and subtract the time for the blank from the time for the sample titration. Calculate the percentage of disulfide in the unknown.

Calculation.

$$\text{Per cent of disulfide} = \frac{(i)\ (t)\ (\text{mol. wt. of unknown})\ (100)}{(\text{sample weight})\ (10)\ (96487)}$$

B. Amperometric Titration of Disulfides

The procedure described in Section A makes use of an oxidative reaction of disulfides with bromine. As mentioned, however, it also is possible to determine disulfides by a reductive procedure.

Disulfides react with sodium sulfite, in alkaline solution, as follows:

$$RSSR + SO_3^{2-} \leftrightarrow RS^- + RSSO_3^-$$

In the presence of mercuric ion, the slightly dissociated mercaptide, $Hg(RS)_2$, is formed and drives the reaction to quantitative completion (5). Titration with a standard mercuric ion solution provides sufficient mercury for this purpose; thus the analysis becomes similar to that for mercapto compounds as described in Chapter 15. Mercaptans, therefore, constitute an interference in the disulfide analysis. However, a separate analysis for mercaptans can be carried out and the results subtracted from the disulfide titration.

A rotating platinum electrode cannot be used for the mercurimetric titration of disulfides; the mercury is complexed in sulfite solution and is not reduced at this type of electrode. A dropping mercury electrode is used instead. An undesirable reaction between the mercury metal which normally collects at the bottom of the cell and mercuric ion in solution can be eliminated by covering the bottom of the cell with chloroform.

338   Dialkyl disulfides

EXPERIMENTAL METHOD [BASED ON THE PROCEDURE OF STRICKS, KOLTHOFF, AND TANAKA (5)]*

Apparatus.   Any polarograph, recording or manual, can serve for this analysis. A dropping mercury indicating electrode (DME) and a saturated calomel reference electrode (6) are used.

The cell may be a small beaker or flask fitted with a stopper which has holes for the DME, reference electrode salt bridge, buret tip, and nitrogen inlet tube. Stirring can be carried out by bubbling nitrogen through the solution.

Reagents.   The titration medium is made up of aliquots of several different solutions. The following stock solutions are needed: 0.1 *M* borax, 4 *M* potassium chloride, 1 *M* sodium sulfite.

The titrant is 0.01 *M* mercuric chloride.

Procedure.   Place 10 ml of 0.1 *M* borax solution and 2.5 ml of 4 *M* potassium chloride in the cell. Add sufficient pure chloroform to cover the bottom of the cell. Then add a disulfide sample of about 0.02 mM to the cell. Add 4 ml of 1 *M* sodium sulfite solution and water to bring the total volume to 20 ml. Bubble nitrogen through the solution for 10 minutes. Introduce the reference electrode salt bridge and the DME into the solution and titrate with the mercuric chloride solution; the potential should be -0.35 V. An atmosphere of nitrogen should be maintained over the mixture during the titration. In a separate experiment determine the residual current at -0.35 V. (The titra-

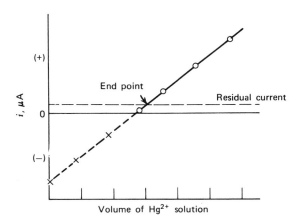

Figure 16–1.   Amperometric titration of disulfides with mercuric chloride.

tion can be carried out quickly, since the current need not be measured until it has become cathodic.) The intersection of the straight part of the excess reagent line with the residual current is the end point. A plot of a typical curve is shown in Figure 16-1.

III. NUCLEAR MAGNETIC RESONANCE METHODS*

The application of nmr to the analysis of disulfides and polysulfides is reported by Martin and Pearce (7). These authors were measuring the effect on the alpha-carbon proton resonances. The dialkyl disulfide chemical shifts were as follows:

	Chemical shift, ppm
$(CH_2{=}CHCH_2)_2S_2$	3.26
$(n\text{-}Amyl)_2S_2$	2.62
$CH_3CH_2)_2S_2$	2.62
$(CH_2{=}CHCH_2)_2S_3$	3.43
$(n\text{-}Amyl)_2S_3$	2.82
$(CH_3CH_2)_2S_3$	2.82
$(CH_2{=}CHCH_2)_2S_4$	3.52
$(n\text{-}Amyl)_2S_4$	2.92
$(Ethyl)_2S_4$	2.92

The benzyl polysulfides were studied in the same manner, and the sulfur content was determined.

Related to this study is the publication of Vineyard (8). The reaction of mercaptan with sulfur yields predominantly the trisulfide. The mercaptans used were isopropyl,

$$2RSH + 2S \rightarrow RS_3R + H_2S$$

secondary butyl, isobutyl, and n-butyl. The chemical shifts of the alpha hydrogen are reported. The tertiary butyl mercaptan formed only the tetrasulfide.

Grant and Van Wazer (9) reported results on similar materials, as well as the selenium analogs.

IV. RADIOCHEMICAL METHODS†

The only radioreagent method presently available for determining the alkyl disulfide group involves fission of the —S—S— bond in the presence of excess sodium sulfite and a labeled organomercury compound (10):

---

* Written by Harry Agahigian.
† Written by D. Campbell.

$$R-S-S-R + SO_3{}^{2-} \rightarrow R-S^- + R-S-SO_3{}^-$$

$$R-S^- + R'-Hg-X \rightarrow R-S-Hg-R' + X^-$$

Both phenylmercuric-^{203}Hg acetate and methyl-^{14}C-mercuric iodide have been employed. The analytical method, which has been applied thus far only to insoluble proteins, involves treatment of the specimen with a buffered (pH 9) water-dimethylformamide solution containing a large excess of sulfite and a smaller excess of radioreagent. Since the samples also contain free mercapto groups, the disulfide content is determined by difference, using this analysis and a separate procedure for thiol content (cf. Chapter 15). Three techniques for estimating —S—S— + —S—H, analogous to those used for determining —S—H alone, were developed; use of the ^{203}Hg-labeled reagent with assay by direct solid scintillation counting was again preferred for sub-micro amounts of sample.

REFERENCES

Absorption Spectrophotometric

1. I. M. Kolthoff, D. R. May, P. Morgan, H. A. Laitinen, and A. S. O'Brian, *Ind. Eng. Chem., Anal. Ed.*, **18**, 442 (1946).
2. S. Siggia and C. R. Stahl, *Anal. Chem.*, **29**, 154 (1957).

Electroanalytical

3. S. Siggia and R. L. Edsberg, *Anal. Chem.*, **20**, 938 (1948).
4. J. W. Sease, C. Neiman, and E. H. Swift, *Anal. Chem.*, **19**, 197 (1947).
5. W. Stricks, I. M. Kolthoff, and N. Tanaka, *Anal. Chem.*, **26**, 299 (1954).
6. I. M. Kolthoff and J. J. Lingane, *Polarography*, 2nd Ed., Interscience, New York, 1952.

Nuclear Magnetic Resonance

7. D. J. Martin and R. H. Pearce, *Anal. Chem.*, **38**, 1604 (1966).
8. B. D. Vineyard, *J. Org. Chem.*, **31**, 601 (1966).
9. D. Grant and J. R. Van Wazer, *J. Am. Chem. Soc.*, **86**, 3012 (1964).

Radiochemical

10. S. J. Leach, A. Meschers, and P. H. Springell, *Anal. Biochem.*, **15**, 18 (1966).

17 Dialkyl sulfides

I. ABSORPTION SPECTROPHOTOMETRIC METHODS*

Measurement after Chemical Reaction

Aliphatic sulfides interact with elemental iodine to produce a complex having an absorbance maximum in the ultraviolet at about 310 mμ:

$$RSR + I_2 \rightleftharpoons (RSR \cdot I)^+ I^-$$

Hastings (1) presented a spectrometric method based on this reaction for the determination of sulfides in hydrocarbons with isooctane as the reaction medium. Hastings and Johnson (2) pointed out the need for the presence in the reaction mixture of a large excess of iodine to force the equilibrium essentially completely to the right.

Drushel and Miller (3) also recognized the necessity for a relatively large excess of iodine in the reaction mixture. These workers used carbon tetrachloride as the solvent.

METHOD FROM DRUSHEL AND MILLER (3)

Apparatus. Beckman Model DU spectrophotometer with a photomultiplier attachment and a pair of 1-cm matched fused-quartz cells.

Reagents. Iodine solution. Dissolve exactly 10.00 grams of iodine in carbon tetrachloride and dilute to 1 liter.

Procedure. Weigh a sample, dissolve in carbon tetrachloride, and dilute to the appropriate volume. Prepare a dilution of this solution to produce absorption of the sample-iodine solution falling in the range of optimum accuracy (an absorptivity between 0.4 and 1.0). For very small samples,

---

* Written by J. Gordon Hanna.

341

such as chromatography fractions, weigh the required quantity of sample on a micro- or semimicrobalance for dilution to 25 ml. A list of appropriate sample weights is as follows:

Aliphatic sulfide, sulfur in sample, %	Approximate weight of sample, mg, diluted to 25 ml to obtain absorptivity between 0.4 and 1.0
0.05	60–140
0.10	30– 70
0.20	15– 35
0.40	8– 18
0.60	5– 12
0.80	4– 9
1.0	3– 7
1.5	2– 5
2.0	1– 3

Dilute 1 ml of the iodine reagent to 10 ml with the above sample solution at the proper concentration. Dilute 1 ml of carbon tetrachloride to 10 ml with the same sample solution and use this solution as the reference for the spectrophotometric measurement. Immediately after preparing the sample blend and the reference solution, obtain the absorbance of the iodine blend at 310 mμ. Make the measurement as soon as the solutions have been placed in the cell compartment.

Obtain a blank value for the iodine reagent by diluting 1 ml of the reagent to 10 ml with carbon tetrachloride and measuring the absorbance of this solution at 310 mμ with carbon tetrachloride as the reference.

Calculation.   Calculate the weight per cent of aliphatic sulfide as follows:

$$\text{Weight per cent of aliphatic sulfide sulfur} = \frac{(A_1 - A_2) \times 100}{400 \times 0.9 \times c}$$

where A_1 is the absorbance of the sample-iodine solution at 310 mμ, A_2 is the absorbance due to iodine (blank), and 0.9 is the dilution factor necessary to correct for diluting 1 ml of iodine to 10 ml with the sample solution, the concentration of which (in grams per liter) is represented by c in the equation.

Interference due to formation of hydrocarbon-iodine complexes were shown to be negligible. Aromatic sulfides show relatively much smaller absorptivities and offer no serious interference. Olefins and mercaptans form, with iodine, complexes of low absorptivities and, unless present in concentrations high enough to consume a large amount of iodine, a large excess of which is needed for complete reaction of the sulfide, are not expected to interfere.

II. GAS CHROMATOGRAPHIC METHODS*

See the discussion of mercaptans in Chapter 15, Section II.

In a number of studies on residues of pesticides containing a $>P{=}S$ group as well as a sulfide linkage (e.g., Fenthion or Phorate), the $P{=}S$ group was found to be oxidized at least partially to $>P{=}O$ and the sulfide linkage to the sulfoxide and sulfone. These compounds were determined by GC after a preliminary separation on an adsorbent column, the sulfoxides being the most difficult to chromatograph (4–8). As many as six compounds had to be determined in a single pesticide residue sample (e.g., sulfide, sulfoxide, and sulfone of both the $P{=}S$ and $P{=}O$ compounds of each pesticide).

For routine analyses, a more rapid method was devised, in which all compounds are oxidized to their highest oxidation state, that is, the sulfone of the $P{=}O$ compound, and this compound is determined (9). This procedure has simplified the determination considerably; although it gives only the total residue, rather than the amounts of the individual compounds, in most residue analyses this information is adequate.

III. ELECTROANALYTICAL METHODS†

Coulometric Titration of Dialkyl Sulfides with Bromine

In Chapter 16 we discussed the oxidation of the disulfide linkage to a sulfonyl bromide group by reaction with bromine. Dialkyl sulfides also react with bromine to produce sulfoxides or sulfones:

$$RSR + Br_2 + H_2O \rightarrow R\overset{\displaystyle O}{\overset{\displaystyle \|}{S}}R + HBr$$

or

$$RSR + 2Br_2 + 2H_2O \rightarrow R\underset{\displaystyle \underset{\displaystyle O}{\|}}{\overset{\displaystyle O}{\overset{\displaystyle \|}{S}}}R + 4HBr$$

In many instances, an excess of bromine tends to carry the oxidation to the sulfone stage. For this reason, the classical volumetric methods (10)

* Written by Morton Beroza and May N. Inscoe.
† Written by Alan F. Krivis.

stipulate that an excess (or at least a large excess) must be avoided; in this way, the reaction for many compounds may be stopped at the sulfoxide stage. There are some compounds, however, which are very readily oxidized to the sulfone, and a specific unknown may be in this class. Therefore, if a bromine oxidation of a sulfide is to be used, it is wise to determine the stoichiometry of the reaction with pure standards before running unknowns.

The advantages of coulometric titrations with bromine, as compared to buret titrations, were discussed in Chapters 1, 12, and 16. These advantages also apply to the determination of dialkyl sulfides by oxidation with bromine.

EXPERIMENTAL METHOD

Apparatus and Reagents. The equipment and materials needed for the coulometric generation of bromine are described in detail in Chapters 1, 12, and 16.

Procedure. The procedure for the coulometric titration of disulfides (Chapter 16, Section II) may be used also for dialkyl sulfides.

Calculations.

$$\text{Per cent of sulfide} = \frac{(i)\ (t)\ (\text{mol. wt. of unknown})\ (100)}{(\text{sample weight})\ (2n^*)\ (96487)}$$

IV. NUCLEAR MAGNETIC RESONANCE METHODS†

The sulfides, disulfides, and polysulfides have been studied by nmr. Aryl thioethers were reported (11), and the effect of oxidation of the sulfides to the sulfones on the ^{1}H chemical shifts was also discussed:

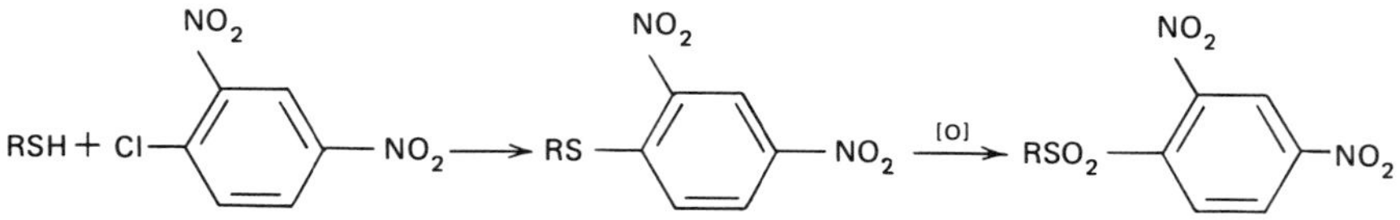

There is little or no effect on the protons of the thiol portion of the ether, but the protons on the ring containing the nitro groups are altered by the conversion of the sulfide to the sulfone. The 2,4-dinitrophenyl thioethers reported were methyl, ethyl, propyl, isoamyl, decyl, benzyl, phenyl, *o*-totyl, *m*-tolyl, and *p*-tolyl. The sulfones reported were ethyl, propyl, isoamyl,

---

* The value of n may be 2 or 4, depending on the stoichiometry of the particular reaction.
† Written by Harry Agahigian.

benzyl, decyl, phenyl, and tolyl. The alkyl sulfides, sulfoxides, and sulfones can be distinguished with ease; once again, however, the protons that are close or adjacent to this heteroatom are measured, so that the method is an indirect one.

REFERENCES

Absorption Spectrophotometric

1. S. H. Hastings, *Anal. Chem.*, **25**, 420 (1953).
2. S. H. Hastings and B. H. Johnson, *Anal. Chem.*, **27**, 564 (1955).
3. H. V. Drushel and J. F. Miller, *Anal. Chem.*, **27**, 495 (1955).

Gas Chromatographic

4. M. C. Bowman and M. Beroza, *J. Agr. Food Chem.*, **16**, 399 (1968).
5. M. C. Bowman and M. Beroza, *J. Assoc. Offic. Anal. Chemists*, **52**, 1054 (1969).
6. M. C. Bowman, M. Beroza, and J. A. Harding, *J. Agr. Food Chem.*, **17**, 138 (1969).
7. M. C. Bowman, M. Beroza, and C. R. Gentry, *J. Assoc. Offic. Anal. Chemists*, **52**, 157 (1969).
8. J. C. Maitlen, L. M. McDonough, and M. Beroza, *J. Agr. Food Chem.*, **16**, 549 (1968).
9. M. C. Bowman and M. Beroza, *J. Assoc. Offic. Anal. Chemists*, **52**, 1231 (1969).

Electroanalytical

10. S. Siggia, *Quantitative Organic Analysis via Functional Groups*, 3rd Ed., John Wiley, New York, 1963, pp. 614–615.

Nuclear Magnetic Resonance

11. H. Agahigian and G. D. Vickers, *J. Org. Chem.*, **27**, 3324 (1962).

18 Sulfonic acids, salts, esters, halides, amides, and imides

I. ABSORPTION SPECTROPHOTOMETRIC METHODS*

A. Direct Infrared Measurement

Toluene and Xylene Sulfonates. Kullbom and Smith (1) developed an infrared method for toluene and xylene sulfonates based on the sulfonate band centered at 1175 cm^{-1} and arising from the sulfur-oxygen vibration. They found that the intensity of the band is not significantly affected by changes in alkyl substitution on the benzene ring. The fact that the method is based on the absorbance of the sulfonate group and is insensitive to the vibrations of the parent molecule suggests that it has potentially more general application.

B. Direct Ultraviolet Measurement

1. Toluenesulfonic Acids. Pinchas and Avinur (2) determined the total amounts of toluenesulfonic acids in the presence of a large excess of sulfuric acid on the basis of their total absorption at 222 mμ. Cerfontain, Duin, and Vollbracht (3) performed ultraviolet determinations of the three isomeric toluenesulfonic acids in excess sulfuric acid by subjecting the absorbances of the unknown mixture and of its constituents, gathered at a large number of wavelengths, to a least-squares treatment by an electronic computer.

2. Toluenesulfonamides. Stewart, Caldwell, and Uelner (4) determined the isomeric distribution in mixed *o*- and *p*-toluenesulfonamides by ultraviolet absorption measurements. The method depends on the ratio of

* Written by J. Gordon Hanna.

346

the absorbance at a wavelength at which the absorptivities are appreciably different (276 mμ, where the absorbance is due predominantly to the ortho isomer) to the absorbance at the isoabsorptive point (256 mμ, where both components have essentially equal absorptivities).

C. Measurement after Chemical Reaction

Alkyl Benzene Sulfonates—Reaction with o-Tolidine and Hypochlorite.   Harris (5) determined alkyl benzene sulfonates on the basis of the blue coloration obtained when the solution of sulfonate is added to a solution of *o*-tolidine and sodium hypochlorite.

METHOD FROM HARRIS (5)

Apparatus.   Cenco-Sheard-Sanford photelometer with a B filter (maximum transmission at 525 mμ) or other type of spectrophotometer for measurement in the range 500 to 600 mμ.

Reagents.   Sodium hypochlorite. Prepare a solution containing 300 ppm of available chlorine.

o-Tolidine solution. Add 1.0 gram of *o*-tolidine to 5 ml of 20% hydrochloric acid and grind to a thin paste. Add 150 to 200 ml of water to produce solution. Transfer to a 1-liter graduate and made to 505 ml with distilled water. Then make to 1 liter with 20% hydrochloric acid. Store in an amber bottle out of direct sunlight.

Procedure.   Transfer 40 ml of distilled water to each of two 50-ml volumetric flasks. Add 1 ml of 300-ppm sodium hypochlorite solution to each and mix. Then add 2 ml of *o*-tolidine solution to each and mix. Dilute the volume of one of the flasks to 50 ml with distilled water and mix for use as a blank. Add the unknown solution of alkyl benzene sulfonate of definite dilution carefully in known amounts to the other flask until a color change just appears. Make up to volume and compare spectrophotometrically with the blank. Read the sulfonate concentration from a calibration curve comparing known concentrations of standard sulfonate with absorbance.

A color change is produced by as little as 0.05 mg of pure sulfonate in 50 ml of solution (equivalent to 1 ppm).

II. GAS CHROMATOGRAPHIC METHODS*

Arylsulfonic acids and their salts are rapidly determined by fusion with KOH at 380 to 400°C under a nitrogen or helium atmosphere; the fusion is

* Written by Morton Beroza and May N. Inscoe.

carried out in a small heated chamber connected directly to a gas chromatograph which is used to determine the amounts and the identities of the phenols evolved when a solution of maleic acid is added to the fusion mixture (6). Isomeric sulfonic acids, ammonium salts, or amides may be converted first to sulfonyl chlorides and then to the sulfonyl fluorides for determination by GC (7, 8).

Pyrolysis gas chromatography can be used for the determination of arylsulfonic acids and their salts; quantification is accomplished by measurement of the sulfur dioxide produced or in some cases the parent hydrocarbon (9).

III.   RADIOCHEMICAL METHODS*

Conversion of Sulfonamides to Derivatives of Radioactive Silver

Sulfonamides which form insoluble silver salts have been determined by using silver-110m nitrate (10) in a radiometric adaptation of a standard volumetric method (11). The sample is dissolved in acetone, magnesium oxide is added, and the solution is titrated; the end point is established by monitoring the supernatant phase for radioactivity due to excess $^{110m}Ag^+$. The error in the radiometric method is $+0.6$ to -0.8%.

REFERENCES

Absorption Spectrophotometric

1. S. D. Kullbom and H. F. Smith, *Anal. Chem.*, **35**, 912 (1963).
2. S. Pinchas and P. Avinur, *Anal. Chem.*, **30**, 2022 (1958).
3. H. Cerfontain, H. G. J. Duin, and L. Vollbracht, *Anal. Chem.*, **35**, 1005 (1963).
4. F. N. Stewart, J. E. Caldwell, and A. F. Uelner, *Anal. Chem.*, **31**, 1806 (1959).
5. J. C. Harris, *Ind. Eng. Chem., Anal. Ed.*, **15**, 254 (1943).

Gas Chromatographic

6. S. Siggia, L. R. Whitlock, and J. C. Tao, *Anal. Chem.*, **41**, 1387 (1969).
7. J. S. Parsons, *J. Gas Chromatog.*, **5**, 254 (1967).
8. A. A. Spryskov and V. A. Kozlov, *Izv. Vysshykh Uchebn. Zavedenii, Khim. i Khim. Tekhnol.*, **11**, 785 (1968).
9. S. Siggia and L. R. Whitlock, *Anal. Chem.*, **42**, 1719 (1970).

Radiochemical

10. P. Bebesel and I. Sirbu, *Riv. Chim. (Bucharest)*, **11**, 288 (1960); *Chem. Abstr.*, **58**:5034a.
11. L. Kum-Tatt, *Analyst*, **82**, 185 (1957).

* Written by D. Campbell.

19 Automated wet chemical analysis: application to the determination of functional groups*

I. INTRODUCTION

A variety of approaches to the automation of wet chemical analyses has been introduced during recent years. However, because of the development of physical methods (e.g., spectroscopy, and chromatography) over the same period, the application of wet chemical principles to functional group analyses has received little attention. In situations in which physical methods are inapplicable, automated wet chemical methods have been actively pursued. This is particularly true of process control and clinical analysis, where frequent tests are required on large numbers of similar samples. The combination of specificity and the low cost of chemical methods with the speed and reproducibility of instrumentation has led to extensive use of automated wet chemical analyses in these fields. In complex systems this chemical selectivity frequently makes wet chemical analysis superior to physical methods. This chapter describes the automation of wet chemical methodology for functional group analyses in terms of philosophy, available equipment, and applications.

"An automated system" in this context denotes the ability to carry out a wet chemical procedure from one step to another without manual manipulation. Reports of applications of automated functional group analysis itself are scattered throughout the literature and hence are difficult to find. In 1969 a symposium was held which included a discussion of automated functional group analysis (1–5); a paper by Mitchell (4) on this subject is included in the proceedings. A partial survey and classification of available literature (6–14) and equipment through 1970 is presented.

* Written by R. A. Hofstader and W. K. Robbins.

Most of the available references to instrumentation and application were obtained by surveying *Analytical Abstracts*, 1959–1970 inclusive; *Chemical Abstracts*, 1957–1970 inclusive; the *Technicon Symposia* 1965–1967 (15–17); and manufacturers' literature (mostly domestic). Because of the great diversity of the fields which required searching, there are many omissions which must be apologized for at the outset. It should also be pointed out that no attempt has been made to discuss automation from a computerization viewpoint. Many excellent reviews are available on this topic (18).

II. PHILOSOPHY

In general, it can be assumed that any wet chemical procedure can be automated. The considerations for automation are strictly economic and are based on cost of equipment, process improvements resulting in greater efficiency, and savings in manpower and time. The same economic justification must be applied to the question of how much to automate—should the entire procedure be automated, or just one part? Automation, like bureaucracy, is not an end unto iteslf. The added expense of instrumentation required for automation can be justified only where it:

Handles a large number of samples.
Increases process rate.
Maintains precision and accuracy.

When these criteria are met, the initial expense is offset by increased efficiency. Thus automation of a procedure must result in an overall reduction in cost.

The need to automate procedures has been found most frequently in the biomedical field (20–22). Today, the clinical laboratory may be equipped with automated techniques for enzyme analyses (23–26), for preparing slides for microscopy, and even for preparing a twelve-point profile of blood samples (27). In the pharmaceutical field many applications have been made (28–29). Although none of the elegant equipment mentioned above has been applied to functional group analysis, much of the rationale in such instrumentation is applicable to this purpose.

Another area which requires the advantages of automation is process control analysis. The complex nature of process systems has made it necessary to automate chemical, as well as physical, analyses. Description of this automation has been presented by several authors (30–33). Although only a few functional groups have been determined in process streams, much of the apparatus utilized has potential application in the analysis of such groups.

Thus, in large part, the automation of wet chemical analysis has evolved from two diverse fields.

Although the basic step in a wet chemical analysis is the reaction, the complete analysis may be considered to consist to a combination of four steps:

Sampling.
Preparation and separation.
Reaction.
Measurement and detection.

A completely automated analysis may be defined as one in which no manual manipulation is required between the sampling and measurement steps. *A partially automated system* is one in which some human intervention or assistance is required between one or more of these steps. For example, in a totally automated system a sample will be automatically carried into a reaction chamber, measured, and recorded. In a partially automated system, however, a diluted aliquot might be placed in a titrator and then the final measurement carried out automatically. Patient (34) presents some alternatives of partial and total automation.

Two approaches have been developed in totally automated systems— *continuous flow analysis* and *discrete sample analysis*. The first of these, as its name implies, employs instrumentation which permits analytical processing of a flowing stream. In discrete sample analysis, on the other hand, a sample is treated on an individual basis in a contained vessel. *Hybrid systems*, commercially available instruments which combine two or more analytical functions, have been developed for both of these approaches.

The basic principle of wet chemical methods is the use of a characteristic chemical reaction for the group being measured. This reaction must be as specific as possible for the functional group of interest. If this reaction is to be automated, it must also be rapid and must involve a reactant or product which is easily measured. Any reaction which produces or consumes measurable amounts of acids, bases, oxidants, metallic ions, precipitates, or colored complexes can be automated with available equipment. Some of these reactions have been discussed in the preceding chapters of this book; others may be found in standard texts.

In some cases the application of automated operations to solution chemistry allows the chemist to use procedures which could not be suitably performed manually. The reproducibility which characterizes automated systems, combined with the use of frequent calibration, makes it possible to obtain accurate, reliable results by means of incomplete reactions. In particular, if analytical procedures are adjusted so that the results are directly proportional to the concentrate of analyte (i.e., if pseudo-first-order kinetics applies), then reactions which do not go to completion may be utilized in automated systems. Within similar limitations, partial separations have been utilized. Some success has even been achieved with the use of only partial recorder response (non-steady-state detection).

III. INSTRUMENTATION

Since both continuous flow and discrete sample analysis have been success-fully applied in automation, each must be judged on its relative merits. The continuous flow analysis, which Ferrari (35) and Skeggs (36) were instru-mental in developing, is based on simple mechanical principles and provides a continuous record of the analysis so that process variations can be rapidly detected. However, this approach consumes large quantities of reagent. In addition, in order that steady-state concentrations of the detected species may be reached, this system requires a relatively large sample. The use of the same system for many samples reduces cost but can introduce problems due to sample diffusion (i.e., broadening and sample carryover).

On the other hand, the discrete sample approach uses small quantities of reagent and sample. Although this method resembles classical procedures, it often requires complex mechanical concepts for total automation. Various parts of the discrete sample approach often serve as the basis of partially automated systems.

In this section, equipment for automated wet chemical analysis is discussed in terms of its availability and use in sampling, preparation and separation, reaction, and measurement. Reviews of automated instrumentation have appeared previously (38–40). Table 19-1 in Section V describes the equip-ment available for use in automated analytical procedures. Most of the equipment included in the table has not yet been applied to the determina-tion of organic functional groups; rather it consists of the commercial equip-ment which is readily available for the automation of wet chemistry. The table is intended to give the reader an initial source for such instrumentation and to familiarize him with the characteristics and availability of the equipment.

Several types of equipment which fit the given definition of automated wet chemical analysis instrumentation are not included in the table. Fraction collectors, heating baths, and similar equipment have been omitted because they are so commonly used in laboratories. Since gas chromatography has been discussed in other chapters of this book, only automatic injectors for gas chromatographs are included. Recent reviews (41, 42) concerning the interfacing of gas chromatography and mass spectral determinations are available; hence such instrumentation also has not been included. For similar reasons, the reader is directed to other reviews for information regarding automated methods of amino acid analysis (43–45).

A. The Analytical System

1. Sampling. "Sampling," in its broadest sense, refers to obtaining a homogeneous portion of material to be analyzed. Very often preparation of this material for analysis is also required. For the sake of simplicity, this discussion will be limited to sampling material which is ready for analysis.

Most samplers used in automated methods are designed to transport *liquids* and are characterized by combination with a pump. In these samplers, vessels containing sample solutions are presented to a dip tube. An aliquot of sample is drawn into the tube and, depending on the pump used, is discharged into a reaction vessel, pumped into a continuous flow system, flushed to waste, or returned to the vials. The aliquot size is determined either by a timer on the sampler dip tube or by the pumping rate. Frequently the pumps are not integral parts of the samplers but either are incorporated in other instruments or are separate units. An example of a dip-tube sampler and separate pump for a continuous flow system is shown in Figure 19-1.

For continuous flow analyses, peristaltic pumps, such as the one shown in Figure 19-2, are most frequently used. Usually, in this class of pumps,* motor-driven rollers pass over elastic tubing, pinching off segments of liquid. (Special organic-resistant tubing is available.) The flow rates of these pumps are controlled by varying tube diameters or by changing motor speeds. Because of their working principle, however, the peristaltic pumps are limited to low-pressure applications.

Slightly higher pressures may be obtained with continuous syringe pumps.† In these, two syringe plungers are driven 180° out of phase in fixed barrels. By varying syringe size and plunger speed, a wide range of pump rates can be obtained. If higher pressures are required, precision metering pumps may be utilized.‡ In these, motorized plungers, cam-driven diaphragms, or pressure-operated bellows produce pulses of flowing liquid at pressures up to 1000 psi. Flow rates can be varied by changing drive motor speed or stroke

---

* Harvard Apparatus, tubing pumps; Glenco, Stal Produkten Model SP-1; AnalTech, Fenet Evenflo Control-O-Rate pump.

† Sage, Inc., unlimited volume pump; Hooke and Tucker, multireagent Dilutor; Bausch & Lomb, automatic diluter pipetter.

‡ Instruments Specialties Co. (ISCO), precision metering pump Model 310; Hach, Synchro-Flow solution pump; Milton Roy, Minipump; Glenco, precision electronic chromatography pump; Chromatix, Inc., Cheminert metering pumps.

Figure 19–1. Dip-tube sampler for continuous flow analyses. Photograph courtesy of Technicon Corp., Tarrytown, N.Y.

length. These pressure pumps are generally constructed of solvent-inert materials such as metal, glass, or Teflon.

For discrete sampling analyses, syringe pumps are generally used. These syringe pumps, however, are of the noncontinuous variety, and the syringes often act only as aspirators, in that they are isolated from the actual sample by glass or plastic tubing. The volume of these pumps is most frequently determined by syringe size and by plunger setting.

Samplers are available which automatically dump solid samples from a turntable into a high-speed mixer which contains the desired solvent or extraction solution (Figure 19-3). The resulting solution is pumped into a flow-through system, where a fraction is resampled.

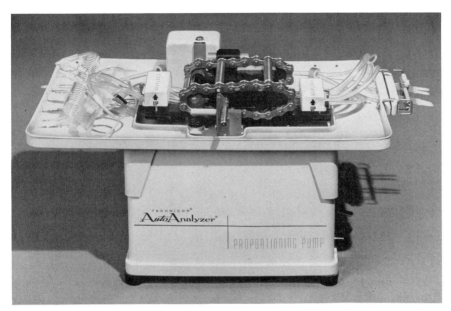

Figure 19–2. Peristaltic pump with 18-tube capacity for continuous flow analyses. Photograph courtesy of Technicon Corp., Tarrytown, N.Y.

For more fully automated systems, automatic recording balances capable of completely automatic weighing of tablets or capsules are available.

Automatic *gas sampling* is accomplished by gas scrubbing systems. The gas enters an absorbing solution through the top of a spiral coil and flows concurrently down the column. The ratio of solution to gas flow can be carefully controlled. At the base of the column, the spent gas is separated and an aliquot automatically withdrawn for further processing.

2. Preparation and Separation. "Preparation" designates the step which prepares the material for the reaction that it will undergo. Since solubilization of solid or gas samples by sampling devices has already been presented, this discussion is limited to dilution, extraction, and/or removal of interferences from the sample solution.

In many systems involving automation, the same pump system used to withdraw the sample serves to introduce the diluent. In a continuous flow system the predetermined quantity of diluent is mixed in a flowing stream with the sample. In a discrete sample analyzer the diluent may be used to flush the sample into the reaction vessel.

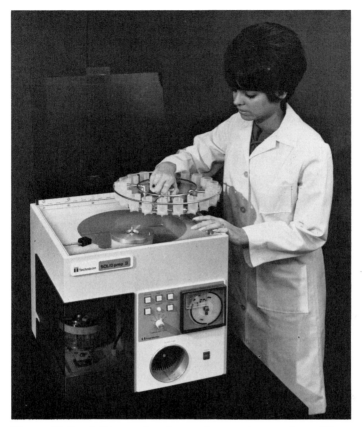

Figure 19–3. Solid sampler for continuous flow analyses. Photograph courtesy of Technicon Corp., Tarrytown, N.Y.

Instrumentation is available which permits the analyte to be separated from interferences automatically. Systems which carry out *extractions* from one phase to another are on the market. In one continuous flow system, fluids pass up a vertical glass coil filled with glass beads, which give a large working surface for phase interaction. At the top of the beaded column the fluid passes into a separator. The two phases are permitted to stratify. The desired phase is continuously aliquoted, and the unwanted phase pumped to waste. The operation of one such system is shown in Figure 19-4.

Separation can also be achieved automatically with the use of a *centrifuge*. Several types capable of being applied to automated systems are available.

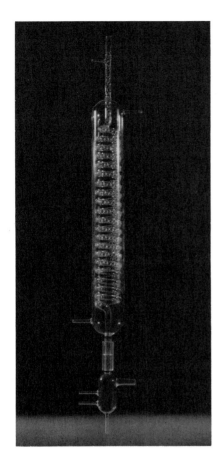

Figure 19–4. Continuous solvent extractor. Photograph courtesy of Technicon Corp., Tarrytown, N.Y.

Their general application has been in removing suspended solids from the system so that the liquid could be further sampled for analysis. However, modified systems have been developed which permit the use of centrifugal forces for solvent separations. The use of special porosity cups make it possible to separate the desired phase, as illustrated in Figure 19-5.

Several systems now under development would carry out complete analyses in a specially designed centrifuge rotor (46). These centrifuge separations are used mostly as parts of discrete sample analytical systems.

For precipitate removal in continuous flow systems, a special filtration apparatus has been designed which operates by having the sample drop onto moving paper continuously drawn over a Teflon platen. Slightly downstream,

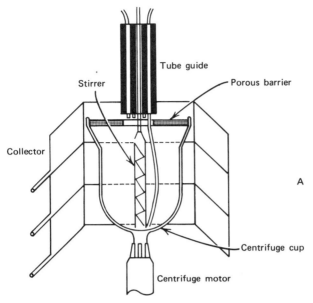

Stirrer

Tube guide

Porous barrier

Collector

A

Centrifuge cup

Centrifuge motor

Figure 19–5. A: Centrifugal solvent extraction and/or filtration system. B: Solvent extraction. (a) Aqueous and organic phases are introduced; (b) Aqueous and organic phases are mixed; (c) Aqueous and organic phases are separated—aqueous phase ejecting. C: Precipitation. (a) Reagents are added, solution is stirred, precipitate forms; (b) Suspension is being centrifuged, supernatant ejects into collector, precipitate remains behind porous barrier; (c) Precipitate is redissolved. Courtesy of Tech. Op., Inc., Burlington, Mass.

tubing attached to the underside of the platen draws the liquid, which is then further processed. Figure 19-6 illustrates the apparatus.

A separation technique which has been used extensively in automated procedures in the clinical field is *dialysis*. A commercially available continuous unit consists of two grooved plates separated by a membrane. The spiral grooves on the plates coincide to form a tube 7 feet long divided lengthwise by the membrane. The dialyzable material in the sample solution passes through the membrane to pick up the stream flowing on the other side. In a recent version of this device a much shorter (7 inch) split channel is used as a pseudo-first-order method (Figure 19-7).

Dialysis procedures are limited to aqueous systems in general. Ultrafiltration, a pressurized dialysis technique (47), may be used with selected organic solvents as a means of concentrating column effluents.

A specially designed microstill permits isolation of *volatiles* in an absorption

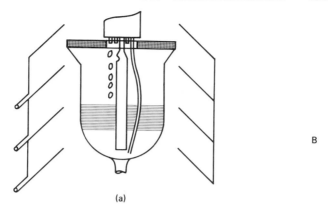

B

(a)

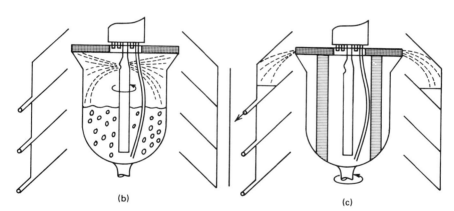

(b)   (c)

Figure 19-5 (Continued)

solution or in a subsequent condensation. The residual absorbate or condensate may then be further processed (Figure 19-8).

Most *solution chromatographic methods* (discussed in other chapters of this book) utilize a sample injection and fraction collector to accomplish the separation of a discrete sample. However, there exist many types of flow-switching devices which would make it possible to use this technique in a continuous flow analysis. As an alternative, a continuous flow system would be coupled with the fraction collector, and the fractions treated as discrete samples for further handling.

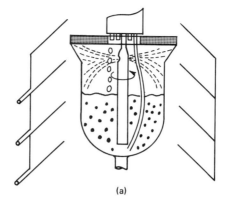

C

(a)

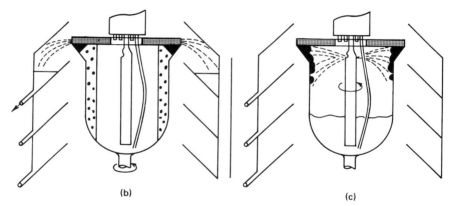

(b)   (c)

Figure 19–5 (Continued)

Several automated determinations of amino acid residues in a peptide hydrolyzate have been developed on the basis of gradient elution ion-exchange methods (43–45). Equipment of this type, designed to produce smooth continuous buffer gradients, has been applied also to sugar residue analysis (48, 49).

3. Reaction.

$$\text{Function} + \text{reactant} \xrightarrow[\substack{\text{specified} \\ \text{conditions}}]{} \text{product}$$

This is the step in an analytical procedure in which the analysis takes place; the function is reacted, producing a species capable of being meas-

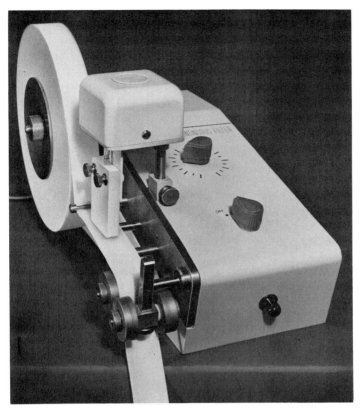

Figure 19–6. Filter apparatus for continuous flow analyses. Photograph courtesy of Technicon Corp., Tarrytown, N.Y.

ured. The reaction is measured by determining either the quantity of product formed or the amount of reactant which is equivalent to the function being measured. In the former case, usually an excessive reactant is added and the resultant product (i.e., intensity of color, change in potential, etc.) is measured. In the latter case, some signal is monitored when the equivalent amount of reactant has been added (i.e., titration).

In continuous flow systems, reactant is pumped and mixed with sample, and the reaction permitted to go to the desired extent in a flowing stream. Since mixing in narrow-bore tubes does not occur easily, air bubbles are introduced into the flowing stream. The air bubbles segment the stream; the small segments are repeated as the stream passes through a coil, and the well-mixed solution may then be debubbled before the next operation is per-

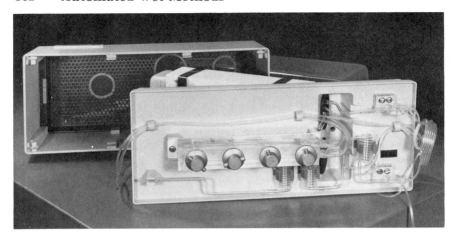

Figure 19–7. Automatic flow-through dialysis unit. Photograph courtesy of Technicon Corp., Tarrytown, N.Y.

formed. In discrete sample analysis the reaction vessel has a much larger diameter. In such a vessel diffusional forces are generally used for mixing; however, care must be exercised in the design of any given analysis to prevent a density gradient from being involved. Thus a concentrated reagent should be added to the top of the dilute sample solution. Often the velocity of the reagent dispenser provides some mixing.

If elevated or reduced temperatures are required for the reaction, the material can be introduced into a constant-temperature bath. In a continuous flow system the mixed solution is pumped through a coil emerged in the bath. For a discrete sample analysis the reaction vessel is drawn mechanically through such a bath. In either case the extent of reaction is adjusted by the bath temperature and the time of heating. Recently an automated commercial instrument has been introduced which is designed to derivatize amino acid residues (in a peptide hydrolyzate) to their volatile N-acetyl, N-propyl* esters and to inject these into a gas chromatograph. Similar procedures may be anticipated for other functional groups which are readily converted to volatile derivatives.

4. Detection. The development of detection systems has reached a point of high sensitivity, low holdup, and rapid response, which makes many satisfactory for use in automated systems. Many of the detectors which have been developed for continuous analysis could be applied equally well to

---

* Vivonex amino acid chemical processor, Model D-3.

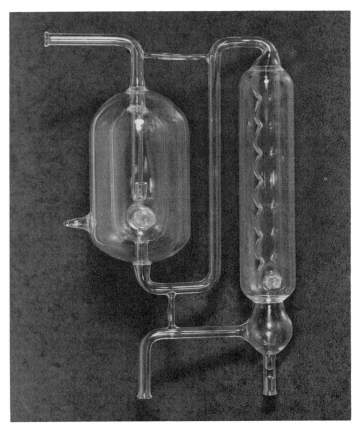

Figure 19–8. Distillation unit for continuous flow analyses. Photograph courtesy of Technicon Corp., Tarrytown, N.Y.

discrete sample techniques. In particular, detectors based on physical measurements, which are used in continuous analysis, could be applied to determine sample fractions for a discrete sample system. For a compilation of these continuous physical sensors, the reader is directed to two recent reviews (31, 50). The list of detectors presented in Table 19-1 includes chiefly instruments which would permit direct quantification of organic functional groups after a wet chemical reaction.

Because of the simplicity and ease of automation, the two areas on which most attention has been focused are titrimetry and colorimetry. A variety of titrators and spectrometers capable of being automated are listed in

Table 19-1. The applications of these techniques to automated functional group analyses are discussed in Section IV.

B. Hybrid Automation Units

In order for several of the analytical steps to be combined into a single system, the various components of the analytical processing must be properly integrated. The coordination of timing and volume is of primary concern.

Combinations of continuous flow modules frequently do not require separate controls. Proper selection of flow rates (tubing diameters) and reaction time (tubing lengths) permits both sequence timing and volume ratios between units. The major drawback in the continuous flow approach is reagent consumption. Although small samples are pumped into the system, the volume of reagent pumped remains constant. The use of smaller-diameter tubing, shorter flow lengths, and miniaturized systems relieves this problem considerably. Figure 19-9 illustrates a combination of continuous flow units for colorimetric analysis.

Perhaps the combination of several units to make a continuous flow system may be best illustrated by an example. Consider a colorimetric analysis which

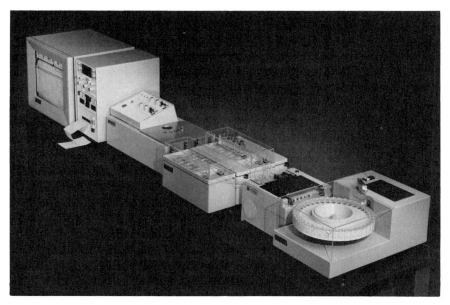

Figure 19–9. Module combination for continuous flow analyses. Photograph courtesy of Technicon Corp., Tarrytown, N.Y.

manually requires dilution, reagent addition, heating, filtration, and an absorption determination. With the continuous flow approach, this analysis might be automated as illustrated in Figure 19-10. In this system an aliquot of sample is pumped from the sampler into an air-segmented stream of reagent and diluent; the resultant stream is pumped through a coil immersed in a heating bath. The solution is dropped onto a moving strip of filter paper, and the recovered filtrate pumped past a recording colorimeter. Such a system may be constructed from commercially available modules such as those produced by Technicon.

The discrete sample approach to our hypothetical analysis is somewhat different. Figure 19-11 illustrates the Beckman DSA-560 (see Table 19-1), which would perform the analysis after it had been programmed to do so (51). In this system, a sample aliquot is pumped by positive displacement, (i.e., motorized syringe) into a reaction vessel, to which a fixed amount of reagent is added. The vessel is then mechanically moved through a temperature bath at a fixed rate. A filter cup is lowered into the sample cup, and the solution aspirated through the filter. The filtrate is transferred to a new sample cup by a syringe pump and pumped into a fixed stop-flow cell. The absorption of the solution is then measured and recorded on a printout tape. The filtrate is thereupon returned to the reaction vessel. Meanwhile, other samples are being similarly processed in separate reaction vessels.

In general, discrete sample analysis requires a controlling device to connect the analytical steps* and the mechanical transfer of the sample through various processes is therefore more complex than in continuous flow systems. The discrete equipment may control the sequence and the size of transfers by the use of an operational program. The program sequence can be determined by a template, by the positioning of various components, or by using a fraction collector. Both electronic and pneumatic controls are used in connection with timers to provide proper advancement of each sample through the sequence. The program sizes as required by the analytical reaction are controlled by selecting appropriate pipet delivery volumes or by allowing procedures to be performed for fixed time periods. When properly coordinated, a discrete sample analyzer uses minimum volumes of sample and reagent to perform the complete analysis.

With the exception of Technicon, most commercially available hybrid units utilize discrete sample methodology. Because of the flexibility of the modular approach used by Technicon, however, continuous flow methods are probably more common in the literature.

---

* Simple independent controllers are available. For example, the Luft Controller 77 can be used to coordinate as many as 52 operations.

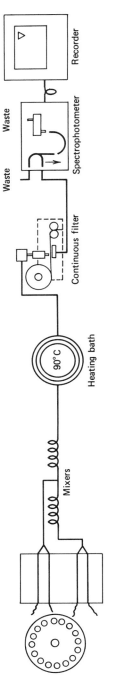

Figure 19–10. Flow-through scheme for analysis.

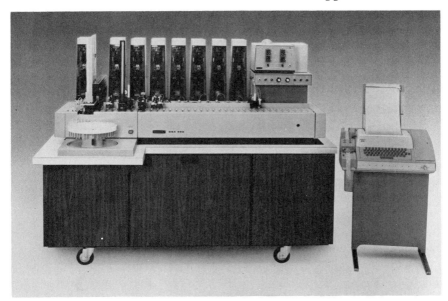

Figure 19–11. The Beckman DSA-560 discrete sample analyzer. Photograph courtesy of Beckman Instrument Co., Fullerton, Calif.

IV. APPLICATIONS

A. Automatic Colorimetric Analysis

The area of automated analysis in which the most extensive application has been made during the past decade is colorimetry. Automated units of both the flow-through and the discrete analyzer types have been applied. The Technicon literature of 1957–1967 lists over 1500 applications of colorimetric analyses to automation (52). Most of these applications are characterized by a large number of similar samples to be analyzed. In clinical analyses the sample matrix is similar, and consequently many applications have been made in this area. A very rapid review of the literature clearly demonstrates that it is possible to automate any colorimetric analysis if there is economic justification for doing so.

The early developments in the field of automatic colorimetric analysis as applied to aqueous systems were reported by Hans Fuhrmann (53) as early as 1952. Sheen and Serfass (54) discuss the early process instrumentation of this type. Their paper describes the history, development, and applications in the chemical industry to 1960, including applications to the determination of dissolved oxygen as well as formaldehyde. In 1967 Lang (55) described a spectrophotometric system built from commercial parts supplemented with an automatic sample changer, a displacement pump, and capillary cells.

The designs of flow-through colorimeters have been described by Epps and Austin (56), Shapira and Wilson (57), Bishop and White (58), Loebl (59), and Gualandi and Morisi (60). The Technicon colorimeter, which has been most widely applied, is a double-beam, narrow-half-band interference type. In this colorimeter, a single light source is viewed simultaneously by right-angled light axes, one by the reference photocell, the other by the sample photocell. Front surface lenses gather and focus the light, passing it through the corresponding interference filter, where it then passes through a small tubular flow cell. The internal volume of this tubular flow cell is 0.1 cm^3.

The colorimeter constitutes part of the electronic network of the recorder, thus forming a ratio-recording system. In this type of network differences in light intensity due to voltage fluctuations or a change in the emittance characteristics of the light bulb are automatically ignored, inasmuch as both photocells see these changes simultaneously. The only factor that changes the output of this system is a reduction in the intensity of the light striking the sample-measuring photocell due to the absorptive characteristics of the solutions flowing past the photocell. Other commercially available equipment is described in Table 19-1.

Specific examples of the colorimetric determination of functional groups have been reported in the literature as far back as 1960. Most applications to functional group analysis have been made using the commercially available Technicon apparatus. Other equipment could be substituted, however, on the basis of specific applications.

Dukes and Hyder (61) reported on the determination of peroxide by automatic colorimetry. Their procedure consists of the oxidation of the leuco base of phenolphthalein by peroxide in the presence of cupric iron. The resultant color is measured at 534 mμ. The instability of the color makes this procedure difficult when carried out manually. Lamy et al. (62) also automated a procedure to determine peroxides but employed the titanyl sulfate procedure.

Ashworth and Walisch (63) described an automatic ultramicromethod for the determination of alkoxyl groups. Kuzel (64) reported an automatic procedure for the determination of tertiary amines in solid samples, utilizing a bromocresol purple procedure. In the same paper he described a novel extraction technique for use with flow-through systems. Atkinson (65) reported a technique for the on-stream automated determination of esters and acids in alcohol.

Duncombe and Shaw (66) applied automated techniques to the deter-

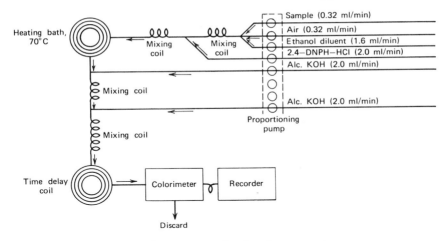

Figure 19–12. Continuous flow analyses of carbonyl compounds.

mination of aldehydes and ketones in liquids. Bartkiewicz and Kenyon (67) reported on the determination of trace quantities of carbonyl in organic solvents. They automated the procedure of Lappin and Clark (68) as shown in Figure 19-12. As can be seen from the flow diagram, sample, air, and diluent are mixed in the first mixing coil. The reagent, dinitrophenylhydrazine hydrochloride, is mixed into the flowing stream and the mixture heated to 70°. Alcoholic KOH is added, the stream is pumped through a flow-through colorimeter, and the response is recorded. Special tubing capable of resisting attack by organic reagents was used in this investigation. This technique makes it possible to analyze twenty samples per hour as opposed by two by the conventional procedure. Moreover, the data shown indicate excellent agreement between the manual and the automated approach.

In a more recent work Friestad, Ott, and Gunther (69) report a procedure for the automatic colorimetric microdetermination of phenol. The procedure is based on the oxidative coupling of the phenol with 3-methyl-2-benzo-thiazolinone hydrazone. The methodology reported in the procedure involves an automated microdistillation of the phenol before the coupling reaction. The method is capable of analyzing twenty samples per hour.

B. Automatic Titrimetric Analysis

Any instrument which records a titration curve or stops a titration at some set end point without manual intervention is considered to be an automatic titrator. The concepts behind these instruments were developed 50 years ago (70), and some commercial autotitrators have been available for more than 20 years (71). During the past two decades, automatic titrations have been applied to the determination of a wide variety of organic functional groups.

Many of the older commercial instruments have been presented in detail by Phillips (72), and he, as well as Tucussel (73) and Zajicek (74), has discussed the principles and operation of automatic titrators. Squirrell (75) presents many applications of automatic titrators to functional group analyses in his book.

1. Automatic Titrators.   Most automatic titrators are designed on the basis of classical titration procedures. Therefore, these instruments must be classified as discrete sample analyzers. Automated continuous titrators, which find use in process control analyses, are included in the review of Blaedel and Laessig (31).

An automatic titrator generally consists of four parts: (*a*) a dispensing unit to introduce the titrant into the reaction vessel, (*b*) a sensing device to determine the end point and to signal a controlling device, (*c*) a controlling unit to regulate the dispensing unit so that the "true" end point is determined, and (*d*) a readout system to present the determined end point. A wide variety of these components are used in the instruments listed in Table 19-1. Each component is functionally described in the following paragraphs.

The *dispensing unit* may be a selenoid-operated valve installed on an ordinary buret or precision pump of some type. Most commercial instruments use a piston buret (motorized syringe) to dispense the titrant because it has low drift with accurate delivery provided by an easily regulated motor drive. Often the syringes are combined with a large reagent bottle in a single unit which permits rapid adaptation of the titrator to many procedures.

The *sensing devices* for most commercial automatic titrators determine the end point either potentiometrically or photometrically. For most potentiometric titrations glass, platinum, or silver electrodes are used as the sensing device. The recent development of specific ion electrodes adds even greater versatility to the potentiometric titration. Automatic photometric titrations use specially balanced photocells equipped with appropriate color and intensity filters to sense the titration end point. Once instrumental corrections are made to blank out sample background, the end point may be reproduced accurately. The light may pass through the entire sample, or it may be guided through a short path by light pipes.

The *controlling device* regulates the rate of titrant introduction into the reaction vessel. In many instruments, signal feedback slows the rate of reagent addition as the end point is approached. This may be accomplished by proportional band controllers, by derivative (incremental) response feedback, or by a built-in time delay. These techniques permit sharp end points to be observed for small potential breaks or for reactions which proceed to completion slowly (76). Figure 19-13 illustrates a photometric titrator with a proportional band controller.

The *readout systems* are quite varied and depend a great deal on the type of end point which is determined. Several instruments plot the titration curve on a recorder which has its drive mechanism synchronized with the syringe motor so that one of the recorder axes is mechanically linked to the reagent volume. In a modification of this approach a linked printer system prints the change in potential for an incremental change in reagent added. Other instruments plot the first derivative, which has been reported to give a sharper end-point determination, directly. A different type works on the dead-stop principle adding the reagent until a preset fixed potential (or color intensity) is reached. In such instruments, only the final total volume used for the sample is printed.

2. Automated Titrations of Organic Functional Groups. Almost every organic functional group may be determined directly or indirectly by a titrimetric procedure, and almost every titration can be automated. For example, organic acids have been titrated automatically in nonaqueous media by both colorimetric (77, 78) and potentiometric (79–81) methods. Such direct automatic titrations of functional groups are supplemented by indirect titrimetric procedures. For example, Squirrell (75) suggests the automatic titration of the free hydrochloric acid produced in the oximation of aldehydes or ketones. Similar techniques have been applied to more functional group analyses than space limitations permit us to discuss. Suffice it to say that almost any functional group may be determined by using an automatic titrator.

However, many of these procedures require several manipulations by a technician before the "automatic" determination is made. Consider the oximation procedure suggested by Squirrell (75). A water-immiscible sample is added to a 100-ml stoppered flask containing 20 ml of 2% (w/v) solution of hydroxylamine hydrochloride adjusted to pH 24. to 4.3 with sodium hydroxide, and the flask is shaken for 30 minutes. Then the solution is washed with 5 ml of water into a beaker, and the aqueous hydrochloric acid layer titrated with 0.1 N sodium hydroxide to an end point determined by a reagent blank. Since manipulations precede the automatic titration, this procedure does not fit our definition of a totally automated analysis.

Johnson (82) has reported totally automated wet chemical functional group analysis using a titrimetric finish. In his procedures, the effluent from a chromatographic column, containing a carboxylic acid, flows continuously through a stirred titration vessel. A titration diluent and indicator are added

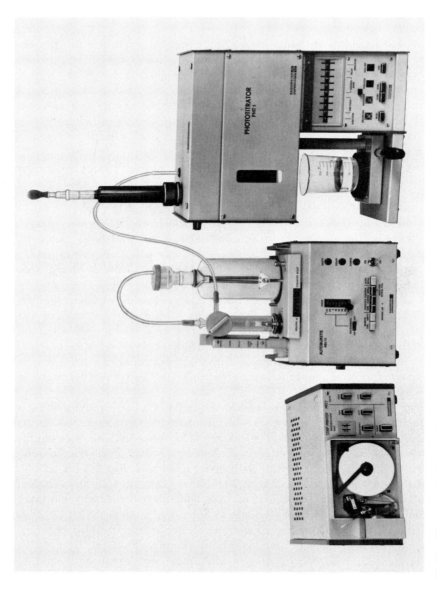

Figure 19–13. Photometric titrator with proportional band controller. Photograph courtesy of the London Company, Cleveland, Ohio.

continuously as well. The acids passing through the cell are titrated to a constant indicator absorption, using electrically generated base. The potential required to create sufficient current to produce the base gives a peak rather than a "staircase" record. Although there are other methods for "continuous titration" of column effluents (83, 84), these procedures, which utilize the addition of an excess of indicator, are better characterized as colorimetric methods.

The methods mentioned illustrate the potential of continuous flow titration procedures. However, as discussed earlier, most commercial automatic titrators are designed to be discrete sample analyzers. Consequently, to date there have been few, if any, totally automated analyses of functional groups using these instruments. Within the last few years, several turntable titrators have been introduced. These instruments could be used to integrate a wet chemical analysis with a titrimetric finish, although this has not been done as yet. In one model, the beakers are rotated under a pair of electrodes for potentiometric titrations or light pipes for photometric titrations. The electrodes are lowered into the beakers, the titration is performed, and the preset end-point volume is printed out. The electrodes or light pipes are then raised, and the turntable rotates so that the next peak is positioned under the sensors for another determination (Figure 19-14).

An alternative device consists of a turntable holding fifty 100-ml beakers,

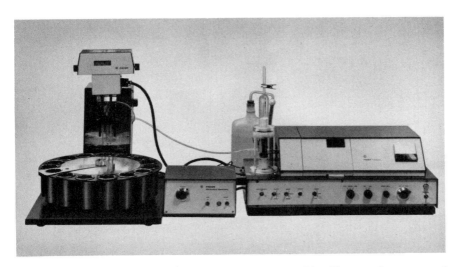

Figure 19–14. Photometric titrator with sample turntable. Photograph courtesy of Fisher Scientific Co., Pittsburgh, Pa.

a timing mechanism, and an elevator. A beaker is raised on the elevator pedestal to present it to the sensor probes (Figure 19-15).

To perceive how these units could be integrated into an automated analysis, consider the oximation-tritation analysis attributed to Squirrell (75). A sample could be mixed with a continuously flowing stream of the pH-adjusted hydroxylamine hydrochloride. After mixing for the required period, the product solution could be collected, using a fraction collection inlet, in the beakers of the turntable. By timing the titration sequence, each sample solution would be collected while a previous sample is being titrated to the set pH determined by a reagent blank. Although hypothetical, this system illustrates the potential of the turntable titrator combination.

Although most of the information presented has concentrated on potentiometric and photometric procedures, we should mention other titrimetric methods which have been automated. For example, mercaptans may be

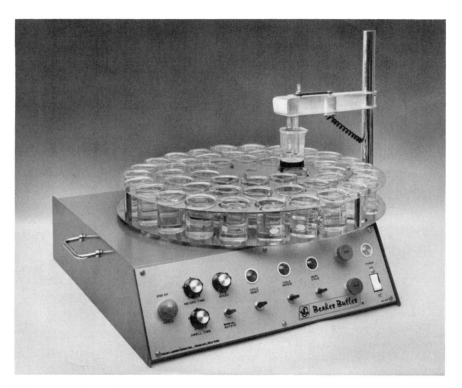

Figure 19–15.   Pedestal turntable apparatus. Photograph courtesy of Vision Laboratories, Oriskany, N.Y.

determined by an amperometric titration with silver nitrate solutions, and some functional groups may be analyzed by thermometric methods (85–87) which are based on automatic titration principles. Another approach which has application to automated methods is coulometric titrations. Bandisch, Beilstein, and Rasch (88) have applied this technique to the determination of double bonds. To be most useful, however, each of these methods must be integrated into a complete analytical system.

V.   COMMERCIALLY AVAILABLE APPARATUS FOR USE IN CONJUNCTION WITH AUTOMATED SYSTEMS

In the preceding sections a considerable amount of attention was given to the equipment which is needed to design a fully or partially automated system for use in wet chemical analysis. Table 19-1 is a compilation of equipment, already commercially available, which has possible application in the design of an automated system. The table includes the manufacturer, model, and basic characteristics of each item of equipment.

Table 19-1. Commercially Available Equipment for Use in Automated Analysis

Item, manufacturer, model	Characteristics
I. Samplers	
A. Liquid sampling	
1. Pye-Unicam, SP-40P	*Capacity:* Circular tray with capacity for 50 samples. *Sample size:* Variable, 1.4–2.0 ml. *Principle of operation:* Built-in metering pump withdraws sample on pneumatic principle—sample returned to container after measurement. *Primary use:* Designed for use with flow cell spectrophotometers.
2. Pye-Unicam, SP-40AU	*Capacity:* Similar to SP-40P. *Sample size:* 3–6 ml. *Principle of operation:* Sample drawn through flow cell by external vacuum source, then flushed to waste. *Primary use:* Similar to SP-40P.
3. Vision Laboratories, Beaker Butler	*Capacity:* Trays of 100. *Sample size:* 50-ml beakers. *Principle of operation:* Beakers rotated over a pedestal which raises the beaker to any desired apparatus for a programmable period (time or detector actuated). Inserts for smaller beakers available (Figure 19-15).
4. Bausch & Lomb, automatic sampler	*Capacity:* 80 samples in prenumbered racks of 5. *Sample size:* 0.5 ml and larger. *Principle of operation:* External aspirator withdraws sample, flushes it to waste. All of the samples in sampler remain submerged in a water bath which maintains a constant temperature. *Primary Use:* Part of Spectronic-400 system.
5. Perkin-Elmer, automatic sample changer	*Sample size:* 2 ml. *Principle of operation:* Programmed aspiration of samples from a fraction collector (not included). Several sequences of sampling possible for either sample return or waste. May be used to activate fraction collector, pipet mechanically lowered into fraction collector tubes. *Primary use:* Used with P-E and Coleman 139 spectrophotometers.

6. Gilford, Model 2443 Rapid Sampler

Capacity: 300 samples/hr. *Sample size:* 0.7 ml. *Principle of operation:* Vacuum. *Primary use:* Used with Gilford spectrophotometer.

7. Evans Electroselenium, Ltd. (EEL), Model 178 Autosampler

Capacity: 48 tubes, 4 quadrants of 12. *Sample size:* 2.5 ml. *Principle of operation:* Sampling probe mechanically lowered into sample tubes; designed for external pump control. *Primary use:* EEL 171, automatic colorimeter.

8. Cecil Instruments, 404-2, automatic sample changer

Capacity: Turntable for 30 samples. *Sample size:* 3-ml volume. *Principle of operation:* Probe lowered mechanically into vials. External vacuum draws sample out. Sample wasted into bulk reservoir held at reduced pressure. *Primary use:* Cecil CE 404 colorimeter.

9. Struers, Samplomat

Capacity: Several turntable types holding up to 300. *Sample size:* 3–4 ml. *Principle of operation:* Integral piston pump draws sample into a cell set at 3 samples/min, returns sample. *Primary use:* Colorimeters, especially Beckman DBG.

10. Technicon, sampler + proportioning pump

Capacity: One hundred 15 × 100 mm test tubes held in a circular tray. *Sample size:* 0.025–5 ml determined by peristaltic tube diameter and sample time. *Principle of operation:* A pivoted dip tube is lowered by a timing mechanism. Peristaltic pump rollers (up to 18) pass over elastic tubing held on a spring-loaded tube. Air introduced for mixing in some channels. Washes possible via gooseneck or as alternate samples. *Primary use:* Used in Technicon AutoAnalyzer assemblies.

11. Technicon, sampler II + proportioning pump III

Capacity: Interchangeable trays of 40 samples (4 sizes of vessels). *Sample size:* Variable, 0.025–5 ml. *Principle of operation:* Dip tube lowered into sample, followed by preset wash cycle. Peristaltic pump passes under 23 tubes held against a spring-loaded table. Separate air inlet channel setup. *Primary use:* AutoAnalyzer assemblies (Figures 19-1 and 19-2).

B. Solid sampling

Technicon, SolidPrep sampler

Solids are dumped from containers into a high-speed blender containing an appropriate solvent; the resulting blend is aliquoted by a peristaltic pump for continuous flow analysis (Figure 19-3).

(Continued)

Table 19-1. (cont.)

Item, manufacturer, model	Characteristics
C. Gas sampling	
Technicon, gas absorption system	Gases are drawn concurrently by a vacuum pump through a beaded helical coil containing a trapping reagent. This trapping solution is then aliquoted by aspiration with a peristalic pump for continuous flow analysis.
D. Recording balances	
Cahn, Model FA automatic recording balance	Samples (capsules, pellets, pills) placed in bowl feeder. Vacuum probe picks up sample and places it on balance (0–2.5 gram capacity). The weight is recorded, and the vacuum probe then picks up the sample and drops it into an exit chute.
II. Separation	
A. Centrifuges	
1. Sorval, KSB tube-type continuous centrifuge	"Szent-Gyorgyi and Blum" KSB tube-type continuous flow system. Eight, four, or two tubes are placed in centrifuge fitted with a special head, which introduces a continuous stream into the middle of the tube. The centripetal forces push the supernatant out of the head through a second tube. Designed as a preparative method of continuous concentrating.
2. Joyce, Loebel, Ltd. (Tech/Ops), Centrichem	Tape-recorder-controlled automatic centrifugal solvent extraction system. Consists of a thistle-shaped vessel surrounded by an annulus-shaped porous barrier. The devise is mounted in the chuck of a laboratory motor while the spray collector, which is activated by a cam, moves vertically to collect different effluents from the centrifuge. Discrete sample, discussed in text (Figure 19-5).

3. Quickfit, Quartz, Ltd., microcentrifuge

Part of Quickfit 617 system. Hollow disk is rotated at 3000 rpm; sample is introduced into central wall and rotation started. Mixture is forced out onto disk, effecting separation. Hollow scoop is fed into inner layer, and decantation takes place along the scoop. When the center layer is reached, a diversion valve is opened, and the heavier phase is taken with vacuum to waste.

B. Continuous filters

Technicon, continuous filter

A roll of filter paper is drawn continuously over a Teflon platen. As it passes, the samples drip onto it. The liquid is aspirated through and continues in the stream. If two liquids mix to give precipitate, they are fed onto a continuous high-speed mixer (an integral part of the unit) before filtration (Figure 19-6).

C. Solvent extractors

Technicon, continuous solvent extraction apparatus

Glass helical coil fitted with glass beads; fluids pass up through the beads. Well-extracted fluid at column top passes into clear section of extraction, where phases separate. The heavy phase is aspirated through the bottom; the lighter phase is drawn (or ejected) through the top. A constant portion of the wanted phase is aliquoted for further treatment (Figure 19-4).

D. Distillation unit

Technicon, continuous distillation apparatus

Samples pass through a heating bath, causing volatiles to rise into a glass distillation head. There they are absorbed by the inflowing reagent stream and carried through the remainder of the analytical system, while the nonvolatiles are wasted.

E. Dialyzers

1. Technicon, dialyzer I

Two grooved plates separated by a cellophane membrane. The spiral grooves coincide to form and effect a tube 7 feet long, divided lengthwise by the membrane. The sample and pickup (reagent) solutions flow concurrently while held at constant temperature in a heating bath.

(Continued)

Table 19-1. (cont.)

Item, manufacturer, model	Characteristics
2. Technicon, dialyzer II	Similar to dialyzer I but smaller ($\sim$7 inch effective tube) and simpler to change. May be temperature controlled by internal heating bath (Figure 19-7).
3. Amicon, CEC-1, on-line column concentrator	Eluant from a pressurized column flows through a thin channel of a Diaflo membrane; concentration ratio is set by a valve. Continuous ultrafiltration. Limited organic suitability for the membranes.
F. Gradient elution apparatus	
1. Instrument Specialties Co., (ISCO), Dialagrad 310	Programmed version of ISCO Model 310 precision metering pump. Program duration, 5 min to 16 days with a total flow rate of 1–3200 ml/hr. From 0 to 100% of two components. Maximum output pressure is 50 psi low pulse system.
2. Technicon, Autograd gardient elution apparatus	Nine chambers containing various buffers are interconnected. Each buffer is drawn toward the apparatus while gradually mixing with the buffer preceding it, thus giving a smooth gradient.
3. Glenco, Radiant elution apparatus	Two chambers of equal diameter are connected by a Teflon valve. One chamber is mixed as the other adds to it by gravity feed. These units may be combined in series for complex gradients.
G. Automatic injectors	
1. Chromatix, Cheminert Valve-CAV	Teflon switching valve, which may be set up with several sample loops so that automatic fill and inject cycles are performed pneumatically. Coupled with Type 20 Cheminert rotary valve—20 samples may be obtained, then injected into a chromatographic system. Maximum operating pressure is 500 psi. Several models are available.

2. Pye-Unicam, automatic solid sample injector	Injector for solid (or fluid) samples for gas chromatography with capacity for 36 glass sample holders (40-μl volume). Samples are released automatically from an injector controller. Glass sample is removed from the column area. Autozero for drift correction.
3. Hewlett-Packard, Model 7670A automatic sampler	For gas chromatography; utilizes syringe techniques by mechanically duplicating manual procedure—pump to fill, fill to volume, rotate 90°, and inject. Handles up to 36 septum-sealed samples on a programmed basis (1–50 μl injections). There may be 1–99 min cycles between injections.
4. Hamilton, automatic sampler	Handles 48 samples in piston-sealed vials, using capillary needles which automatically flush. Samples as small as 0.01 μl. Piston-operated, much like the syringe in the Hewlett-Packard unit.

III. Detection

A. Colorimeters

| 1. Evans, Electroselenium
Ltd. (EEL),
Model 171 automatic colorimeters | 400–700 nm at 30-nm half-width interference filters; accommodation for single filters between 347 and 700 nm at 10-nm half-width; 2.5-ml sample presented in pump-read-return cycles of 15 sec. Two-stage hour-glass cell filled and emptied with air flushing by a fluid-isolated syringe pump. Electronics for autostandardize, autozero, and first-order Beer's law correction. Results presented on digital printout. |
| 2. Cecil Instruments,
CE 404 colorimeter system | 400–700 nm at 20-nm half-band width, accommodation for 6 filters in a selection disk; 3.0-ml sample (10-mm cell, 2-ml volume) used in a pump-read-waste cycle of 20 sec (using Cecil CE 404-2 sample changer). Flow cell measurements, 2.5 ml/min volume. Prewash by excess sample minimizes cross-contamination. Outlet provided for 10-mV recorder. |

(Continued)

381

Table 19-1. (cont.)

Item, manufacturer, model	Characteristics
3. Technicon, colorimeter I	Dual-beam, 340–700 nm, recorder acts as the null balancing indicator yielding per cent transmittance. Narrow half-width band-pass interference filters select wavelength. 15-mm small-bore tubular flow cell used with flowing sample stream. Paper available to permit absorbance valves to be read directly.
4. Technicon, colorimeter II	Similar to colorimeter I, except that 3 flow cells utilize the same lamp, housing, etc., permitting automatic blank correction or inverse colorimetric tests in a differential mode. Covers 340–700 nm and reads directly in absorbance units.
B. Spectrophotometers	
1. Cecil Instruments, CE 303 spectrophotometers	Diffraction grating monochrometer for 325–1000 nm, band width 10 nm. Pump-read-waste cycle, 3/min (with CE 404 autosample changer). 3-ml sample; automatic stray light filter. Options for 10-mV recorder or 1-V digital output. Four cell lengths, up to 40 mm.
2. Pye-Unicam, SP-1800 spectrophotometer	190–700 or 190–850 nm ranges, Ebert 8/10 monochrometer. Four result presentations: 0–2, 0–1, 0–0.5, and 0–0.2 A. Linear absorbance or concentration 10-mV recorder or 2-V printer outputs. Resolution, 3 mm/mm slit width. SP-40P or SP-40-AU sample changers determine pump-read-return or pump-read-waste. Cycles using 1.4–2.0 or 3–6 ml, respectively, at cycles of 45 sec/sample.
3. Pye-Unicam, SP-3000 spectrophotometer	More fully integrated version of SP-1800: automated wavelength selection, sampling sequence printout, etc.
4. Bausch & Lomb, Spectronic 100 spectrophotometer	335–925 nm with 8.0-nm band pass. Used in Spectronic 400 system. Sample (0.5 ml/min) drawn into microflow through cell, then wasted. The cell is evacuated and purged between samples. Automatic sampling working on a time basis (6–48

5. Technicon,
   UV spectrophotometer

sec/sample) may be read %T, linear A, or concentrations when interfaced to a data processer (analog, 2 V).

Beckman DBG spectrophotometer, modified to accommodate AutoAnalyzer inlet and outlet flows. Square flow cell used, rather than Technicon tubular-type cell.

6. Instrument Specialties Co. (ISCO), Model SRS

Spectroradiometer and programmed recorder designed to cover 380–750 nm at 25-nm band width. Wedge-interference filter system plots watts/cm² versus millimeters over 8 intensity ranges. Uses a Teflon diffusing screen and precision planar photodiode detection system. With a fiber optic head and program scanner, spectra can be scanned over any range for 24 hr at 15-min intervals. 10-mV output.

C. Fluoriometers

1. Technicon,
   fluorimeter

Designed for use in continuous flow. 60 samples/hr; response time, $\frac{1}{2}$ sec. Excitation from 250 to 600 nm via 85-W hp lamp. Interference and glass gelatin filters pass light onto dual photomultipliers (minimizing drift). Intensity ranges adjusted by selecting 1 of 6 apertures varying by a factor of 4.

2. Farrand Optical Co. (FOCI), automated turret spectrophotometer

Range in 200–700 mμ, batches of 8 or 16 samples, strip chart recorder; high-intensity mercury lamp excitation. Dual monochrometer system. Constant-temperature sample chamber.

3. Farrand Optical Co. (FOCI), manual spectrophotometer

Similar to the ATS, high-pressure, 150-W dc xenon-arc source $f/3.5$ optical system. Blocking filters may be combined with dual monochrometers.

D. Automatic polarimeter

Bendix,
automatic spectropolarimeter

Spectropolarimeter with ±500-min rotation, effective range ±50° for 1-mm cell (flow through), spectral range 450–750. Filters: Hg green (546 nm), Na$_D$ (589 nm); others on request. Sensitivity, 0.1 millidegrees arc. High-intensity tungsten bulb. Output up to 1 V. Useful as column monitor.

(Continued)

Table 19-1. (cont.)

Item, manufacturer, model	Characteristics
E. Ultraviolet absorption detectors	
1. Varian, absorption detector 4000-1	Microcell, 8 μl volume. Absorbance, 0.005–0.64. A full-scale fixed analysis at 254 nm dual beam, designed for use with high pressure (1000 psi) for liquid chromatography.
2. Instrument Specialties Co. (ISCO), UA-2 ultraviolet analyzer	Quartz flow-through cell (0.5–0.1 cc volume), source lamp, and shutter for setting zero on one side, filter and photomultiplier on the other. Used to monitor 254 nm in a flowing stream. Linear absorbance (0–0.5 or 0–2.5 full scale) suitable for external recordings. Can be used to control fraction collector. Several spectral range detectors may be used interchangeably in the unit. *254 single beam*—filtered 254 line from a low-pressure mercury lamp. *Visible single beam*—wedge interference filter with half-band width 25 nm over 410–700 range. *Near IR*—wedge interference filter with 40-nm half width over 700–950 nm. *Dual beam*—254 or 280 nm may be selected; the latter source is a fluorescence crystal with 17-nm half-band width. *Model 660*—seven mercury lines may be selected for observing flowing streams (254, 313, 364, 405, 435, 546, 679). Wavelengths selected by snap-in filters.
F. Potentiometric titrators	
1. Precision, Precision Dow Recordomatic	0–14 pH units in 10-unit increments; −1500 to +1500 mV in 1000-mV increments. Glass syringe pumps. Stops automatically at end point while recorder responds; plots millivolts (pH) versus milliliters added. Chart drive and pump mechanism linked mechanically. Feed rate adjustable, $1-3\frac{1}{8}$ ml/min for 50-ml pump. Scales may be expanded 2 to 4 times. Two pump assemblies mounted on opposite ends share recorder, permitting alternate runs or two different setups. Feed pumps must be refilled manually. 10–20 ml samples/250-ml beaker. Microadaptions (5- or 1-ml pumps) are available.

2. Radiometer (London Co.), titration equipment

Modular approach combines 4 units: a tritrator, a titrant delivery unit, a titration assembly, and a recorder (20 modules available). Titrator controls reagent addition via proportional band controller. Includes calibration, range, and sensitivities. Titrant delivery units—mounted piston burets (0.25–50 ml) and reagent bottles readily interchanged with drive unit of continuous variable speed (30–200% syringe volume), digital readout. Titration assemblies—paddle-stirred assemblies containing sample, electrodes, and titrant delivery units. Special vessels required (1–500 ml). Recorder—mechanically linked to syringe pump. Plots milliliters versus potential. Various assemblies may be used to obtain dead-stop or titration curves.

3. Fisher, Fisher Titralyzer

Two measuring units containing the controls, an automatic buret (45 ml, constant-temperature bath). Titration stand consists of a turntable (sixteen 200-ml tall form beakers or twelve 400-ml beakers) and a mechanism for raising and lowering the buret tip and electrodes into the vessel. Magnetic stirring. Titration to preset end-point with a fixed time delay to avoid "anticipation." Final titration volume printed out. A temperature bath available to permit titration of solutions in beakers up to 90°C (Figure 19-14).

4. Fisher Model 35 Titrimeter

Manually prepared sample presented to electrodes. Standard buret connected to selenoid valve, which controls titration to dead-stop end point. Technician reads initial and final volumes.

5. Metrohm (Brinkman), Titroprint E-425

Electrodes lowered into standard beaker, magnetically stirred. Feed from interchangeable piston burets (5, 10, 20, 50 ml). Printer records milliliters and the resulting millivolts continuously, using fixed incremental addition. Time delays may be introduced for slow reactions. Stepwise addition may be limited to regions near end point, i.e., program a slug followed by increments. Printout includes sample number (1–99). Wide set of potentials or pH ranges may be selected for end-point detection.

(Continued)

Table 19-1. (cont.)

Item, manufacturer, model	Characteristics
6. Metrohm (Brinkman), Potentiograph	Precursor to the Titroprint, with similar feed mechanism and range. Readout, however, is on a recorder plotting potential versus volume.
7. Metrohm (Brinkman), Mutititration E-440	Designed for process control analyses. Automatic buret sampling aspirates sample from stream into titration vessel. Programmed pin matrix logic. Contains 4 burets for reagent titration; 2 large and 2 small piston burets (5–50 ml) may be used for multiple titrations or back-titration. Autocalibration possible, magnetically stirred. Proportional band dead-stop measurement may be made for 2 dead-stop end points on one sample. When finished, system is flushed before next analysis.
8. Sargent-Welch, Model DG	Two titration stands share single recorder; separate controls. Standard beakers (up to 600 ml) present sample to electrodes with magnetic stirring. Delivery from buret pumps at constant rates (from 50-ml buret) or at rate inversely proportional to the slope of the titration curve (10-ml buret also available). Multiple potential ranges (200–1000 mV) can be adopted to handle micro samples of 5 ml with combination electrodes. Chart drive and piston buret synchronized. Dead-stop titration possible by a microswitch on the recorder. Refilling of buret is automatic.
G. Photometric titrators	
1. Fisher, Photometric Titralyzer	Similar to mechanism for potentiometric titrations except for its detection system. 15×20 ml tall form beakers are held in light-tight shields in the turntable. Two Lucite "light pipes" are lowered into the beaker with a special bellows which forms a dark box for the titration. Titration to set intensity wavelength over 400–700 nm in 33-nm band with variable wedge interference filter. A reference detector balances the system for lamp instability. Results printed as before (Figure 19-14).

2. Radiometer (London), Phototitration, Type PMT1

Sample in standard beaker is placed on a Lucite pedestal containing 2 sets of 2 filters and 2 photodetectors. Paired filters permit "background" correction at wavelength other than end. Sample raised by hand into box. Light shines down through solution. Titrator adds preset amount of indicator as a controller for an autoburet (see potentiometric Radiometer). Each titration is coded to control end point, delay of shutoff, control band, direction of titration, and amount of indicator to be added (Figure 19-13).

H. Hybrid analytical systems

1. Polimak (Labline), Clinomak Mark II

Discrete sample, single-module colorimetric system. Samples held in polyethylene cups (0.4 ml) in a circular tray (capacity, 40). Outside the sample cups 90 reaction vessels (3 ml, glass) are arranged. Sample (20–200 ml) is transferred by aspiration through nonwet tip via minipump. Sample cycle time, 12 sec to 43 min. Reagents (up to 5) added by gravity-fed electromagnetic valves. Jet from tip provides mixing after desired delay. Results read (12 sec/sample) by grating colorimeter (400–700 nm). May be programmed for batches of 15; switching to alternate procedures rapid. Special curvette washer recommended. No provision for separation procedures. Useful as preparative for flow-through UV, fluorimetric, or photometric methods.

2. Pye-Unicam, chemical processing unit A-60

Chemical preparations unit for discrete colorimetric methods may be used with any spectrophotometer equipped with a flow cell. No provision included for separation steps. Chain in an air-thermostated (20–60 ± 5°C) rectangular track carries 120 units, each consisting of a sample cup and separate 5-ml test tube as reaction vessel. Sample (10–300 ml) and reagent volumes are controlled by pneumatically actuated precision glass syringes, whose throw is governed by up to 2.0 ml of diluent or reagent held at 20–60°C at each station (1 pickup station, 2 addition stations with stirrers, 1 transfer station can be switched around the track in multiples of 30-sec intervals). The reaction mixture (4.0 to 4.5 ml) is transferred by a pneumatically actuated dip tube to a low residual volume spectrophotometric cell (5 mm) and returned to its reaction vessel after measurement.

(Continued)

Table 19-1. (cont.)

Item, manufacturer, model	Characteristics
3. Bausch & Lomb, Spectronic 400	Modular approach; discrete samples analyzed colorimetrically. Four modules—sample preparation, incubation bath-sampler, Spectronic 100 spectrophotometer, controller. Sampling and reagent addition carried out by *manually* activated syringe pumps. Remainder of the system is described in Section III.B.4.
4. Joyce, Loebel, Ltd. (Tech/Ops), Microlab laboratory analysis system	Modular approach to discrete samples—racks of 15 or 40 are manually transferred between modules. All module operations are set at 15-sec intervals. Units are sampler, reagent addition, centrifuge, supernatant, and autocolorimeter. The sampler holds racks of 15 or 40 plastic vials from which 0.01 to 3.5 ml samples are aspirated, then flushed into glass reaction tubes with 0.2 to 5 ml of diluent. Two other reagents may be added. Reagent addition similar but lacks transfer capacity. Centrifuge (3000 rpm for 3–5 min) uses angled racks. Supernatant unit macro (3.5–4.0 ml) sampler with two-reagent addition if required. Autocolorimeter-double-beam null balance with optical wedge and color filters. Quartz halogen source and quartz optics to facilitate UV methods. Built-in recorder linear in concentration. Analog converter printer available. All pumping actions via syringe pumps. Flame photometer attachment available.
5. Vickers, (Medi-Computer), M-300 multichannel, automated analysis system	Computer-controlled discrete sample analysis to handle 20 chemistries at the rate of 20 tests/300 samples/hr. Each console may be used as a free-standing dual-channel system. Each consists of a transfer diluter, which withdraws a sample and then dispenses it with a diluent into a reaction cavity. Reagents are added. Alloy tray of 120 reaction tubes may be heated. Laundry systems. Flame photometer interchangeable; continuous result printout. Double-beam colorimeter.
6. Beckman, DSA-560, discrete sample analyzer	Discrete sample approach in single pneumatically controlled unit. Samples (5–50 μl) in rotary tray for 40 tubes (Microfuge® tubes, plastic cups, or glass test tubes) are picked up by probe unit, which moves mechanically over a Q-cup

(a plastic tray with 5 compartments) and is flushed into the compartment. Each cycle moves cups along an incubator held at a desired temperature by a closed fluid circulating system. At each station reagent or reagents (50μ liters to 1 ml) may be added. When filtration is required, a porous cup is dropped over one compartment. The fluid is aspirated through the cup, then discharged to a separate compartment. After color development, the capsule contents are pumped into a double-beam filter colorimeter with high sensitivity. May be operated as dual-channel instrument. Range, 340–700 nm. Results may be presented in concentration units as traces, teletype, or printer. May be used for kinetic measurement. Rate 35–120 samples/hr (Figure 19-11).

7. Technicon, AutoAnalyzer

Modular approach, continuous flow analysis. Most modules described in preceding sections. Basic approach given in detail in text. Extremely large range of capabilities with colorimetric or other finishes. Uses tubing which could be a problem for some organics. Heating baths preset for 37°C or 95°C are available for analyses requiring incubation.

8. Technicon, AutoAnalyzer II

Resembles AutoAnalyzer but is greatly miniaturized. Separate prepackage modules available for some (nonorganic) tests. Package may contain reagents, mixing coils, dialyzer, heating bath, time delay, and phasing steps. Uses fewer intramodule connections and manifold pumps (usually 5). Uses much smaller reagent volumes (Figure 19-12).

9. Quickfit, Quartz, Ltd., Quickfit 617 automatic sampler

Stacked double turntable with microcentrifuge (see II.A) mounted between; specimens loaded into vessels in upper tray. Samples (96/hr) are removed with reagent via syringe. Both are ejected into a collection vessel. A second sampler aliquots the reaction mixture and transfers it to separator (centrifuge). Separator transfers it to collector in the lower tray. A third probe transfers this output with reagent to a reaction vessel suspended in water incubation bath. After fixed number of positions, an aliquot is removed by the colorimeter transfer probe.

REFERENCES

1. W. Koch, *Pure Appl. Chem.*, **18**, Nos. 1–2, 1 (1969).
2. H. Malissa, *Pure Appl. Chem.*, **18**, Nos. 1–2, 17 (1969).
3. L. Mietes, *Pure Appl. Chem.*, **18**, Nos. 1–2, 35 (1969).
4. J. Mitchell, Jr., *Pure Appl. Chem.*, **18**, Nos. 1–2, 81 (1969).
5. W. Simon, *Pure Appl. Chem.*, **18**, Nos. 1–2, 97 (1969).
6. Z. Zagorski, *Chem. Anal.*, **3**, 313 (1958).
7. A. Dijkstra, *Chem. Weekblad*, **64**(40) (1968).
8. H. Malissa, *Ind. Chim. Belge*, **33**, 220 (1968).
9. L. Erdey and F. Szabadvary, *Pure Appl. Chem.*, **13**, 437 (1966).
10. C. J. van Nieuwenburg, *Chim. Anal. (Paris)*, **38**, 423 (1956).
11. R. Muller, *Ann. N.Y. Acad. Sci.*, **87**, 611 (1960).
12. G. D. Patterson, *Anal. Chem.*, **29**, 605 (1957).
13. M. H. Pattison, *Am. Lab*, June 1969, p. 37.
14. W. J. Blaedel and C. L. Olson, *J. Chem. Educ.*, **40**, A549 (1963).
15. *Technicon Symposia 1965*, Mediad, Inc., White Plains, N.Y., 1966.
16. *Technicon Symposia 1966*, Vols. I and II, 1967.
17. *Technicon Symposia 1967*, Vols. I and II, 1968.
18. D. E. Smith, Jr., *Anal. Chem.*, **52**, 206 (1969).
19. C. H. Orr and J. A. Norris, ed., *Progress in Analytical Chemistry*, No. 4, Plenum, New York, 1970.
20. G. R. Kingsley, *Anal Chem.* (Annual Reviews), **41**, (5) 14R (1969).
21. G. L. Wied, ed., *Introduction to Quantitative Cytochemistry*, Academic, New York, 1966.
22. G. T. Skeggs, Jr., and H. Hockstrasser, *Clin. Chem.*, **10**, 918 (1964).
23. G. Guilbault, *Enzymatic Methods of Analysis*, Pergamon, Oxford, England, 1970.
24. M. K. Schwartz and O. Bodansky, *Technicon Symposia 1966*, pp. 489–493.
25. H. U. Bergmeyer, ed., *Methods of Enzymatic Analysis*, 2nd Ed., Verlag Chemie, Weinheim, 1965.
26. W. J. Smythe et al., *Technicon Symposia 1967*, Vol. I, pp. 105–114.
27. L. P. Sinotte, ed., *Ann. N.Y. Acad. Sci.*, **153**(2), 389 (1968).
28. J. R. Gerke and A. Ferrari, *Technicon Symposia 1967*, Vol. I, pp. 531–540.
29. *Chem. Eng. News*, **44**, Oct. 24, 1966, p. 22.
30. T. C. Wherry, *Oil Gas J.*, **54**, No. 56, 125, 129 (1956).
31. W. J. Blaedel and R. H. Laessig, *Advan. Anal. Chem. Instr.*, **5**, 69 (1966).
32. H. Maier and H. Claudy, *Ann. N.Y. Acad. Sci.*, **87**, 864 (1960).
33. M. H. Adelman, *Industrial Analysis* (Proceedings of the Technicon International Congress, 1969), Vol. 2, Mediad Inc., White Plains, N.Y., 1970.
34. D. Patient, *Ann. N.Y. Acad. Sci.*, **87**, 830 (1960).
35. A. Ferrari, G. Kessler, F. Russo-Alesi, J. Kelly, C. Vanderwende, and L. van Petten, *Ann. N.Y. Acad. Sci.*, **87**, 729 (1960).
36. Leonard T. Skeggs, Jr., *Chem. Eng. News*, Aug. 10, 1970, p. 54.
37. M. Hermansky and M. Vondracek, *Cesk Farm.*, **17**(4), 204 (1968).

38. S. Siggia, *Continuous Analysis of Chemical Process Systems*, John Wiley, New York, 1959.

39. H. Noebels, *Ann. N.Y. Acad. Sci.*, **87**, 934 (1960).

40. R. Hagstrom and I. Capuano, "Factors Influencing Selection and Use of Automatic Analytical Instruments," paper presented at ASTM Meeting, Atlantic City, N.J., 1968.

41. L. S. Ettre and W. H. McFadden, *Auxilliary Techniques of Gas Chromatography*, Interscience, New York, 1969.

42. A. B. Littlewood, *Chromatographia*, **1**, 37 (1968).

43. J. H. Peters et al., *Anal. Biochem.*, **23**, 459 (1968).

44. H. W. Kunz, C. F. Bernard, and T. J. Gill, III, *J. Chromatog.*, **32**, 786 (1968).

45. K. Woods and R. Engle, *Ann. N.Y. Acad. Sci.*, **87**, 764 (1960).

46. *Aminco Lab. News*, **26**(3 & 4), 4 (1970).

47. P. Schratter, *Am. Lab.*, Oct. 1969, p. 21.

48. Y. C. Lee, J. F. McKelvy, and D. Lang, *Anal. Biochem.*, **27**, 567 (1969).

49. K. Brendel et al., *Anal. Biochem.*, **18**, 147 (1967).

50. J. Polesuk, *Am. Lab.*, May 1970, p. 27; *Am. Lab.*, June 1970, p. 47.

51. J. D. McCallum, *Am. Lab.*, December 1968, p. 48.

52. "Technicon® AutoAnalyzer® Bibliography 1957/1967," Technicon Corp., New York, 1968.

53. H. Fuhrmann, *Chem.-Ingr.-Tech.*, **26**(7), 401 (1954).

54. R. T. Sheen and E. J. Serfass, *Ann. N.Y. Acad. Sci.*, **87**, 1960.

55. W. Lang, *Mikrochim. Acta*, **1**, 52 (1967).

56. E. A. Epps, Jr., and H. C. Austin, Jr., *J. Assoc. Offic. Anal. Chemists*, **50**(4), 981 (1967).

57. R. Shapira and A. M. Wilson, *Anal. Chem.*, **38**, 1803 (1966).

58. J. F. Bishop and R. S. White, *Ind. Eng. Chem.*, **46**, 1432 (1954).

59. H. Loebl, "Improvements in Automatically Operating [Spectro]photometers," Brit. Pat. 1,149,384; appl. 7/5/65.

60. G. Gualandi and G. Morisi, *Ann. Ist. Sup. Sanita*, **3**(5), 589 (1967).

61. E. K. Dukes and M. L. Hyder, *Anal. Chem.*, **36**(8), 1689 (1964).

62. J. N. Lamy, J. Lamy-Provansal, J. de Russe, and J. D. Weill, *Bull. Soc. Chim. Biol.*, **49**(8–9), 1167 (1967).

63. R. F. Ashworth and W. Walisch, International Technicon Symposium on Automation in Analytical Chemistry, Technicon GmbH, Frankfort, Germany, 1964, p. 628.

64. N. R. Kuzel, *Technicon Symposia 1966*, Vol. I, Mediad, Inc., White Plains, N.Y., 1967, pp. 218–221.

65. T. W. L. Atkinson, *Chem. Process.*, **28**(4), 2 (1965).

66. R. E. Duncombe and W. H. C. Shaw, *Technicon Symposia 1966*, Vol. II, Mediad Inc., White Plains, N.Y., 1967, pp. 15–18.

67. S. A. Bartkiewicz and L. C. Kenyon, Jr., *Anal. Chem.*, **35**, 422 (1963).

68. R. Lappen and L. C. Clark, *Anal. Chem.*, **23**, 541 (1951).

69. H. O. Friestad, D. E. Ott, and F. A. Gunther, *Anal. Chem.*, **41**, 1950 (1969).

70. H. Ziegel, *Z. Anal. Chem.*, **53**, 755 (1914).

71. Precision Dow Dual Recordomatic Titrator (1948), based on the design of H. A. Robinson, *Trans. Electrochem. Soc.*, **92**, 445 (1947).

72. J. P. Philipps, *Automatic Titrators*, Academic, New York, 1959.

73. J. Tucussel, *Bull. Soc. Chim. France*, 1115 (1964).

74. O. T. Zajicek, *J. Chem. Educ.*, **41**, 554 (1964).

75. D. C. M. Squirrell, *Automatic Methods in Volumetric Analysis*, Van Nostrand, Princeton, N.J., 1964.

76. R. Bryden and A. Dean, *Ann. N.Y. Acad. Sci.*, **87**, 813 (1970).

77. T. F. Kelley, *Anal. Chem.*, **37**, 1078 (1965).

78. R. P. Noble, *J. Lipid Res.*, **7**, 745 (1966).

79. H. V. Malmstadt and D. A. Vassallo, *Anal. Chem.*, **31**(5), 862 (1959).

80. I. Shain and G. R. Svoboda, *Anal. Chem.*, **31**(11), 1857 (November 1969).

81. G. R. Svoboda, *Anal. Chem.*, **33**(12), 1638 (November 1961).

82. M. J. Johnson, *Anal. Chem.*, **38**, 1579 (1968).

83. J. Charles, *J. Chromatog.*, **35**, 158 (1968).

84. L. Kesner and E. Muntwyler, *Anal. Chem.*, **38**, 1164 (1966).

85. V. J. Vaygand et al., *Talanta*, **15**, 699 (1969).

86. P. T. Priestly, W. S. Sebborn and R. F. W. Selman, *Analyst*, **88**, 797 (1963).

87. P. T. Priestly, W. S. Sebborn, and R. F. W. Selman, *Analyst*, **90**, 589 (1965).

88. J. Bandisch, G. Beilstein, and E. Rasch, *Z. Anal. Chem.*, **231**, 137 (1967).

Appendix

This Appendix is meant to give the reader some basic, general information and/or references to basic, general information regarding the technique behind the measuring approach used. Needless to say, it would be inappropriate to go into all details in this book. For convenience, the Appendix is divided into the same subsections as are the various chapters: absorption spectrophotometry, gas chromatography, electroanalytical methods, nuclear magnetic resonance, and radiochemical methods.

More detail is included in regard to gas chromatography than to other methods. There are general techniques in this area especially geared to functional group analysis, and it was considered desirable to present here a unified treatment of all these general GC techniques.

I. ABSORPTION SPECTROPHOTOMETRIC METHODS*

Visible and Ultraviolet

1. G. L. Clark, ed., *The Encyclopedia of Spectroscopy*, Reinhold, New York, 1960.
2. I. M. Kolthoff and P. J. Elving, eds., *Treatise on Analytical Chemistry*, Part I, Vol. 5, Interscience, New York, 1964.
    (a) E. J. Meehan, Chapter 54, "Fundamentals of Spectroscopy," and Chapter 55, "Spectroscopic Apparatus and Measurements."
    (b) A. A. Schilt and B. Jaselakis, Chapter 58, Part 1, "Ultraviolet and Visible Spectrophotometry—Principles and Applications."
    (c) B. Jaselakis, Chapter 58, Part 2, "Ultraviolet and Visible Spectrophotometry—Instrumentation and Measurement."
3. L. N. Ferguson, "Organic Analysis with Ultraviolet-Visible Absorption Spectroscopy," *in* C. N. Reilley, ed., *Advances in Analytical Chemistry and Instrumentation*, Vol. 4, Interscience, New York, 1965, p. 411.
4. N. R. Rao, *Ultraviolet and Visible Spectroscopy*, Butterworth, London, 1961.

* Supplied by J. Gordon Hanna.

Infrared

5. H. A. Szymanski, *IR Theory and Practice of Infrared Spectroscopy*, Plenum, New York, 1964.

6. A. Smith, "Infrared Spectroscopy," *in* I. M. Kolthoff and P. L. Elving, eds., *Treatise on Analytical Chemistry*, Part I, Vol. 6, Interscience, New York, 1965, Chapter 66.

7. R. F. Goddu, "Near-Infrared Spectrophotometry," *in* C. N. Reilly, ed., *Advances in Analytical Chemistry and Instrumentation*, Vol. 1, Interscience, New York, 1960, p. 347.

8. C. N. R. Rao, *Chemical Applications of Infrared Spectroscopy*, Academic, New York, 1963.

9. N. B. Colthup, L. H. Daly, and S. E. Wiberley, *Introduction to Infrared and Raman Spectroscopy*, Academic, New York, 1964.

II. GAS CHROMATOGRAPHIC METHODS*

Gas chromatography (GC) has greatly expanded our capabilities in the realm of quantitative functional group analysis of organic chemicals. Classically, in a given procedure, quantitative functional group analysis has been applied mostly to one or at most a few compounds to determine chemical structure, usually by measuring the reaction product of one particular functional group. With GC, qualitative and quantitative data of reaction products can readily be obtained simultaneously; less material, sometimes 1000-fold less, is required for analysis; and the determination of many compounds rather than one or two is routine. These capabilities and the ease of conducting analyses by GC have led to the phenomenally widespread use of this technique today.

The acquisition of reliable analytical data rests, as always, on the competence of the analyst, and knowledge of good GC practice is therefore essential. A detailed discussion is necessarily outside the scope of this book, however, and one or more of the many texts on the subject (e.g., L. S. Ettre and A. Zlatkis, eds., *The Practice of Gas Chromatography*, Interscience, New York, 1967) should be consulted. A brief description is included here to familiarize the reader with the basic nature and capabilities of the processes involved in GC. To avoid needless repetition in presenting methods for individual functional groups, the description of the GC process is followed by discussions of the scope and type of information presented and some factors applicable to functional group analyses in general, for example, derivatization and types of detectors. Throughout the text, frequent reference will be made to this general information.

A. A Brief Description of Gas Chromatography

Gas chromatography is a technique for separating and determining volatile substances qualitatively and quantitatively. A sample injected into the gas

* Written by Morton Beroza and May N. Inscoe.

chromatograph is volatilized (if not already a gas) into an inert gas stream (carrier gas) flowing at a constant rate through a tube containing a packing (the column). In the most widely used arrangement (gas-liquid chromatography), the packing consists of a liquid (or stationary) phase coated on a uniform-sized solid support. Components of the sample are partitioned between the gas and the liquid phases and migrate through the column at different rates, mainly as a consequence of their different partition coefficients under the conditions employed. The detector registers its response on a recorder as the components emerge from the column. Compounds not separated on one column may often be readily separated on another that has a different liquid phase. Packings may also be solid adsorbents (gas-solid chromatography) with no liquid phase used.

An important consideration in quantitative gas-liquid chromatographic analysis is the solid support. Silanization of the support is widely practiced to reduce adsorption (usually manifested by tailing), especially of polar compounds, and also allows for better spreading of nonpolar liquid phases on the support. Glass and metal tubes may also be silanized to reduce adsorption and catalytic effects. Another means of overcoming adsorption effects is to use Teflon supports. A support may also be treated with acid or base to repress ionization of the acids or bases being analyzed. (This same objective is sometimes accomplished by injecting some volatile acid or base with the sample.)

Figure A-1 is a schematic representation of a GC system, and Figure A-2 illustrates a chromatogram with the major peak parameters identified. The retention time, t_R, is the time from injection to the emergence of the peak maximum. (The corrected t_R is measured from the air peak and sometimes from the solvent front to the peak maximum.) The t_R can be used to identify compounds because with a fixed set of conditions each substance has its own characteristic t_R. Pure compounds produce symmetrical peaks in a good GC system. Impurities are readily detected by distortions from symmetry, although one should bear in mind that unsymmetrical peaks can also arise from other causes, such as decomposition or adsorption on the column support. The amount of substance in an analysis is most often determined from the peak area, that is, the area enclosed by the peak and the base line (this may be measured with a planimeter or approximated by multiplying peak height by width at half peak height); the peak area is normally proportional to the concentration of solute. The response per unit weight for each compound is determined, usually with standards, for quantitative analysis. Integrators of a variety of types are available to give areas of peaks, thereby facilitating quantitative analyses.

One of the important advantages of GC analysis over classical methodology is the capability of simultaneously separating and determining many com-

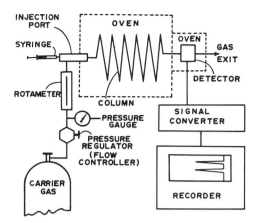

Figure A-1.   Diagram of a gas chromatograph.

pounds in a single analysis rather than only one or two. This inherent feature of GC of separating closely related compounds confers upon an analysis an enhanced specificity and therefore a greater quantitative validity and reliability than are normally obtainable by classical procedures. Peak identities are more secure (but not certain) because of the characteristic t_R of each compound.

The amount and the sophistication of GC equipment have increased considerably because of the need for more precise control and greater versatility in analysis. The column may be maintained at one temperature (isothermal) or may be temperature programmed. With the latter technique the temperature of the column is gradually increased to allow the determination of compounds with a wide range of volatilities in a reasonable time and with good sensitivity. (Unlike isothermal analyses, with temperature programming later-eluting peaks do not spread but remain sharp.) A differential flow controller and high tank pressure are required to maintain a constant flow rate with temperature programming because the rising temperature causes the carrier gas to expand. Dual-column arrangements aid in multidetector operation and in correcting for the bleeding of liquid phase, which causes unstable base lines at elevated temperatures. Open tubular columns (capillary tubes with inner walls coated with liquid phase) that are 50 to 1000 feet in length are used to achieve exceptionally good separations. Ancillary techniques, such as chemical modification of a compound, are often needed to accomplish difficult analyses (1).

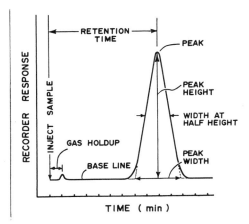

Figure A–2. Nomenclature to a gas chromatogram.

B. Scope of This Presentation and General Comments on Gas Chromatography

The widespread use of gas chromatography in quantitative functional group analysis precludes the presentation of a complete listing of pertinent literature. Instead, an effort has been made in the preceding text to select for mention typical methodology and techniques from current practice and to bypass old procedures when newer ones appear suitable or superior. Nor has it been possible within the space allowed, except in a very few instances, to provide more than the briefest description or mention of analyses. The reader *must* consult the original articles to learn enough detail to carry out procedures and to apprise himself of appropriate limitations and interferences; original articles will also be useful in providing references to earlier relevant methodology, including reports by the originators of the various methods.

The requirement for quantitative methods has eliminated many procedures from consideration. However, the choice was not always clear-cut because many GC methods deal with a number of compounds, only some of which can be determined quantitatively. Furthermore, some methods advanced as qualitative are potentially quantitative or are sufficiently reproducible to provide quantitative data upon the application of appropriate correction factors, or they provide the desired structural information without being quantitative. For example, quantitative data are not required for the location of double bonds in a compound, and the presence or number of a given group in a molecule may often be determined more easily by t_R than by quantitative analysis of the group, considerably less compound being required for such an analysis. In this connection a discussion of carbon-skeleton chromatography (2) and retention indices, has been included separately, as the use of these techniques appears to be applicable to the analyses for most functional groups. Finally, some methods which may not be considered particularly important are included because they complete the range of available methodology or they may be suitable for further development.

With GC facilitating the quantitative analysis of multicomponent mixtures, the concept of functional group analysis has been broadened considerably in recent years. For example, several decades ago, only methoxyl or ethoxyl groups could be determined; today C_1 to C_4 and higher alkoxy groups, individually or as a unit, can readily be determined. Functional group analyses can also be used for mixtures of known or unknown compounds. For example, we may know the fatty acids present in a glyceride oil, but the amount of each may have to be determined. Most important in the conduct of this type of analysis is the preliminary treatment of the sample to secure the fraction containing the compounds with the functional group of interest. Preliminary treatment can consist of separation of the acid, basic, or neutral fractions, or it may comprise a separation of extraneous material by thin-layer chromatography or other chromatographic means, the determination being made on the portion containing the compounds with the functional group of interest. (See Section C, "Preparation of Samples for Gas Chromatography.") These preliminary procedures are generally not given.

Of course, when the type of compound with which one is dealing is known, for example, by history of the sample or through the use of other instrumental techniques such as infrared spectrometry, functional group analysis is an obvious recourse. In connection with this point, it should be mentioned that virtually every form of analytical instrumentation has been combined with GC (because of its separating power) in a stepwise or integrated arrangement to facilitate analyses. However, references to these arrangements—although they are of great value—are few because they usually provide a qualitative rather that a quantitative answer; they are not, therefore, within the purview of this text. Mention must be made, however, of the GC-mass spectrometer combinations, which are now available commercially, more specifically, the double-focusing, high-resolution spectrometers. In many instances, these instruments provide molecular weights and fragments of a compound (actually mass/charge) to the millimass units (e.g., 182.015) (3). Such accurate data allow the analyst to determine the empirical formula of a compound unambiguously in most cases to a molecular weight of at least 500. The need to perform elemental analysis is thereby eliminated, and the results signal the kind of functional groups to look for. For example, if nitrogen is present, groups such as amino, nitroso, nitro, or amide are possibilities to consider. The gas chromatograph also serves another purpose—as an inlet system for the mass spectrometer. Besides separating compounds, it facilitates the introduction of small amounts of compound as dilute solutions by removing the solvent and allowing the compound to enter the spectrometer in concentrated and (hopefully) pure form.

Of further interest is a review of methods of elemental analysis utilizing GC (4). New commercial instrumentation for elemental analysis, most of which utilizes GC, has speeded analyses and made possible determinations on as little as 0.2 mg of compound.

Many of the older functional group analyses were based on the quantity of reactive reagent consumed rather than on the amount of product produced; for example, in the determination of OH content, the amount of acetic anhydride consumed was measured rather than the amount of ester produced. This approach, although still used on occasion, has been largely supplanted by direct determination of products because with GC such determination is far more convenient, rapid, and specific; furthermore, the amounts of several products may be determined in a single analysis. The methodology had to be changed to accommodate this approach. In esterifications, for instance, large excesses of reagent are added to drive the reaction to completion rapidly and thereby speed analyses. In classical procedures a 50% excess of reagent might be used and a long time allowed for the reaction to go to completion, simply because a good answer could not be obtained if too much reagent had to be back-titrated. Since the analyst using GC can measure the products of the reaction and can judge their purity by the symmetry or shape of the peaks, the need for the highly precise results demanded in classical determinations has been somewhat relaxed.

Remarkable progress in the life sciences has been possible through the application of GC because this technique can be used to determine quantitatively very small amounts of chemicals or, more specifically, physiological levels of chemicals in biological materials. Again, preliminary separations and/or cleanup are usually required to isolate the fraction containing the compounds with the target functional group. When functional groups or elements that are sensed by one of the highly specific detectors (see Section D) are present in the chemical being determined, the determination of remarkably small amounts (nanograms or picograms) of the chemical often becomes possible; at the same time, the specificity of the analysis is much improved because the detector responds feebly or not at all to extraneous material in the sample. The highly specific detectors are also invaluable in improving the sensitivity of analysis through the formation of derivatives of given functional groups; these derivatives are chosen to contain elements or groups to which the specific detectors respond with great intensity. (See Section C.3.)

Calculations of results have not usually been included because they are generally straightforward. As already noted, quantification of known compounds is based on peak areas (or heights) by comparison with standards analyzed in exactly the same way. For more precise quantification, an internal standard is employed; usually it is a compound with properties similar to the components being analyzed but with a t_R sufficiently different to avoid interfering with the peaks of interest. In adding an internal standard the analyst can correct for inaccuracies in manipulating solutions and in injecting a given volume into the gas chromatograph. (See a standard GC text for further details.) Usually peak area will be proportional to concentration, and with small (usually submicrogram) amounts peak height will also often be proportional to concentration; response factors (peak area or height per unit weight) may then be calculated. When concentration is not proportional to response, calibration curves must be constructed from a series of dilutions of the compounds being determined. These nonlinear calibration curves are often linear on a semilogarithmic or log-log plot; for example, the response of the flame photometric detector to sulfur compounds is nonlinear, but a log-log plot of its concentration versus response is linear.

In light of these considerations, the use of GC for functional group analyses has much to recommend it. However, since many of the methods are new and have not been thoroughly tested with a wide range of materials, the analyst must exercise caution. Confirmation of the identity of compounds by means other than the t_R on a single column (e.g., by derivatization and analysis) is most desirable. Consistent results in the chromatography of a mixture on a polar and a nonpolar liquid phase or on several liquid phases of different polarities will enhance the validity of an analysis. The more information brought to bear, the more reliable will a particular analysis become.

C. Preparation of Samples for Gas Chromatography

Representative sampling and proper sample storage (expecially of biological materials), although not within the purview of this book, must be given careful consideration if the degradation of compounds of interest is to be avoided.

1. Extraction.   Whereas some samples are provided ready for analysis, others must be extracted from a substrate before they can be analyzed. Liquids can be extracted with an immiscible solvent. Solids (such as plant material) usually have to be chopped, ground, pulverized, or otherwise comminuted to allow complete extraction of the essential ingredients with a solvent. The solvent that completely removes the ingredients of interest with the least amount of interference and of coextractives is to be preferred. Such a solvent is generally the poorest one that will quantitatively extract the substances of interest; it should be easily removed (low-boiling) without loss of the essential ingredients; and it should be free of interferences despite considerable concentration.

2. Cleanup.   The amount of cleanup needed is directly dependent on the GC detector employed. Many samples are not clean enough for direct injection into the gas chromatograph. Interfering impurities must be removed, as well as excessive amounts of nonvolatile matter, which accumulate on the column after repeated injections and cause shifts in t_R, poor separations, unstable base lines, and low recoveries. Material of low volatility can also cause difficulty, even though it may not register on the detector (e.g., a highly specific one), by accumulating in the GC detector and gradually diminishing its response; detectors with a radioactive ionizing source that cannot be operated well above the column temperature are especially vulnerable to this source of contamination. Accordingly, the removal of unwanted material (cleanup) is often a prerequisite to satisfacory analysis.

Cleanup should be as simple and rapid as possible. Partitioning the sample between a pair of immiscible solvents is frequently used for this purpose; for example, fats are almost completely removed from extracts of biological material by partitioning between the hexane-acetonitrile pair; the fat remains in the hexane layer, while the desired compound is in the acetonitrile layer. Group separations (e.g., separation of acid, basic, and neutral fractions by extraction) are often helpful. Passing the sample through a column of adsorbent with solvents of increasing eluting power (column chromatography) is used to remove extraneous material and often to make gross separations of the ingredients of interest. Thin-layer chromatography serves a similar function. In many instances chemical treatment of an extract that does not affect the compounds being analyzed will remove an interference (e.g., removal of fats by saponification). Sometimes chemical treatment will be needed to release the substance(s) being determined; an example is the release of amino acids by hydrolysis of a protein. Although these preliminary separations increase the specificity of the analysis, they also introduce the possibility of losses in purification.

3. Derivatization.   Many compounds, even if pure, cannot be quantitatively determined by gas chromatography because they are thermally unstable, insufficiently volatile, or highly polar. Polar functional groups cause tailing or irreversible adsorption of compounds on many column packings; they also depress the volatility of compounds and contribute to their thermal instability. Since compounds with polar groups are readily derivatized to much less polar compounds, derivatization has been widely adopted to overcome the aforementioned impediments to analysis, and many compounds previously considered not amenable to analysis by GC are now readily determined as derivatives. Indeed, the key to many functional group analyses is found in the formation of an appropriate derivative.

Derivatization should be as rapid and specific as possible for the target functional group. Rapid derivatization is aided by the fact that very little of a compound is needed for GC analyses. As already indicated, comparatively large amounts of derivatizing agent are normally permissible, and reactions are driven to completion very quickly. In a well-designed analysis either the derivatizing agent does not interfere (i.e., it has a different t_R) or it is separated beforehand. Ideally, the derivative will form only from compounds with the functional group of interest, will form quantitatively, will chromatograph well, and will appear on the chromatogram free of interference. If interference occurs, a different chromatographic column may eliminate this difficulty. Alternatively, other derivatives can be prepared to find one that gives a peak sufficiently free of the interference.

Finding a suitable derivative is facilitated by the great variety that is available. Fluorinated derivatives tend to be more volatile than nonfluorinated ones and therefore have smaller t_R's or can be chromatographed at temperatures lower than usual. Fluorinated derivatives have made possible the GC of a variety of compounds, including high-molecular-weight substances and polyhydroxy compounds, and are therefore of considerable importance.

Halogenated derivatives, including fluorinated ones, are frequently employed to improve the sensitivity of an analysis; for example, the electron capture detector responds with great sensitivity to the heptafluorobutyrate derivatives of alcohols. The procedure also gains specificity in this way, as a positive response indicates that the compound contains the functional group being derivatized. However, derivatives may be formed from compounds other than the ones of interest if they have the same functional group or another functional group that reacts with the derivatizing agent; these unwanted by-products of the derivatization reaction can cause troublesome background interference, which is likely to vary in magnitude with each sample. It follows that a compound already having a group that can be determined with great sensitivity (such as a halogen) should not be converted to a derivative with a similar group because background interference will be introduced and the specificity will be decreased.

Because GC separates as well as measures the products of reaction, quantitative analysis is greatly facilitated; at the same time the specificity of the analysis is improved markedly, because a given compound must appear at a definite t_R. Response factors must, of course, be determined for each component of a mixture to make a quantitative determination of each one.

The fact that several compounds of a mixture may be determined in a single GC analysis complicates matters because the analyst may be dealing with compounds possessing a variety of functional groups. As an example, it may be necessary to make the determination of alcohols as acetates more specific by ascertaining that the ingredients of the mixture are neutral (not acidic or basic), thus ruling out the presence of amines and phenols that would otherwise be derivatized also. Should amines, enols, and phenols be present, the analyst may be able to remove these interferences by extractions with acid and base or by use of ion-exchange resins. Such separations are frequently necessary before derivatization to eliminate interference in an analysis.

D. GC Detectors

Although little of the great array of equipment available can be mentioned, the important detectors used in gas chromatography, especially those available commercially, deserve attention. They are listed in Table A-1, along with some of their characteristics, which must be considered approximations because of the great variety of equipment available. The need to determine many types of compounds at levels ranging from milligrams to picograms in a multitude of different materials has led to the development of many useful detectors. [Several recent reviews describing GC detectors (see refs. 5–8) are worth consulting for further information.] Detectors may be highly specific or nonspecific in their response, and linearity of response is important in quantitative analysis. By sequential placement, stream splitting, or other arrangement, two detectors which provide different information may be used to monitor a single GC effluent.

1. Nonspecific Detectors. The most widely used nonspecific (or general) detectors are the thermal conductivity and flame ionization detectors. The thermal conductivity detector (hot-wire or thermistor type) measures the difference in thermal conductivity of the pure carrier gas and a mixture of a solute in the carrier gas stream. Most substances have thermal conductivities much lower than that of the helium or hydrogen normally used as the carrier gas and are therefore readily detected. This type of detector is sensitive to flow rate and temperature changes and consequently requires good control of these parameters. Careful calibration of the detector with standards (determination of response factors) and operation within the linear range are desirable for quantitative analysis. The thermal conductivity detector is

Table A-1.   Summary of Characteristics of Important GC Detectors

Detector	Type	Substances detected[a]	Linearity range	Limit of detection[b]
Thermal conductivity	Nonspecific	All	10^3	0.1–1 μg
Flame ionization	Nonspecific	All organic	10^6	1–100 ng
Gas density balance	Nonspecific	All	10^3	1 μg
Cross section	Nonspecific	All	10^4	1 μg
Mass spectrometric	Nonspecific	All	...	0.01–1 μg
Electron capture	Specific	[c]	5×10^2	pg level
Thermionic	Specific	P-, N-, hal-containing	Narrow	pg to ng
Flame photometric	Specific	P-, S-, hal-containing	$>10^3$	0.1, 3, 20 ng for P, S, hal compounds, respectively
Coulometric	Specific	Hal-, S-, N-containing	Narrow	Several ng
Electrolytic conductivity	Specific	Hal-, S-, N-containing	Narrow	Several ng

[a] N = nitrogen, S = sulfur, P = phosphorus, hal = halogen.
[b] μg = microgram (10^{-6} gram), ng = nanogram (10^{-9} gram), pg = picogram (10^{-12} gram), generally based on amount of compound giving a response of twice the noise level of the detector.
[c] See text.

rugged and usually nondestructive, that is, separated solutes may be recovered intact after passing the detector.

In a flame ionization detector hydrogen is added to the carrier gas as it emerges from the column; the carrier gas is usually nitrogen or helium, and the carrier gas and hydrogen are normally mixed in equal proportions. The mixture burns at a jet in an air or oxygen atmosphere. Ions formed in the burning of organic solutes cause the electrical resistance of the flame to decrease in proportion to the amount of solute emerging. Under the influence of a potential difference (ca. 100 to 300 V), the resistance drop between the jet and an electrode above or beside the flame causes a flow of current which is displayed on a strip chart recorder after suitable amplification. Its high sensitivity, excellent linearity over a great range (10^6), rapid response, excellent stability, ability to sense all organic compounds, insensitivity to inorganic compounds and to small changes in temperature and flow rate, and simplicity of operation have made the flame ionization unit one of the most popular, if not the most popular, of the GC detectors. For accurate quantitative analysis, individual response factors must be determined for each component of interest.

The other nonspecific detectors of Table A-1, although commercially available, are not widely used. A special application of the gas density balance detector is the determination of molecular weights of components in a sample. Mention must again be made of the increasingly popular GC-mass spectrometer combinations, which have become invaluable in identifying components of mixtures and in determining their purity. Since a symmetrical peak of one compound may conceal another having the same t_R, confirmation of purity is frequently essential in establishing the quantitative validity of an analysis.

2. *Specific Detectors.* The highly specific detectors are very selective in their response and are invaluable for the determination of trace amounts of compounds containing certain elements or groupings. They have become especially useful in the analysis of pesticides, pharmaceuticals, petrochemicals, and biochemical substances that contain halogen, phosphorus, sulfur, or nitrogen. The great advantage of these detectors is that cleanup (removal of interfering matter), which is the laborious and time-consuming stage of most analyses, is held to a minimum because the detector is simply blind to the extraneous material.

The electron capture detector is probably the least specific of the so-called highly specific detectors in widespread use, since it responds to many types of compounds, such as halides, certain sulfur-containing compounds, conjugated carbonyls, metal organics, nitro compounds, and nitriles. Its high sensitivity to chlorine- and sulfur-containing pesticides has made it invaluable in the quantitative analysis of traces of these materials. In fact, the use of electron-capturing derivatives to increase the sensitivity of analyses has made the detector generally serviceable in virtually all fields of endeavor. In this type of detector an ionizable carrier gas (usually N_2) is conducted through a cell containing a radioactive source (such as tritium, strontium, radium, or ^{63}Ni), and a potential just sufficient to collect all free electrons produced is applied to the two electrodes. The current is constant until an electron-capturing compound is introduced; then a decrease in ion current that is proportional to the concentration of the electron-capturing solute results. After amplification, the loss of current is displayed on a recorder. (Use of a pulsed rather than a dc potential minimizes certain anomalous responses.) The electron capture detector is inexpensive and is easy to operate. Its linearity range is very narrow, and it is sensitive to water (the carrier gas must be dry). Unclean extracts, even when they do not interfere, may gradually contaminate the detector and cause loss in sensitivity. This difficulty is encountered mostly with detectors that cannot operate above 225°C, such as those with a tritium ionization source. Use of an ionization source capable of functioning at a higher temperature (e.g., up to 400°C with ^{63}Ni) can eliminate this problem.

The flame photometric detector of Brody and Chaney (9), available from Tracor, Inc., Austin, Tex., is a highly specific, modified flame ionization detector in which the light emitted by the burning GC effluent is monitored by a phototube that registers its response on a recorder. With a 526-nm interference filter, the detector responds to compounds containing phosphorus with high sensitivity (at least 0.1 ng) and unusually high specificity; the detector is very stable, and its response is linear with concentration. With a 394-nm interference filter, the detector responds with good specificity to as little as a few nanograms of compound containing sulfur. Although the response of the sulfur detector is nonlinear with respect to concentration (roughly, the response is proportional to the square root of the concentration), it is usually linear in a log-log plot. With a dual version of the detector (having two phototubes, one with a 526- and the other with a 394-nm filter; available from Tracor), phosphorus and sulfur compounds can be determined simultaneously (10), and the P/S ratio in the molecule may be calculated from the ratio of the responses.

The first thermionic (or alkali-flame) detector was a flame ionization detector having a platinum coil coated with sodium sulfate in the vicinity of the flame (11–13). With it, nanogram and subnanogram amounts of compounds containing phosphorus could be determined with high specificity, although the detector also responded to halogen compounds, but with much less sensitivity. (Precise gas-flow control was necessary.) In one modification of the detector, the response of halogenated compounds was virtually eliminated by using potassium chloride as the salt (14). In another (Varian-Aerograph), a plug of CsBr was molded to the tip of the burner (15). In a third modification, the phosphorus response was depressed and the response to nitrogen-containing compounds greatly enhanced by precise placement of the electrode (16, 17).

The major advantages of the thermionic and flame photometric detectors over the electron-capture detector are freedom from fouling and therefore longer detector life, greater specificity in response, ability to operate at high temperature, greater linearity range, and insensitivity to water.

With some detection systems, the organic material in the GC effluent is subjected to combustion in an oxygen stream. The halide or sulfur dioxide generated from halogen- and sulfur-containing compounds is determined by automatic titration in a microcoulometric cell (18, 19) or by an electrolytic conductivity detector (20) with a recorder providing the usual chromatogram of the compounds. Organic material may also be "burned" under reductive conditions. Martin (21) used coulometry, and Coulson (22) employed electrolytic conductivity to determine ammonia catalytically formed (with nickel) from nitrogen-containing compounds. The detectors respond to as little as a few nanograms of compound. With appropriate subtraction agents, the specificity of these detectors is excellent.

E. Carbon-Skeleton Chromatography

In carbon-skeleton chromatography, the functional groups of a compound are stripped off and double and triple bonds are saturated; the product(s) helps in identifying the carbon skeleton. The technique, covered in a recent review (23), has been used in the analysis of a wide variety of compounds— acids, alcohols, aldehydes, anhydrides, ethers, epoxides, ketones, esters, amines, amides, aliphatic and aromatic hydrocarbons, nitriles, sulfides, halides, olefins, and other types. Because of its broad applicability, this method of analysis is discussed here in this general section rather than under the individual functional groups. The analysis, although qualitative, can also be useful for quantitative determinations. However, its major utility is in structure determination, for which quantitative functional group analyses are most frequently needed. The use of the retention indices of the carbon skeleton and of functional groups constitutes a valuable approach to the elucidation of chemical structure and is discussed in the next section.

Figure A-3 is a schematic representation of the apparatus. A hot catalyst-containing tube is introduced into the GC pathway just ahead of the injection port of the usual flame ionization gas chromatograph, and hydrogen is used as the carrier gas. The catalyst (palladium or platinum on a GC support) is held at elevated temperatures, usually between 200 and 350°C. As the injected compound passes the hot catalyst, the compound is chemically degraded to its carbon skeleton or to its next lower homolog. The hydrocarbon products pass into the gas chromatograph and are identified by their retention times. If more sophisticated instrumentation is available, such as a mass spectrometer, identification can be more certain. As little as 1 μg can suffice for analysis.

An apparatus to conduct carbon-skeleton chromatography is available commercially (National Instruments Laboratory, Rockville, Md.). It holds a 9-inch length of the catalyst in an aluminum tube having a $\frac{3}{16}$-inch bore, actually about 2 grams of catalyst. The catalyst is maintained at an elevated temperature by means of a heater and variable transformer. The so-called neutral catalyst (24) is prepared by evaporating to dryness a solution

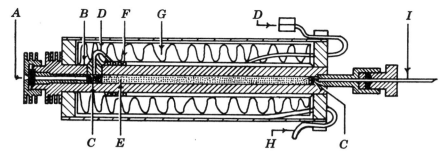

Figure A–3. Cross section of catalyst assembly for carbon-skeleton chromatography. A = Injection port (fins keep septum cool); B = aluminum tube; C = glass wool; D = hydrogen inlet (wrapped around F); E = catalyst; F = heating jacket; G = insulation; H = heating jacket electric cord. Needle stock, I, attaches assembly to GC injection port.

$PdCl_2$ in 5% acetic acid in contact with the support and sufficient alkali to neutralize the HCl that forms when the $PdCl_2$ is reduced ($PdCl_2 + H_2 \rightarrow Pd + 2HCl$). The catalyst is activated by gradually heating it to the analysis temperature in a slow stream of hydrogen.

In carbon-skeleton chromatography, hydrogenation, dehydrogenation, and hydrogenolysis may occur. Hydrogenation (saturation of multiple bonds) greatly reduces the number of possible compounds in determining the carbon skeleton. Dehydrogenation (abstraction of hydrogen) takes place with compounds containing cyclohexane rings and gives aromatic structures. Elevated catalyst temperatures (e.g., 330°C) favor such aromatizations; lower temperatures, between 200 and 250°C, favor hydrogenation over dehydrogenation (25). [Dehydrogenations may be used in a manner similar to the classical zinc or selenium dehydrogenations to determine the identity of ring structures (26).] Finally, hydrogenolysis involves the cleavage of functional groups from a molecule, and the addition of a hydrogen atom to each of the cleaved ends.

In the typical reactions that follow (1% Pd neutral catalyst at 300° C and an H_2 flow rate of 20 ml/ minute were used), the parent hydrocarbon is obtained from halides, sulfides, and compounds with secondary oxygen (or nitrogen) functions:

$$CH_3CH_2CH_2CH_2CH_2 \vdots Cl \rightarrow CH_3CH_2CH_2CH_2CH_3$$

$$CH_3CH_2CH_2CH_2CH_2 \vdots SH \rightarrow CH_3CH_2CH_2CH_2CH_3$$

$$\begin{array}{c} OH \\ ---\vert--- \\ CH_3CH_2CH_2CHCH_3 \end{array} \qquad \rightarrow CH_3CH_2CH_2CH_2CH_3$$

When an oxygen or nitrogen function is on the end carbon atom (aldehyde, primary alcohol, ester, ether, amine, amide, carboxylic acid), the next lower homolog of the parent hydrocarbon is obtained, although the parent compound may be produced concurrently:

$$CH_3(CH_2)_{10} | CHO \qquad \rightarrow CH_3(CH_2)_9 CH_3$$

$$CH_3(CH_2)_{15} | CH_2 OH \qquad \rightarrow CH_3(CH_2)_{14} CH_3$$

$$CH_3(CH_2)_{16} | CH_2 | \overset{\overset{\displaystyle O}{\|}}{OCCH_3} \rightarrow CH_3(CH_2)_{16}CH_3 + CH_3(CH_2)_{15}CH_3$$

Reactions of carbon-skeleton chromatography are summarized in Figure A-4.

Temperature programming of the analytical column is used frequently to show both the small and the large hydrocarbon products as sharp peaks. (The catalyst temperature remains constant.) Temperature programming is also useful in clearing the analytical column for the next run; should any

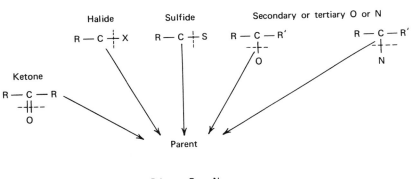

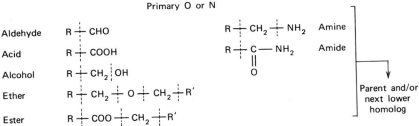

(Unsaturates are saturated)

Figure A–4. Reactions observed with carbon-skeleton chromatography.

unchanged compound pass the catalyst, it will be eliminated when the column temperature is raised.

The limitations of carbon-skeleton chromatography should be recognized. Thus products from compounds with several polar groups or those with 20 or more carbon atoms in their molecules come through in low yield and sometimes not at all. However, the proper products are readily obtained from polar and high-molecular-weight compounds by using shorter catalyst lengths; a 4-inch length of catalyst was found suitable for compounds in the C_{12} to C_{30} range (27). When a short length of catalyst (e.g., a 3-inch length) is sufficient, it is often possible to put the catalyst in the injection port, and no special apparatus is required. This arrangement (28), as shown in Figure A-5, is particularly convenient with a Hewlett-Packard Model 810 gas chromatograph. The insert liner of the injection port is removed, filled with catalyst, and reinserted. With hydrogen as the carrier gas and the injection port heater at the appropriate temperature, carbon-skeleton chromatography is readily carried out. A stainless steel adaptor holds the septum a small distance from the hot zone to keep the septum cooler and to reduce septum bleed. Some compounds, such as acids or their esters, react slowly and need to be held up in a cold trap between the catalyst and the injection port and then released by rapid heating to secure reliable retention times. The testing of model compounds before proceeding with unknowns is standard procedure to be certain that the apparatus is functioning properly.

Carbon-skeleton chromatography has been used to determine the chemical structures of insect sex attractants, sterols, odor chemicals, alkaloid fragments, and other substances. Brownlee and Silverstein (29) devised an integrated system for collecting a sample from a gas chromatograph and reintroducing it through a carbon-skeleton apparatus.

F. Retention Indices in Gas Chromatography

One of the major applications of quantitative functional group analysis is to determine the number of certain functional groups in a compound.

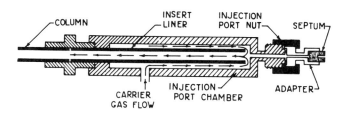

Figure A–5. Cross-sectional view of injection port arrangement for carbon-skeleton chromatography. Catalyst is held in insert liner, and adapter keeps the septum cooler than the injection port.

This information can often be deduced from retention time (t_R) data of similar compounds or from t_R data of the carbon skeleton plus those of functional groups. Although each compound has a characteristic t_R in a given GC system, this t_R is not generally reproducible between laboratories and possibly not even with the same system a month later. However, the ratio of the t_R of a compound compared to that of a standard substance (referred to as the relative t_R) is highly reproducible. In determining relative t_R data, the sample components and the standard are analyzed under identical conditions, and the relative t_R data can be calculated directly from the chromatogram. These data are invaluable in functional group analysis.

A number of systems have been devised to record relative t_R data, the major differences being the standard(s) used. The t_R data are *adjusted* (then designated t_R'); that is, gas holdup must be subtracted. (See Figure A-2.) The Kovats retention index system is based on the t_R' of a compound in relation to the t_R' values of the *n*-alkanes eluting before and after the compound. For example, in isothermal chromatography, the indices (I) of the *n*-alkanes = number of carbon atoms $\times$ 100 (e.g., $I_{\text{decane}} = 1000$, $I_{\text{undecane}} = 1100$), and a compound with an index of 1050 has a t_R' half-way between those of decane (1000) and of undecane (1100) *on a logarithmic scale*. [For a lucid discussion of the Kovats retention index system, see L. S. Ettre, *Anal. Chem.*, *36* (8), 31A (1964).]

A major feature of the system is that the indices of the carbon skeleton and the index increments for the functional groups in a compound are additive. For example, the index of 2-bromo-5-chlorodecane, found to be 1280 with a given column, can be derived from the sum of the index of the alkane, I_{decane} (1000), and the individual index increments, ∂I_{Cl} (120) and ∂I_{Br} (160) (∂I = differences of indices *on the same column*). A compilation of the retention indices of a large variety of compounds with many GC columns has been published (W. O. McReynolds, *Gas Chromatographic Retention Data*, Preston Technical Abstracts, Evanston, Ill., 1966). A further feature of the system is that the difference in indices (ΔI) of a given functional group *on two given columns*, usually polar (P) and nonpolar (NP)—that is, $\partial I^P - \partial I^{NP}$—has a characteristic value, and such differences are likewise additive for several groups; for example, if $\Delta I_{\text{Cl}} = 20$ and $\Delta I_{\text{Br}} = 40$, the ΔI of the compound having the two functional groups is 60. The index of the carbon skeleton remains essentially constant regardless of stationary phase and is therefore excluded when the indices of a compound on two columns are compared in this way.

Carbon-skeleton chromatography frequently offers a good starting point for structural analysis through the use of retention indices. Even when the carbon skeleton of a monofunctional compound is not known, if its t_R can be determined the analyst may be able to narrow the number of functional group possibilities to a few and possibly even to one from the difference between the t_R' of the unknown and the t_R' of the carbon skeleton. By using a second column having a different polarity or selectivity, he may be able to identify the functional group or to confirm its presence. This same approach is applicable to compounds with several functional groups when all but one of the groups are known.

The foregoing exposition is meant only to convey the concept of using t_R' data in functional group analysis, and must be considered a gross oversimplification. For detailed discussions, see E. Sz. Kovats, *Advan. Chromatog.*, **1**, 230 (1965), and G. Schomburg, *ibid.*, **6**, 211 (1968). Retention indices [of compounds (I) and functional groups (∂I)] can vary with temperature, support, stationary phase, position of the functional group, proximity of a neighboring group, interaction of the functional groups, stereochemistry of the functional groups, steric effects, nature of the carbon skeleton or compound nucleus; undoubtedly other more subtle factors also come into play. However, the retention indices of "nonpolar" substances remain almost constant for any kind of stationary phase, and the retention indices of a given substance are about the same for all "nonpolar" (truly nonpolar) stationary phases. Also, similar substitution in similarly constituted compounds increases the retention indices by the same amount for a given stationary phase.

On the basis of the homology principle, it has often been either more convenient or more accurate to use standards other than the n-alkanes for more polar or for complex compounds. Woodford and Van Gent (30) used methyl esters of the straight-chain fatty acids as standards and referred to the number of carbon atoms in the acid chain as the "carbon number"; the t_R' values of other fatty acid methyl esters (branched or unsaturated) were then characterized in relation to the standard esters, as in the Kovats system. Later the term "equivalent chain length" was suggested in place of "carbon number" by Miwa et al. (31) for the same compounds. Retention time indices for functional groups on steroids have also been advanced for use with steroid carbon skeletons (32), and these values, called "steroid numbers," have been found to be most useful.

It follows that index increments of functional groups can be added to either the t_R index of the carbon skeleton or the t_R index of a compound similar to the one of interest, and the sum of the indices can be compared with the t_R index of an unknown compound to determine the functional group content of the molecule. With appropriate model compounds on hand, it is a simple matter to determine the t_R indices needed for such analyses, and the effect of varying a structural feature on the index of a compound of known structure is used to derive the structures of analogous compounds.

G. Reporting of Gas Chromatographic Data (33)

The following guidelines refer to parameters and terminology used in reporting GC methods and will serve as a check list of items that should be taken into account. Many papers have failed to give sufficient pertinent information and consequently have lost much of their effectiveness.

As a general guide, American Society for Testing and Materials' *Recommended Practice for Gas Chromatography Terms and Relationships*, which is based on International Union of Pure and Applied Chemistry's recommendations, is used. *Only what is essential for satisfactory performance is described.* Conditions for the starred items below are always stated as well as any other items that the author deems necessary.

Apparatus

Instrument
Detector Type.* Thermal conductivity, flame ionization, electron capture, etc.
Recorder Range
Detector Voltage
Bridge Current

Column*

Length* and Diameter.* Inside diameter is preferred but outside diameter with wall thickness is acceptable; both may be given
Material.* Glass, copper, stainless steel
Packing.* Weight per cent of liquid phase on support (give mesh size and pretreatment of support—e.g., silanization)
Capillary.* No packing, only liquid phase; support-coated liquid phase; or other type column

Temperatures,* °C.

Injection Port*
Detector*
Column* (oven). Isothermal* or Temperature Programmed.* Give initial and final temperature and rate of temperature rise
Specify other arrangements if made

Flow Rate of Gases. In milliliters per minute (ml. per min.) at exit port

Carrier Gas*
Other Gases

Sample.   Volume injected in microliters (μl.), milliliters (ml.) if gas, concentration of solution

Solvent Used

Retention Time of Compounds.* Minutes

Relative Retention Time. Standard component(s)—e.g., Kovats Index, etc.

Quantitation

Methods.* Peak area, peak height, integrator, planimeter, etc.

Precision. Repeatability

Recovery of Added Amounts. Accuracy. Correction, if made

Weight. Expressed as weight, volume, or mole per cent. State conditions on which calculation is based—e.g., dry weight, sample as received, etc.

Minimum Detectable Limits. Basis, $2\times$ noise level or $2\times$ interference

Standards. Internal standard used for calibration

Interference of Substrate. Baseline correction

Calculations

Corrections, if made

Mathematics

Chromatogram

Typical Analysis. May be very useful

Retention Time (Horizontal) *vs.* Recorder Response (Vertical)

Labels. Peaks, attenuation, temperature program, title

Gas Holdup. Air peak, solvent front

Special

Column conditioning procedure

Adequate extraction (exhaustive) of compound(s) from sample

Specificity of analysis

Speed of analysis

Interfering compounds related to those analyzed

Dual-column operation, background subtraction

Reaction gas chromatography—e.g., hydrogenation, pyrolysis, etc.

Backflush technique

Modifications. Valves, injection devices, stream splitters, trapping, flow controllers, equipment for transfer to other instruments (e.g., spectrometers). Full description of material or apparatus not available commercially

Technique for selecting representative samples

Collaborative results

Errors to be avoided
How to keep apparatus clean and functional

References

Book of ASTM Standards, Part 30, pp. 1071–8 (1967)
International Union of Pure and Applied Chemistry [*Pure Appl. Chem.* **1**, 177–86 (1960)]

H. References

1. L. S. Ettre and W. H. McFadden, eds., *Ancillary Techniques of Gas Chromatography*, Interscience, New York, 1969.

2. M. Beroza, *Accounts Chem. Res.*, **3**, 33 (1970).

3. J. T. Watson and K. Biemann, *Anal. Chem.*, **37**, 844 (1965).

4. M. Beroza and R. A. Coad, *J. Gas Chromatog.*, **4**, 199 (1966).

5. W. E. Westlake and F. A. Gunther, *Residue Rev.*, **18**, 175 (1967).

6. T. A. Gough and E. A. Walker, *Analyst*, **95**, 1 (1970).

7. B. J. Gudzinowicz, *in* L. S. Ettre and A. Zlatkis, eds., *Practice of Gas Chromatography*, Interscience, New York, 1967, pp. 239–332.

8. C. T. Malone and W. H. McFadden, *in* L. S. Ettre and W. H. McFadden, eds., *Ancillary Techniques of Gas Chromatography*, Interscience, New York, 1969, pp. 341–372.

9. S. S. Brody and J. E. Chaney, *J. Gas Chromatog.*, **4**, 42 (1966).

10. M. C. Bowman and M. Beroza, *Anal. Chem.*, **40**, 1448 (1968).

11. A. Karmen and L. Giuffrida, *Nature*, **201**, 1204 (1964).

12. A. Karmen, *Anal. Chem.*, **36**, 1416 (1964).

13. L. Giuffrida, *J. Assoc. Offic. Agr. Chemists*, **47**, 293 (1964).

14. L. Giuffrida, N. F. Ives, and D. C. Bostwick, *J. Assoc. Offic. Anal. Chemists*, **49**, 8 (1966).

15. C. H. Hartmann, *Bull. Environ. Contam. Toxicol.*, **1**, 159 (1966).

16. W. A. Aue, C. W. Gehrke, R. C. Tindle, D. L. Stalling, and C. D. Ruyle, *J. Gas Chromatog.*, **5**, 381 (1967).

17. C. H. Hartmann, *J. Chromatog. Sci.*, **7**, 163 (1969).

18. D. M. Coulson and L. A. Cavanagh, *Anal. Chem.*, **32**, 1245 (1960).

19. L. Giuffrida and N. F. Ives, *J. Assoc. Offic. Anal. Chemists*, **52**, 541 (1969).

20. D. M. Coulson, *J. Gas Chromatog.*, **3**, 134 (1965).

21. R. L. Martin, *Anal. Chem.*, **38**, 1209 (1966).

22. D. M. Coulson, *J. Gas Chromatog.*, **4**, 285 (1966).

23. M. Beroza and M. N. Inscoe, *in* L. S. Ettre and W. H. McFadden, eds., *Ancillary Techniques of Gas Chromatography*, Wiley-Interscience, New York, 1969, pp. 89–144.

24. M. Beroza and R. Sarmiento, *Anal. Chem.*, **35**, 1353 (1963).

25. M. Beroza and R. Sarmiento, *Anal. Chem.*, **36**, 1744 (1964).

26. M. Beroza and F. Acree, Jr., *J. Assoc. Offic. Agr. Chemists*, **47**, 1 (1964).

27. M. Beroza and R. Sarmiento, *Anal. Chem.*, **37**, 1040 (1965).

28. B. A. Bierl, M. Beroza, and W. T. Ashton, *Mikrochim. Acta*, **1969**, 637.

29. R. G. Brownlee and R. M. Silverstein, *Anal. Chem.*, **40**, 2077 (1968).

30. F. P. Woodford and C. M. Van Gent, *J. Lipid Res.*, **1**, 188 (1960).

31. T. K. Miwa, K. L. Mikolajczak, F. R. Earle, and I. A. Wolff, *Anal. Chem.*, **32**, 1739 (1960).

32. E. C. Horning, W. J. A. VandenHeuvel, and B. G. Creech, *in* D. Glick, ed., *Methods of Biochemical Analysis*, Vol. XI, Interscience, New York, 1963, pp. 69–147.

33. M. Beroza and I. Hornstein, *J. Agr. and Food Chem.*, **17**, 160 (1969).

III.   ELECTROANALYTICAL METHODS*

A.   General Coverage

1. M. J. Allen, *Organic Electrode Processes*, Reinhold, New York, 1958. Not really an analytical text, but it discusses many organic electrolytic reactions.

2. M. R. F. Ashworth, *Titrimetric Organic Analysis*, Parts I and II, Interscience, New York, 1964 and 1965. Tabulation of references and very brief sketches of methods for the titrimetric determination of a comprehensive list of organic compounds. Many of the titrations are based on electroanalytical end-point detection. (See also refs B.3 and B.8.)

3. G. Charlot, ed., *Modern Electroanalytical Methods*, Elsevier, Amsterdam, 1958. Papers on a variety of electroanalytical subjects which were delivered at an international symposium.

4. G. Charlot, J. Badoz-Lambling, and B. Tremillon, *Electrochemical Reactions, Electrochemical Methods of Analysis*, Elsevier, Amsterdam, 1962. English translation of the French text discussing several electroanalytical methods.

5. P. Delahay, *New Instrumental Methods in Electrochemistry*, Interscience, New York, 1954. Theoretical treatment of all kinds of electrochemical methods.

6. F. Fichter, *Organische Elektrochemie*, Verlag Theodor Steinkopff, Dresden and Leipzig, 1942. A very complete tabulation and discussion of all the organic electrolytic reactions in the literature up to the date of publication. This book is often listed as a reference in publications, but not many institutional or personal libraries actually possess a copy.

7. I. M. Kolthoff and P. J. Elving, eds., *Treatise on Analytical Chemistry*, Part 1, Vol. 4, Interscience, New York, 1963. Eleven chapters on various topics by experts in the individual fields.

8. J. J. Lingane, *Electroanalytical Chemistry*, 2nd Ed., Interscience, New York, 1958. Excellent coverage, both theoretical and practical, of a wide range of topics in electroanalytical chemistry, with an emphasis on inorganic analysis.

9. D. A. MacInnes, *The Principles of Electrochemistry*, Reinhold, New York, 1939, and Dover, New York, 1961. Inexpensive reprint of a classic textbook on several topics of interest.

10. J. Mitchell, Jr., I. M. Kolthoff, E. S. Proskauer, and A. Weissberger, eds., *Organic Analysis*, Interscience, New York. Several volumes in this series contain specific chapters on electroanalytical subjects. Many of the other chapters also include discussions of electroanalytical approaches to the particular topics under discussion.

11. A. Weissberger, ed., *Technique of Organic Chemistry*, 3rd Ed., Interscience, New York, 1960. A number of chapters in various volumes discuss all of the different aspects of organic electroanalytical chemistry. Very comprehensive coverage.

* Written by Alan F. Krivis.

B. Potentiometry

1. R. G. Bates, *Electrometric pH Determinations*, John Wiley, New York, 1954. Treats pH measurement from all aspects in a very comprehensive manner.

2. A. H. Beckett and E. H. Tinley, *Titration in Nonaqueous Solvents*, British Drug Houses, Poole, 1962. Booklet outlining the essentials of nonaqueous titrations, including potentiometric measurements.

3. N. D. Cheronis and T. S. Ma, *Organic Functional Group Analysis*, Interscience, New York, 1964. Very useful compendium of organic analytical references and methods, many of which are based on potentiometric titrations.

4. W. M. Clark, *Oxidation-Reduction Potentials of Organic Systems*, Williams and Wilkins, Baltimore, 1960. Detailed discussion of the theory and technique for measuring organic redox systems and the significance of the data.

5. J. S. Fritz, *Acid-Base Titrations in Nonaqueous Solvents*, G. F. Smith Chemical Co., Columbus, 1952. Similar in scope io ref. B.2.

6. W. Huber, *Titrations in Nonaqueous Solvents*, Academic, New York, 1967. English translation of the German text dealing with acid-base titrations in nonaqueous media.

7. I. M. Kolthoff and H. A. Laitinen, *pH and Electro-Titrations*, John Wiley, New York, 1941. Short volume on several topics, including potentiometric measurements.

8. S. Siggia, *Quantitative Organic Analysis via Functional Groups*, 3rd Ed., John Wiley, New York, 1963. The intent of this very popular book is to offer, *within its pages*, detailed methods suitable for most organic analytical problems. As also indicated for ref. B.3, many of these methods involve potentiometric titrations.

C. Voltammetry

1. M. Brezina and P. Zuman, *Polarography in Medicine, Biochemistry, and Pharmacy*, Interscience, New York, 1958. A useful collection of methods for the polarographic determination of compounds of interest in the fields mentioned.

2. I. M. Kolthoff, and J. J. Lingane, *Polarography*, 2nd Ed., Interscience, New York, 1952. The bible for polarographic and voltammetric work.

3. L. Meites, *Polarographic Techniques*, 2nd Ed., Interscience, New York, 1965. Comprehensive one-volume monograph on polarography and closely related techniques.

4. G. W. C. Milner, *The Principles and Applications of Polarography and Other Electroanalytical Processes*, Longmans, Green and Co., New York, 1957. Excellent discussion of the common and not so common variations of voltammetry.

5. E. H. Sargent and Co., *Bibliography of Polarographic Literature*, 1969. Expanded and updated edition of the very valuable bibliographies on polarographic and voltammetric papers published by Sargent from 1937 to 1956.

6. K. Schwabe, *Polarographie und Chemische Konstitution Organischer Verbindungen*, Akademie-Verlag, Berlin, 1957. Tables of half-wave potentials of organic compounds.

7. J. T. Stock, *Amperometric Titrations*, Interscience, New York, 1965. A complete discussion and description of amperometric titrations of both inorganic and organic species.

8. P. Zuman, *Organic Polarographic Analysis*, Pergamon, London, 1964. Analysis of organic compounds, direct and indirect, by polarographic means.

D. Coulometry

1. K. Abresch and I. Claasen, *Coulometric Analysis*, Chapman & Hall, London, 1964. Translation into English of the German text covering various aspects of practical coulometry.

2. G. Patriarche, *Contributions a l'Analyse Coulometrique*, Arscia, Brussels, 1964. Coulometric analysis with an emphasis on pharmaceutical materials.

E. Conductivity

1. H. T. S. Britton, *Conductometric Analysis*, Van Nostrand, Princeton, N J., 1934. Classic discussion of the field.

2. R. M. Fuoss and F. Accascina, *Electrolytic Conductance*, Interscience, New York, 1961. Discussion of the theory of electrolytes by one of the authorities in the field.

3. R. A. Robinson and R. H. Stokes, *Electrolyte Solutions*, 2nd Rev. Ed., Butterworths, London, 1965. Theoretical and practical aspects of several phases of electrochemistry with an emphasis on conductance measurements.

F. Dielectric Constant

1. H. Fröhlich, *Theory of Dielectrics*, Oxford University Press, London, 1949.

2. E. Pungor, *Oscillometry and Conductometry*, Pergamon, London, 1965. Combination of conductance and dielectric constant coverage with many practical applications.

3. C. P. Smyth, *Dielectric Behavior and Structure*, McGraw-Hill, New York, 1955. Comprehensive discussion of the theory and measurement of dielectric constant and dipole moment.

IV. NUCLEAR MAGNETIC RESONANCE METHODS*

The spectroscopic technique described as nuclear magnetic resonance is exactly that. The experiment measures precisely the resonance frequency of nuclei that have a nonzero spin. Many excellent texts that deal with this subject at length, from both the theoretical and the application approach, are available.

The data from the nmr spectrum are quite fundamental and yield a great deal of structural as well as quantitative information. Most nmr analyses are carried out on systems containing hydrogen. Other frequently studied nuclei are ^{19}F, ^{11}B, ^{31}P, ^{14}N, and ^{13}C.

A. NMR Experimental Method

The nmr experiment is accomplished as follows:

1. The sample must be a liquid or a solid that can be placed in solution. Liquids can be analyzed as they are, but care should be taken with the integrator data.

* Written by Harry Agahigian.

2. The usual solvents are deuteriochloroform, deuterioacetone, deuterium oxide, trifluoroacetic acid, deuteriated dimethyl sulfoxide, and deuteriated dimethylformamide. The solution is usually in the 10 to 20% range for optimum nmr operation. Although exact weights and sample volumes may be used, the concentration is not critical unless the chemical aspect of the experiment calls for it. For example, the level of unsaturation with respect to saturation can be followed from sample to sample without the solutions being exactly the same concentration.

3. Once the compound is in solution, the solution is transferred to a nmr tube. A level of 1 to 1.5 inches is sufficient so that the liquid level is in the receiver coil and vortexing as the sample is spun is not a problem. The standard used is tetramethylsilane.

4. The spinning of the sample cannot be emphasized enough, as this is critical in the nmr experiment. In some cases the spin rate has to be adjusted quite carefully to obtain good resolution and signal stability. The design of current nmr spectrometers is such that stability (and even resolution) is no longer a problem and in most cases is proportional to the ability of the operator rather than reflecting a limitation in the instrumentation.

5. Once the spinning rate has been set and the spectrum is viewed on either the oscilloscope or a recorder, adjustments are made, usually to the Y gradient. After symmetrical ringing has been achieved on the standard peak, tetramethylsilane, and maximum peak height has been obtained, the spectrum can be scanned and there is reasonable assurance that a good-quality spectrum will result. The usual sweep time is 200 to 250 seconds.

6. Once the scan is completed, an integral is obtained. Adjustments have to be made on the spectrometer, and the instrument is switched to the integration mode. The scan rate is usually increased, and a balancing of the integrator has to be accomplished. This is most critical, as is the phase adjustment. Otherwise, good integrals are not obtained. Care in this adjustment is important, as the best spectrum is of no value with a poor integral.

7. The spectrum and integral thus being obtained, the analysis of the spectrum follows. The chemical shift is the most important measurement to consider. This is, quite simply, the distance in hertz (cycles per second) from a standard such as tetramethylsilane to a certain peak in the spectrum. A problem arises in reporting hertz or cycles per second because the radio-frequency may be 40, 60, or 100 MHz, although 60 MHz is the routine frequency at this time. As a result, parts per million is used. This is obtained by dividing the difference from the tetramethylsilane standard to the peak in question by the radiofrequency used and multiplying by 10^6. Parts per million does not designate concentration. If a value of 1.2 ppm from TMS is reported and the radiofrequency is 60 MHz, this means that there is a

resonance or a group of lines centered 72 Hz from tetramethylsilane (60 × 1.2 = 72).

8. The integral values do not really mean much until the line or multiplet of lines is analyzed. This is the other important parameter, the coupling constant, that is, the effect of different spin states of the adjacent nuclei on the lines of the nucleus measured. In a naive way, if there is a single proton on an adjacent carbon, the adjacent methyl (e.g., acetaldehyde) "sees" two spin states and the methyl appears as a doublet. Not all spectra are so simple. With a little experience, however, those which are not first order are easily recognized and in truth give more information than simple single lines. The coupling constants vary according to proximity and stereochemistry and provide details as to not only the adjacent nuclei but also the geometry.

9. The groups of lines are now considered in the evaluation of the integrals. In functional group analysis it is easy to look at one convenient resonance in the spectrum and to follow its change as a function of time. For example, the change of an acetaldehyde to an acetal may be followed by observing either the methyl resonance of the CH. One can be used to cross-check the other, although it suffices to observe one resonance or nucleus.

Subjects such as spin-spin decoupling, deuterium exchange, and computer analysis are outside the scope of this presentation. These approaches to nmr spectroscopy can be applied when the nature of the problem warrants their use.

B.   Recommended Texts

1. J. A. Pople, W. G. Schneider, and H. J. Bernstein, *High Resolution Nuclear Magnetic Resonance*, McGraw-Hill, New York, 1959.

2. L. M. Jackman, *Applications of Nuclear Magnetic Resonance Spectroscopy in Organic Chemistry*, Pergamon, New York, 1959.

3. R. H. Bible, *Interpretation of NMR spectra*, Plenum, New York, 1965.

4. H. Conroy, *Advances in Organic Chemistry*, Vol. II, Interscience, New York, 1960, pp. 265–325.

5. F. G. Nachod, *Ann. N.Y. Acad. Sci.*, **70,** 763 (1958).

6. J. R. Dyer, *Applications of Absorption Spectroscopy of Organic Compounds*, Prentice-Hall, Englewood Cliffs, N.J., 1965.

7. S. Siggia, *Survey of Analytical Chemistry*, McGraw, New York, 1968, pp. 102–114.

8. D. Pasto and C. Johnson, *Organic Structure Determinations*, Prentice-Hall, Englewood Cliffs, N.J., 1969.

9. J. W. Emsley, J. Feeney, and L. H. Sutcliffe, *High Resolution Nuclear Magnetic Resonance Spectroscopy*, Pergamon, London, Vol. I, 1965; Vol. II, 1966.

10. J. W. Emsley, J. Feeney, and L. H. Sutcliffe, *Progress in Nuclear Magnetic Resonance Spectroscopy*, Pergamon, London, Vols. I–V, 1965–1969.

V.   RADIOCHEMICAL METHODS*

The following books are important references dealing with radioactivity measurements and methods:

1. C. G. Beil, Jr., and F. N. Hayes, eds., *Liquid Scintillation Counting*, Pergamon, London, 1958.
2. G. D. Chase and J. L. Rabinowitz, *Principles of Radioisotope Methodology*, 3rd Ed., Burgess, Minn., 1967.
3. D. A. Lambie, *Techniques for the Use of Radioisotopes in Analysis*, D. Van Nostrand, London, 1964.
4. R. T. Overman and H. M. Clark, *Radioisotope Techniques*, McGraw-Hill, New York, 1960.
5. V. F. Raaen, G. A. Ropp, and H. P. Raaen, *Carbon*-14, McGraw-Hill, New York, 1968.
6. E. D. Branscome, Jr., *The Current Status of Liquid Scintillation Counting*, Grune & Stratton, New York, 1970.

* Supplied by D. Campbell.

Index